ROUTLEDGE LIBRARY EDITIONS:
INTERNATIONAL SECURITY STUDIES

Volume 19

THE STRUCTURE OF THE DEFENSE INDUSTRY

THE STRUCTURE OF THE DEFENSE INDUSTRY

An International Survey

Edited by
NICOLE BALL AND MILTON LEITENBERG

LONDON AND NEW YORK

First published in 1983 by Croom Helm Ltd

This edition first published in 2021
by Routledge
2 Park Square, Milton Park, Abingdon, Oxon OX14 4RN

and by Routledge
52 Vanderbilt Avenue, New York, NY 10017

Routledge is an imprint of the Taylor & Francis Group, an informa business

British Library Cataloguing in Publication Data
A catalogue record for this book is available from the British Library

ISBN: 978-0-367-68499-0 (Set)
ISBN: 978-1-00-316169-1 (Set) (ebk)
ISBN: 978-0-367-70121-5 (Volume 19) (hbk)
ISBN: 978-0-367-70125-3 (Volume 19) (pbk)
ISBN: 978-1-00-314467-0 (Volume 19) (ebk)

Publisher's Note
The publisher has gone to great lengths to ensure the quality of this reprint but points out that some imperfections in the original copies may be apparent.

Disclaimer
The publisher has made every effort to trace copyright holders and would welcome correspondence from those they have been unable to trace.

The Structure of the Defense Industry

AN INTERNATIONAL SURVEY

Edited by Nicole Ball and Milton Leitenberg

CROOM HELM
London & Canberra

Croom Helm Ltd, Provident House, Burrell Row,
Beckenham, Kent BR3 1AT

British Library Cataloguing in Publication Data

The structure of the defense industry.
1. Munitions
I. Ball, Nicole II. Leitenberg, Milton
382'.45'6234 HD9743.A2

ISBN 0-7099-1611-6

Printed and bound in Great Britain by
Biddles Ltd, Guildford and King's Lynn

CONTENTS

TABLES

FIGURES

MAP

PREFACE

Milton Leitenberg and Nicole Ball

The 1970s were designated as the 'First Disarmament Decade' by the United Nations. No disarmament took place during that time. The four years between the first United Nations Special Session on Disarmament in 1978 and the second Special Session in 1982 did not bring any disarmament either. Quite the opposite. Efforts by the US and the USSR to achieve arms control were in fact suspended for the most part between 1980 and 1982. While the Strategic Arms Limitation Talks (SALT) negotiations were in progress throughout the 1970s, the two negotiating parties, the US and the USSR, each introduced new generations of strategic missile systems. Both sides continued to make qualitative improvements in their strategic nuclear systems, and the numbers of warheads for these, which were essentially unconstrained by the SALT agreements, continued to rise. Though unratified by the US Senate, the 1979 SALT II agreement was still being observed in 1982 by both nations, except for the provisions on reductions that would have applied to the forces of the USSR. Conventional arms transfers rose in the four years, as they had continuously since the early 1970s. By the 1980s, Soviet arms transfers had reached levels equivalent to those of the US.

In the United States and in the West, there is a constant undercurrent of public opposition to armament policies, which periodically reaches levels substantial enough to be politically effective. Western intra-alliance disagreements over priorities can also be strong enough to alter military planning. There has been an upsurge of public opposition and fear in the West during this period, in large part due to the failure of the SALT process and the continued pace of US and USSR nuclear weapon deployments. In the USSR and the states of the Warsaw Pact, however, strategic and other major weapon deployments follow one another without significant public or political opposition.

In the area of nuclear weapons proliferation, there were renewed fears that development programmes currently in progress might lead to the appearance of several new nuclear-weapon states in the 1980s. Though the greatest danger to the world for a long time to come will in all likelihood remain the enormous deployment of nuclear weapons that presently exist in the military forces of the US and the USSR, the coming years may see a sizeable expansion in the number of nuclear weapon states. The outcome of such a situation is unpredictable but is certainly likely to be dangerous.

Global military expenditures have also continued to rise. The 1981 estimate of these outlays, at current prices, was $600-650 billion per year, rising in volume at the rate of roughly 3 per cent per year. The rate of increase of global military expenditure in recent years has been higher than that of global economic growth. It is a common suggestion that some portion of these funds could be more usefully applied to other serious problems world-wide, but it is no less true for that. A very large number of wars and conflicts have taken place since the end of World War II – numbering in the hundreds, depending on various criteria of definition. A half dozen active wars were being fought as the second United Nations Special Session met in 1982, and military coups d'etat had continued apace during the previous decade. In short, it was an unruly period, and all of the major indicators of military activity world-wide remained at high rates, or rose even higher.

The purpose of this book is to examine the role of defence industries in the industrial and economic structure of modern states, regardless of their political complexion, in the period since the end of World War II. Its aim is to help define the degree to which institutional factors play a role in the armament of states. Examples of such factors would be the domestic political considerations which support the maintenance of employment funded by government expenditure on defence procurement, or the role of the defence sector in the pattern of national industrial production.

This volume presents a comparative study which attempts to establish the economic parameters of defence industries in ten different states in as much detail as evidence permits. It thus deals with a group of considerations which has been almost totally neglected in previous systematic research on weapons acquisition. This book therefore does not deal with the international political considerations which have been the primary determinants of weapons acquisition in the postwar period.[1] Nor is there space to examine the bureaucratic and military-service pressures that are also major driving forces behind weapons acquisition and succession. A large, informative literature already exists on both of these other subjects, with case studies for the most part based on the experience of the development of major weapons in the United States. This book was undertaken precisely because there appeared to be no comparative assessment of the economic parameters of the defence industries of several countries. This is so despite the fact that questions regarding economic considerations frequently are raised when arms control is discussed. In addition, assumptions regarding economic aspects are often heard, even though they are usually vague and rarely based on any empirical evidence.

Two nations, the US and the USSR, are responsible for by far the overwhelming portion of post-World War II military expenditure, weapons accumulation and arms transfers. We are convinced that the reasons for this are political, a result of conscious and explicit policy decisions on the part of successive administrations of both nations. In the same way, it is choices and decisions primarily based on political considerations – not economic pressures relating to domestic employment or export earnings – that are the chief determinants of military expenditure and arms-procurement policies in most other states, both developed and less developed. Nevertheless, in order for military expenditure, force levels and weapon procurement to be reduced, a reorientation of the defence-industrial capacity of arms producing countries would be necessary.

Proponents of arms control and disarmament are often confronted with the argument that domestic economic considerations are the driving forces in arms competition and that reductions in military-related production would lead to economic hardships. The studies in this volume confirm the view that such reductions would cause relatively minor economic dislocations in nearly all cases, and that even these would be mitigated by the ability of economies to adapt to changing patterns of production.

Any realistic assessment of the effects of reductions in defence industries – that is, the conversion of military industries to civilian uses – must take as its starting point the role played by these industries in present-day national economies. Comparative studies such as this one are therefore essential in order to establish exactly what these economic relationships are. As with most defence-related topics, there is a considerable amount of information available for the United States, less for the major European countries, and very much less for East-bloc countries, China and the developing nations. This volume has been designed to draw together such data as do exist on the defence industries of an important sample of major weapon-producing countries. The accent is on comparability of information. Each chapter takes as its core a number of topics which we feel ought to illuminate the role played by defence industries in the industrial and economic structure of modern nations. These include:

1. Some comments on the historical context of the defence industry, its origin and any recent transformations;
2. Employment in the defence industry as compared to national totals, and within particular industrial sectors of importance to the defence industry;
3. Measures of defence industrial output, compared to national totals and to output of particular industrial sectors;
4. Capital value of the defence industry;
5. Amount of government input of varying kinds: funding, subsidies, organization and control;
6. Research and development (R & D) in the defence sector, compared to other industrial sectors and to civilian portions of constituent industries;
7. The internal organization of the major producers, proportions of management, technical and blue-collar workers, compared to other industrial sectors;
8. Profitability of the defence industry;
9. Concentration and competition in the defence industry;
10. Arms exports as a proportion of the nation's defence industrial output, total export and 'heavy industrial' export;
11. Regional employment concentration in the defence industry;
12. Degree of reliance of particular industrial firms or sectors on defence contracting;
13. Amount of raw materials used by the defence industry, amount of national R & D manpower absorbed by defence industry;
14. The role of the defence industry as a 'leading edge' in maintaining national technological capabilities.

Although data and information on each of these questions are not available for all countries, we consider that this volume provides a good and useful summary of the defence industries in the US, the USSR, France, the Federal Republic of Germany, Sweden, Czechoslovakia, Italy, China and Israel. The original plans for this volume also included a chapter on the United Kingdom. Two different sets of authors failed, however, to produce this chapter and it was necessary to substitute a statistical appendix on the United Kingdom, prepared by one of the editors, for a full chapter. In addition to the individual country studies, there is a chapter examining the development of military industries in the Third World and an introductory chapter assessing the case studies as a group and additionally surveying the effect of expenditure in the defence-industrial sector on three major economic variables not dealt with by the individual case studies: growth, employment and inflation.

Acknowledgements

A number of people assisted in preparing the manuscript for publication. Harry Dean helped to assemble the statistical appendix on the United Kingdom. Alberto Izquierdo did the graphics for the volume. Elsa Vingren typed the manuscript. Ulf Modin and Ingvar Lindblom translated the chapter on Czechoslovakia from German into English. One of the editors (Milton Leitenberg) was supported by the Berghof Foundation for Peace and Conflict Research during a portion of the period that work on this volume was under way. The Swedish Institute for International Affairs provided support for typing of the manuscript. We wish to express our appreciation to all of these for their assistance. In addition, we would like to thank David Croom for his interest in the book.

Stockholm

Note

1. John L. Gaddis, *Strategies of Containment. A Critical Appraisal of Postwar American Security Policy*, Oxford: Oxford University Press, 1981. Thomas W. Wolfe, *Soviet Power and Europe, 1945-1970*, Baltimore, Md.: The Johns Hopkins University Press, 1970. Thomas W. Wolfe, *The SALT Experience: Its Impact on US and Soviet Strategic Policy and Decisionmaking*, R-1686-PR, Santa Monica: RAND, 1975. David Holloway, *The Soviet Union and the Arms Race*, New Haven, Conn.: Yale University Press, 1983.

INTRODUCTION: THE MILITARY SECTOR AND THE ECONOMY

Frank Blackaby

The studies in this book describe the industrial structure of the military weapons industries in the main industrial countries of the world – in much the same way as one might describe any major industry, such as the steel industry in those countries. The studies add considerably to our knowledge of the structure of these industries. Yet our knowledge of this industrial sector is still patchy – indeed in some ways fragmentary. One way to see this is to compare the material available about this industry (or group of industries) with the material available about one of their civil counterparts.

Suppose these studies were about, for example, the road vehicle industry (or industries) in the countries surveyed here. It would be possible – indeed necessary – to provide tables which showed, year by year, the quantities produced in each country of the main products of that industry. There would, further, be input/output tables showing imports, production, and exports for each main item; and there would be trade matrices showing the exporters and their markets. Such material does of course exist for the main civil industries of the world. We know precisely who produces cars of different engine sizes, the share of imports in final purchases, and we have a detailed knowledge of the patterns of trade – who exports to whom.

By contrast, our picture of the weapons industries in these countries is fragmentary. There is no single military product for which we can construct a world table of production and trade, with country detail, year by year. The picture of trade is very thin, with some information about quantities exported by some countries, and some information about values, but with very little official information about the country destination of exports.

Yet it is important to begin the attempt – as these studies do – to analyze the world weapons industry as if it were any other industry. The major manufacturers treat it as such. They have their forecasts of world demand for fighter aircraft, with detailed studies of obsolescence and replacement demands. In this matter, as in other military matters, it is not good enough that knowledge should be restricted to practitioners.

The Macro-economy

This chapter does not attempt a world view of the industrial structure of the weapons industry. It is concerned with another question – the macro-economic consequences of the military sector: that is, the effects of the existence, and of the changes, of the military weapons industry (and indeed of military expenditure in general) on the economic performance of the countries in this study.

It is proper to start with a warning. There are many critics of the vast scale of military expenditure in the industrial countries of the world; indeed, the proportion of national product devoted to this sector in the past twenty years has no peacetime precedent. There is no precedent for the enormous armouries on either side of the dividing line in Europe. Nor is there any precedent for the very rapid rate of military technological advance by the two great powers. The critics of this scale of military expenditure are under the temptation to look to this factor as a cause of many other economic and social ills. The industrial world, they observe, has a high level of military expenditure. It also, in the past decade, has been troubled by a chronically high rate of inflation and by rising unemployment. There is a strong temptation to link the poor performance of both Western and socialist economies in recent years with the size of their military budgets. There is always an attraction in simple, single explanations for a miscellany of troubles.

How far, then, can the military expenditure of the countries studied in this book be held responsible for the worsening economic performance of the industrial world since 1973? There is no doubt that economic performance has worsened, by all the standard criteria; and there is also little doubt that the break came in the early 1970s.

First, economic growth. From 1950 to the early 1970s, the growth rate in the Western industrial world – that is, the countries belonging to the Organization for Economic Cooperation and Development (OECD) – was extraordinarily rapid; their aggregate and national production was rising at an average annual rate of 4.5-5.0 per cent a year, in real terms. There is no precedent to be found in the historical statistics for the countries in this area, for so rapid a rate of growth for over twenty years. The break came in 1973. From 1973 to 1982 – a period of nine years – the growth rate halved.

It is the same story for the other main criteria of relative economic success or failure – unemployment and the rate of price increase. Over the long period of economic success – up to 1973 – the standardized unemployment percentage in the OECD was under 3 per cent. By the first quarter of 1982 it had more than doubled, to 7.7 per cent. Finally, the average annual rate of increase in consumer prices in the OECD area, over the whole period from 1950 to 1973, was also under 3 per cent. Again, by the first quarter of 1982 – inspite of the doubling of unemployment – consumer prices were still 9 per cent higher than a year earlier.

How far is it possible to invoke changes in military expenditure as an explanation for this worsening in economic performance? Quite briefly, it is not. For military expenditure to be used as an explanation of the economic developments, it would have to show a sharp acceleration in its upward trend around the beginning of the 1970s. Nothing like this appears in the figures. In the long period of economic success – from 1959-62 to 1971-73 – military spending in the OECD area was rising, in real terms, at an annual rate of around 2.2 per cent. In the nine years of relative economic failure – comparing 1971-73 with 1979-81 – the average annual rise was only 0.7 per cent a year. Further, as a share of the national product of the OECD countries, military spending was lower in the period of economic failure. The figures, however manipulated, simply do not support any hypothesis of a connection between military expenditure and the worsening economic performance of the Western industrial world.

For the group of socialist countries the absence of firm figures for military expenditure makes it very difficult to examine the hypothesis of a connection between the trend of military spending and a slowing-down in the rate of economic growth. Estimates of the rate of growth of the USSR's national product also show a decline, from 6 per cent in the 1950s to 3 per cent in the period 1976-9. Western estimates of Soviet military spending tend to put it on a steadily rising trend with little variation – though the actual figure for that steadily rising trend varies a good deal (see pp. 57-60). There is just insufficient evidence to link the behaviour of the military sector with the slowing-down of the Soviet economy's growth rate. However, whereas with the Western industrial world the figures strongly suggest that such a hypothesis is false, with socialist countries the verdict has to be more agnostic.

So far, the question has simply been about the connection, or absence of connection, between the movements in military spending and the general worsening of the industrial world's economic performance in the last decade. The argument is that, certainly in the Western industrial world, it was not military expenditure which caused that deterioration, since the trend in that expenditure was decelerating rather than accelerating at the time. That is, of course, not an argument that there can never be a connection between military spending and economic performance; the next sections consider the circumstances under which this might be so.

Theoretical Problems

It is, in general, not usual to advance complex hypotheses about the macro-economic consequences of investment in particular industries (or groups of industries), or of separate sectors of government expenditure. To take an industrial example, we do not in general ask whether the steel industry has caused inflation or unemployment. To take a government expenditure example, we do not ask that question separately about health expenditure or educational expenditure. (It is, of course, a question which is frequently asked about government expenditure in general.) There is, therefore, a certain initial oddity about the amount of discussion of the connection between military expenditure and inflation and unemployment.

The theoretical problems of any analysis of this question arise, of course, because there is no consensus about the determinants of either inflation or unemployment. This is not the place for an extended discussion of Keynesianism or monetarism: but there is no way of avoiding the problem that Keynesians and monetarists would analyze the consequences of military expenditure on inflation and employment in rather different ways. Many modern-day Keynesians link the problem of inflation in the Western industrial world with the wage-bargaining process, and with the way in which annual increases in money incomes are settled. The monetarist approach looks rather at the figures for the money supply, and for the consequences for the money supply of budgetary deficits.

This section considers in turn: (a) military expenditure as a cause of excess demand; (b) bottleneck inflation; (c) inflation arising simply from budget deficits; (d) military expenditure and employment; (e) structural consequences of high military spending in industrial countries; (f) structural consequences of high military spending in developing countries; (g) the problems of conversion.

Military Expenditure as a Cause of Excess Demand

Both Keynesians and monetarists would agree that excess pressure of demand can bring about inflation – though they might disagree about the extent to which it was excess pressure in the goods market or the labour market. A very big rise in military expenditure can, of course, bring about excess demand in the economy. It is war which is the prime example of such a phenomenon. In wartime, the inflationary effects of shortages of goods and labour are normally reduced by price and wage controls, and the effects of inflation are seen in the movement of black market prices. However, there are examples other than those of the two world wars of excess demand resulting from sharp upswings in military expenditure. The Korean War is one such instance – with a particularly strong effect from stockpiling on the prices of commodities and raw materials. The rapid rate of inflation in Israel can be considered, at least partly, as a wartime phenomenon (see pp. 289-90). So also can the acceleration in the rate of price increase in the United States at the time of the Vietnam War. Between 1965 and 1969, unemployment in the United States fell from 4.5 to 3.5 per cent, and the rate of increase in consumer prices moved up from 2.0 per cent to 5.5 per cent a year.

It is quite clear, therefore, that military expenditure can cause excess demand and lead to inflation in that way. However, given the high figure of unemployment in recent years in the Western industrial world, it is not easy to label the inflation of the late 1970s as an excess-demand inflation.

Military Expenditure and 'Bottleneck' Inflation

Even when unemployment is relatively high, and the general pressure of demand in the economy is relatively low, a sharp upswing in military spending – particularly if it is concentrated on procurement – can cause some 'bottleneck' inflation. This can arise from specific shortages of materials of skilled personnel needed for the new weapons programmes. It has been argued that the upswing in military procurement in the United States in 1980 and thereafter would bring about an inflation of this kind, for in the previous period when weapons procurement in the United States stagnated, many subcontractors who had previously been largely engaged in military production turned to civil production (see p. 34). When military orders increased, in the second half of 1980, bottlenecks began to appear, and delivery times for such items as aluminium forgings and aircraft landing gear lengthened. However, with the deepening recession in the civil economy which developed during 1981 and the first half of 1982, these bottlenecks became less noticeable.

Developing countries would seem to be particularly prone to this problem, since so many of them lack a solid manpower and material resource base for heavy industrial production of any kind. Chapter 10 has estimated a 'potential arms production capacity' for 34 developing countries and has concluded that a number of them (notably Israel, Pakistan and the Philippines) 'have apparently overburdened their industrial and manpower bases with ambitious arms-production programs' (p. 325). (Israel's industrial base is presumably better developed than that of most Third World countries but its level of arms production has been extremely high.) In addition, a number of developing countries have ambitious plans for investment in arms production which, if implemented, are likely to absorb more manpower and other resources than their economies can reasonably be expected to provide (see p. 322).

Military Expenditure, Budget Deficits and Inflation

By definition, military expenditure – since it is government expenditure – is a component part of any central government budget deficit. Do budget deficits inevitably lead to inflation – and if not inevitably, under what circumstances does it happen? This is an area of controversy. It is linked with the Keynesian questions: is the economy automatically self-righting or is it not? Can the government, by demand management, maintain a reasonable degree of full employment? The question of budget deficits is linked with this issue, because budget deficits tend automatically to increase when unemployment rises and the economy moves into recession. So a rising budget deficit may not be a signal that the government should take anti-inflationary action; it may be a signal that it should take action to stimulate the economy.

In sum, if there is a central government budget deficit, then inevitably military expenditure is part of that deficit. However, if the economy is at less than full employment (however defined), it cannot be categorically stated that the deficit is necessarily a cause of inflation.

Military Expenditure, Unemployment and Employment

Can a government, by increasing military expenditure, increase employment or increase unemployment? This is, of course, the obverse of the previous question. The issue is debated, not in terms of military expenditure alone, but in terms of government expenditure in general. Nonetheless one of the more famous 'returns of full employment' was at the time of the Second World War; hence the working-class saying of the time: 'Your peace is our war, and your war is our peace'.

It is not easy to question the proposition that in wartime the government can create full employment. However, it could be argued that this needs an array of price and wage controls – to prevent the inflationary consequences – which are unacceptable and unworkable in peacetime. There is, therefore, at the moment no consensus either among economists or politicians about whether increased government expenditure (military or other) would set in motion forces which would reduce unemployment. It is clearly difficult, in a free trading world, for one nation to move on its own to re-establish full employment: it will tend to get into difficulties on its balance of payments and will either lose foreign reserves or find its exchange-rate falling. Further, there is also the danger that government reflationary action will act as a signal to trade unions to press for higher money-wage increases.

In the arguments conducted in 1981 and 1982 in the United States on the consequences of the rise in the military budget in that country, we find some strange alliances. Liberal, or Keynesian, economists tend to doubt whether budget deficits are necessarily very damaging, but they tend also to be aligned with those who dislike big military budgets. Conservative, or monetarist, economists have long been arguing for balanced budgets, but they in turn will generally be on the side of those who want military spending. We can also find the same ambivalence in government pronouncements in both the United States and the United Kingdom. Treasury ministers deny that increased government spending could reduce unemployment. At the same time, defence ministers offer figures for the numbers of new jobs created for each new weapon programme.

It is not for this book to resolve current economic controversies: there is no consensus about the efficacy of peacetime reflationary action, through government expenditure, as a way of reducing unemployment. However, events of the last two years – with unemployment continuing to increase in the Western industrial world – have certainly put in question the proposition that these economies would right themselves without such reflationary action. Those who advocate such action would agree, however, that government military expenditure has no advantage for that purpose over civil expenditure. It probably creates rather fewer jobs than the equivalent amount of civil spending.

The popular conception in all major arms producing countries certainly is that one result of less military expenditure, particularly in the form of arms procurement, is higher levels of unemployment. This notion is fostered by those who have political or economic reasons for wanting to see arms expenditure and arms production receive an important share of state resources. This volume, however, provides evidence that military expenditure is neither a particularly important source of employment in most countries nor the best means of increasing employment. In West Germany, the employment effects of military expenditure began to be used to justify particular procurement proposals during the economic downturn of the late 1970s. In pure economic terms, however, Chapter 4 argues that the German government would have been better off funding other programmes since, among other things, employment in the arms industry is a small proportion of total industrial employment and fewer jobs are created through expenditure on arms production than by investment in other sectors of the German economy such as construction, health and social services (see p. 134).

Similar calculations have been made for the employment effects of arms production in the United States. A study prepared for the US Arms Control and Disarmament Agency concluded that 'there is nothing unique about the capacity of military spending to generate jobs'. A detailed analysis prepared by the US Bureau of Labor Statistics in 1976 estimated the employment generated by a billion dollars of expenditure for 'national defense' for five types of construction and for state and local government purchases for education and health. It was estimated that state and local spending for health and education would generate a significantly higher number of jobs per unit of expenditure than would defence spending, while expenditures on construction would generate fewer jobs (Table 1).

What is true, however, is that different geographical regions, industrial sectors and occupational groups would be affected to different degrees by reductions in military expenditure. In the United States, for example, reduced arms procurement would affect scientists and engineers most severely. Nonetheless, the negative consequences for this group – and indeed for any group – of affected workers could be mitigated by careful planning and retraining programmes (see pp. 45-46). In Italy, although total defence-related employment is estimated to be under 2 per cent of total employment in manufacturing industries, particular regions of the country are more dependent on defence-related employment than others and would require special assistance were cuts in defence procurement to occur (see pp. 221, 238-43).

Structural Consequences of High Military Spending: Industrial Countries

It has been argued that the course of military spending cannot explain the general slowdown in economic growth in the Western industrial

Table 01.1: Comparison of the Employment Effects of Optimal Department of Defense Budgets and Offsetting Government Actions, by Industry, Fiscal Year 1973 (thousands of jobs)

Industry	Estimated employment generated by 1973 defence budget	Employment change generated by:			Employment change generated by:		
		Higher defence budget option	Offsetting tax increase and monetary restraint[a]	Offsetting reductions in other government expenditures[b]	Lower defence budget option	Offsetting tax cut and monetary case[a]	Offsetting increases in other government expenditures[b]
Agriculture	45	+ 1	- 14	- 64	- 3	+ 30	+135
Mining	25	+ 4	- 3	- 5	- 7	+ 6	+ 10
Construction	40	+ 5	- 25	- 64	- 9	+ 54	+135
Manufacturing	1,315	+166	-109	-112	-337	+230	+238
Services	595	+ 60	-285	-191	-114	+600	+402
Department of Defense (military and civilian)	3,400	+200	0	0	-450	0	0
Total	5,420	+436	-436	-436	-920	+920	+920

Note: a. Assumed to produce an across-the-board change in private spending, almost all in consumption.
b. Assumed to involve an across-the-board change in all non-defence government programmes.
Sources: US Arms Control and Disarmament Agency, *The Economic Impact of Reductions in Defense Spending*, Publication 64, Washington, DC: July 1972, 31 pp. The first column is adapted primarily from Richard P. Oliver, 'Employment Effects of Reduced Defence Spending', *Monthly Labor Review* 94 (December 1971): 3-11. All other estimates are adapted from Bernard Udis (ed.), *Adjustments of the US Economy to Reductions in Military Spending*, ACDA/E-156, Washington, DC: December 1970.

world. However, there is a wholly separate question: whether high military spending in individual countries puts them at a 'growth disadvantage' compared with countries where military spending is lower. It is very difficult to test this proposition. This is because we are dealing with observations of a relatively small number of countries and there is a very long list of possible determinants of those growth-rate differences. The most common argument here is to cite the comparison between West Germany and Japan on the one hand, and the United States and Britain on the other. Japan and West Germany have been high-growth countries through the postwar period. The United States and Britain have been low-growth countries. Japan devoted less than 1 per cent of its national product to military expenditure in 1981, and West Germany 3.5 per cent. The figures for Britain and the United States in 1981 were 5 per cent and 5.8 per cent respectively. These differences have been of long duration. For example, United States military expenditure as a percent of national product has been falling from much higher levels constantly since the mid-1950s.

Further, both Britain and the United States devote a much larger proportion of their total research and development expenditure to military uses than either Japan or West Germany. In the US, for example, US military-related R&D expenditures accounted for some 50 per cent of total US R&D outlays and about 85 per cent of government-financed R&D during the 1950s and 1960s. This situation changed somewhat during the 1970s and military-related R&D in the US currently accounts for about 25 per cent of total US R&D outlays and about 50 per cent of government-financed R&D (see p. 34). During the 1960s and 1970s, defence-related R&D expenditures in Britain as a proportion of government-financed R&D ranged between 64.8 per cent (1961) and 41.1 per cent (1969). At the beginning of the 1980s, defence-related R&D accounted for nearly 56 per cent of government-financed R&D in Britain (see p. 356). In West Germany in the early 1960s a rather high proportion of government-financed R&D was military related (nearly 70 per cent). Government funding for R&D was, however, a relatively small element in total R&D expenditures and military-related R&D in 1963 accounted for only just over 10 per cent of total West German expenditure on research and development. By 1978, only 17.3 per cent of government-financed R&D went to military purposes and for total West German R&D the proportion was 5.7 per cent (see p. 121).

However, there are of course very many other differences between

West Germany and Japan on the one hand and the United States and Britain on the other – too many differences for it to be possible to quantify the relative importance of any one of them. When other countries are brought into the comparison, the apparent relationship between high military expenditure and slow growth becomes less clear. France, for example, has been a high growth country. It has, however, devoted a considerable volume of resources to military research and development – some 25-30 per cent of public expenditure on R&D was defence-related in the 1970s. This is roughly half the British figure but 50-75 per cent more than the West German one.

There is a strong prima facie case for saying that if the resources devoted to military research and development were devoted to civil research and development, this would accelerate the economic growth-rate of the civil sector of the economy. Undoubtedly there is some spinoff from military work but it is hard to believe that this spinoff produces as much, in civil applications, as would be produced if the sums spent on military research were used to attack civil problems directly. (Comparisons of growth-rates of civil output are, of course, not the same as comparisons of growth-rates of national product: national product includes military output.) However, there is always the additional question whether governments which cut back on military research would expand civil research commensurately.

Structural Consequences of High Military Spending: Developing Countries

As in the study of industrial countries, we cannot expect much help from statistical exercises comparing military expenditure – either its level or its rate of growth – with figures for the economic growth of various Third World countries. First of all, the statistics for both military expenditure and real national product are highly suspect for many countries. Wherever a large part of production and consumption are outside the monetary economy as they are in many developing countries, national product figures are hazardous. Second, there is the problem of the direction of causation. Although national accounts statisticians may classify military expenditure as a regrettable necessity, many governments treat it rather as some kind of common good: the country grows richer, so it can afford a larger military sector. Relatively fast growth in the economy may therefore lead to relatively fast growth in military expenditure as well. There is no adequate statistical technique for disentangling the direction of causation.

In the absence of statistical demonstration, there are nonetheless certain points which can validly be made about the probable consequences of military spending. Where imports of weapons compete (as they often do) with imports of capital goods, it is reasonable to assume that development will be delayed. Further, the skill requirements of the maintenance of weapon systems must be a heavy drain on a limited pool of skilled manpower. With regard to expenditure on military industries in particular, Chapter 10 demonstrates that many of the expectations regarding arms production in Third World countries have by and large not been fulfilled. Dependence on imports has not been reduced in most cases. Production costs remain high and foreign exchange savings have for the most part not materialized to the anticipated extent. The number of jobs created by arms industries are relatively few in terms of the employment requirements of Third World countries. Self-sufficiency in arms production remains elusive.

It is often argued that there are positive effects of military expenditure. It is possible that in some cases conscription into military forces can be used as a unifying force in a country beset by ethnic or religious differences. Unfortunately, the opposite is more likely to occur as military recruitment is used to reinforce the dominant position of the major ethnic group(s) in divided societies. This phenomenon can be found in the post-independence militaries of Indonesia, Burma, Pakistan, Ghana, Uganda and Nigeria to name but a few examples. The military can also be used for civil construction or other domestic purposes and serves as an avenue of training which might not otherwise be available. However, this is really saying that military expenditure in such cases approximates more closely civil expenditure and should perhaps be classed as such.

Another argument which is often heard is that military-led industrialization can set developing countries on the road to economic modernity and development. Chapter 10, however, demonstrates that to move beyond the stage of assembly and simple production is extremely difficult for most Third World countries and that dependence on foreign sources of technology and capital increases, rather than decreases, as countries seek to produce more sophisticated items. Furthermore, to the extent that technical knowledge is transferred through the arms production process, it tends not to be the sort of technology that will be most useful to the development of Third World countries which really need labour-intensive processes to produce consumer goods and the ability to produce basic heavy industrial goods. Military technology tends to be capital-intensive and to require many specialized industrial products (see pp. 322-23). In general, it is only the largest Third World countries which can expect to benefit from military-led industrialization but even here, as the example of China demonstrates, there can be difficulties (see Chapter 8).

Conversion

The final economic question concerning military expenditure is about the economic problem which might be posed if there were at any time substantial disarmament. Here again, the common assumption is that there is a formidable economic problem involved. This is probably not the case, as most of the chapters in this volume make clear. Nonetheless, some countries would find economic adjustment to disarmament and the reduction of their defence-industrial sector more difficult than others. In the 1970s, for example, France sought to offset increased oil imports by expanding French exports, especially exports of armaments. In 1977, one-quarter of France's oil import bill was paid for by arms exports and in recent years French arms exports have been more dynamic than exports of civilian capital goods (see pp. 92-96). Just over 7 per cent of all French exports of machinery and transport goods in 1978 were defence-related. Great Britain is another country which has a relatively high dependence on arms exports: in 1978, nearly 7 per cent of all exports of machinery and transport goods were military in nature. For the USSR, available figures indicate that arms trade accounts for 12-15 per cent of all exports for the years 1974-8 (see pp. 67-69). This means that USSR arms exports, as a fraction of its machinery and transport exports, would be very much higher since major sectors of USSR exports are oil and other raw materials – and this figure very probably rose further after 1978. These figures tell one that in some cases arms trade may be an important source of export earnings for certain nations. This can have important implications for the overall balance of trade and hence the stability of a nation's currency but it tells one very little about the technical feasability of defence industrial conversion.

Adjustment would not in fact be impossible. The following are some of the relevant considerations which suggest that adjustment to the conversion of defence industries would on the contrary be quite manageable. First, there is a great deal of flux in modern industrial economies under normal conditions. Industries rise and decline; old processes become obsolete. In many industrial countries, labour turnover in the typical manufacturing plant is about 30 per cent a year – one worker in three leaves his or her existing job in the course of a year. Disarmament would add only a small additional fraction to the amount of industrial change occcurring in any case. In countries such as West Germany or Italy where the proportion of the total workforce engaged in military-related production is quite small, the disarmament-connected increment would be very small indeed.

Second, we have examples of very substantial reductions in military expenditure which have been accommodated with very little distress. Of course, the major transformations have been the changes at the end of major wars – and here the experience after the Second World War showed how military expenditure could fall from 40 per cent to 3 per cent of the national product within the space of two or three years, with no significant rise in unemployment. Peacetime examples are perhaps more helpful. In the United Kingdom, the government cut the military budget by one-third in real terms between the years 1953 and 1957. The simplest assessment of this substantial change is to say that virtually nobody noticed that it had taken place. In 1953, unemployment in Britain stood at 1.5 per cent; in 1957, at 1.4 per cent. Of course, at present the problem appears more formidable because the general level of unemployment is so much higher than it was then. But the problem is the problem of the general level of unemployment, not the problem of accommodating the resources displaced by reductions in military expenditure.

Third, military expenditure is wholly within the government domain. That means that the government could, if it wished, continue to pay full pay to ex-soldiers or displaced factory workers. The balance of the budget would be no worse than before. So, extremely generous transition payments could be afforded with no difficulty and could be combined with any retraining programmes necessitated by conversion from military to civilian production, thereby strengthening the economy in the long run.

Conclusion

The main economic point to make about military expenditure is a very simple one: it uses up resources which might alternatively be employed to provide consumer satisfactions – either in the provision of private or of collective goods and services. In particular, if the skill and ingenuity devoted to weapons development were diverted to civil objectives, the process of technological advance in the civil field could be appreciably accelerated.

However, except when there are major changes in trend in military expenditure, it is a mistake to consider that the military sector is responsible for such macro-economic developments as upswings in prices or in unemployment. In particular, the worsening economic performance in the industrial economies during the last decade cannot properly be attributed to changes in military spending.

1 THE UNITED STATES

Judith Reppy

Introduction

Billions of dollars are spent each year by the US on the development and production of weapons, making the defence industry clearly an important sector of the economy. Yet, even though military hardware has been produced in the United States since the earliest days of the republic, the emergence of a full-scale defence industry in the private sector of the economy did not occur until World War II. Throughout the nineteenth century the manufacture of guns and ships for the government was shared between government arsenals and private firms, but only during actual war fighting was there a significant level of arms production. Indeed, the dominant feature of arms procurement up to World War II was its episodic quality. Typically, the start of a war found the US military unprepared, and US industry engaged in the manufacture of civilian goods. During the fighting there would be a strenuous effort to increase production of war material followed by sharp cutbacks in defence spending and production when the war ended.

Thus, during World War I the United States sent two million men to Europe, but they had to fight mostly with French and British-made weapons: only a small number of US weapons were shipped to Europe before the war ended. The exception was the US output of ammunition and explosives, but in other categories the volume of production was very small. For example, only 145 field guns and sixteen tanks were shipped to France before the Armistice (although if the war had continued longer, US production would have become important).[1] When the war ended most of the firms that had converted from civilian to wartime production either went out of business or returned to production for the civilian market.

Nor was private industry heavily involved in developing new military technology. The main source for new weapons design in the nineteenth century was the arsenal system of the military departments, the naval shipyards and War Department armories. Military technology tended to mirror the general level of civilian technological development, with specialized applications for military use being developed in Army or

Navy research establishments that were co-located with the arsenals themselves. Earlier, the arsenals had been an important source of innovation for civilian industry, for example, in the development of machine tools, but their significance in this role faded as the civilian economy became industrialized, and low military budgets forced cutbacks in arsenal activity.[2]

Recognition that this system might not be adequate, even in peacetime, came with the increasing importance of new technology and new weapons, particularly the introduction of aircraft. By the 1920s and 1930s, the Army Air Force was contracting with private firms to develop a succession of aircraft prototypes. These designs advanced the state of the art, even though none of them was procured in large numbers, and this military support was important for the very survival of the industry during the Depression.[3] Among the firms that supplied aircraft to the military services during this period – for example, Boeing, Chance-Vought, Pratt & Whitney and Grumman – one can recognize the future major defence contractors of the postwar period. The groundwork was laid for major governmental funding and close relationships between private contractors and military services in developing new weapons, characteristics that became dominant features of the defence industry in the United States after 1945.

World War II was a watershed in the history of the US defence industry. From the early build-up in production for European needs to the high point of wartime production, spending for defence jumped from less than 2 per cent of gross national product (GNP) to nearly 40 per cent. Hundreds of firms were mobilized to produce for the war effort; new governmental bodies for organizing and controlling the production and distribution of war material were instituted; and new institutional arrangements for performing military-related research and development were devised.

At the end of the war there was a short period during which it seemed that the traditional pattern of cutbacks in spending to low peacetime levels would be repeated. But two major influences intervened to alter the historical pattern. Peace gave way to the Cold War, as the United States and the Soviet Union confronted each other in Europe and elsewhere. The Soviet threat to US interests was seemingly confirmed by the Berlin Blockade of 1948 and the first Russian atomic bomb tests in the following year. The start of the Korean War in 1950 unleashed a surge of defence spending by the United States that was only partially related to the war itself. Defence budgets jumped to over $40 billion, establishing a level for 'peacetime' military spending that in real terms (that is, corrected for inflation) has been nearly constant, except for the bulge caused by the Vietnam War, until the current upward trend beginning in 1976. This high and relatively steady level of spending formed the basis for the establishment of a 'permanent' defence industry.

The second major influence on the evolution of a defence industry in the private sector was the significant increase in the rate of technological change in weapons and associated systems during and after World War II. The new technologies associated with nuclear weapons, jet aircraft, missiles, radar, satellites and nuclear-powered submarines, to name just a few examples, did not have a well-established base in the existing arsenal system. Indeed, there was a general perception that the in-house laboratories of the military services were not well suited to advance technology in these rapidly changing areas because of their limited flexibility in terms of such things as salary levels and hiring practices under the civil service regulations. Technology and a political preference for private enterprise combined to shift resources from the in-house establishments to outside contractors.

The consequence of these influences, plus others such as interservice rivalries over new weapons and their associated missions, was the development of a large-scale peacetime defence establishment in the United States. Weapons procurement came to be characterized by heavy reliance on contracts with private firms for both research and development (R&D) and production of weapons, and by the emphasis on new technology. The arsenals and in-house laboratories that had earlier dominated the development and production of military hardware became only a relatively small part of the total economic activity supported by the Department of Defense (DOD); currently government arsenals still retain a significant role only in the production of munitions. Overall, 70 per cent of DOD's programme for research and development is now performed by private industry, and the figure for procurement approaches 100 per cent.

Table 1 shows the total DOD budget, the part allocated to weapons development and procurement for selected years from 1960 to 1981, and the ratio of defence spending to the federal budget and to gross national product. These data reveal the very large fraction of the budget that is spent on research and development. Compared to civilian industry, where R&D is, on the average, around 3 per cent of sales, the DOD has spent 10 per cent or more of its budget in the category Research, Development, Test and Evaluation (RDT&E) for the past twenty years. In some years RDT&E has been as much as 40 per cent of the total funds available for investment in new hardware (that is, RDT&E plus procurement funds), reflecting the emphasis that the military have put on new technology rather than purchasing weapons in large quantities.

Table 1.1: Department of Defense Budget: Total, and Selected Components, and as a Percentage of Federal Budget and Gross National Product, Selected Years

Financial Year	Total Obligational Authority ($ million)			Budget as percentage of	
	Total	Procurement	RDT&E[a]	Federal budget	GNP
1950	14,337	4,176	553	27.4	4.4
1955	33,790	8,917	2,621	51.3	9.2
1960	40,257	11,137	5,476	45.0	8.3
1965	49,561	14,112	6,433	38.7	7.0
1968	74,965	22,528	7,263	43.2	9.3
1970	75,517	19,161	7,399	39.2	8.0
1972	76,502	18,526	7,584	32.4	6.7
1975	86,176	17,320	8,632	26.0	5.8
1978	116,494	30,346	11,474	22.9	5.0
1980	139,343	35,792	13,517	22.6	5.1

Note: a. Research, development, testing and evaluation.
Source: US Department of Defense, OASD (Comptroller), 'National Defense Budget Estimates for FY 1981', Washington, DC: mimeo, pp. 69-71, 100.

It is worth noting that one component of US military spending lies largely outside the DOD budget. For historical reasons the development and production of nuclear warheads is under the aegis of the Department of Energy (DOE), which has inherited virtually intact the weapons programmes that were initiated by the Atomic Energy Commission and its successor agencies. Close liaison between DOE and DOD is necessary, since the warheads must conform to the delivery vehicles, for example, the missile frames and propulsion units, that are designed and produced under DOD contracts. In the case of naval nuclear reactors for submarines, the same man, Admiral Rickover, served as both the commanding Navy officer and, under various titles, as the chief of the energy agency's office for the naval reactor programme for many years, until his retirement in January 1982.

Because of the special safety problems posed by nuclear weapons, private industry plays a much smaller role in the production of nuclear warheads than it does in other weapons. The major nuclear weapons laboratories at Los Alamos and Livermore are responsible for warhead design, and production is carried out in a number of government-owned, company-operated facilities such as the Pantex plant at Amarillo, Texas. The DOE budget for defence-related atomic energy activities in the financial year 1981 which includes a number of programmes in addition to the weapons programme, was $3,658 million, of which $1,460.3 million was classified as Research, Development and Testing.[4]

Another element of military-related spending outside the defence budget is the space programme. The projects of the National Aeronautics and Space Agency (NASA) are obstensibly civilian, but they are closely related to the satellite programmes of the military services. A substantial fraction of the payloads scheduled for the space shuttle will be military packages, and military support for the shuttle was important in gaining political support for increased funding to cover cost overruns in the shuttle project. Thus, some portion of NASA's FY 1981 budget outlays of $5.3 billion should be added to the $156.6 billion for DOD and $3.6 billion of military-related programmes in DOE in estimating US spending for defence.

These dollar figures convey the size of the military programme of the United States. In 1980 defence spending was 23 per cent of the total federal budget, 5.2 per cent of gross national product and included 24 per cent of the resources allocated to research and development in the economy as a whole. There has been a secular decline in these percentage figures since the 1950s, but the recent increases in military spending have reversed the trend. Under the proposed Reagan plan, spending for defence is expected to rise to 37 per cent of the federal budget and 6 per cent of GNP by 1984.

Characteristics of the US Defence Industry

The defence industry is not defined in the usual way by its product; rather it consists of those firms from a number of different industries that sell to the Department of Defense. The nature of this governmental customer is the single most important determinant of the characteristics of the industry, because the DOD's monopsony power permits it to set directly the specifications of the products sold in the market and the rules under which business is transacted. This is not to argue that the private defence firms have no market power, but only that they exercise their power in the context of the rules set by the government.

Important features of the defence market that can be attributed to the role of government demand are:

1. the emphasis on high technology with performance of new weapons receiving more emphasis than cost;
2. the acceptance by the government of most of the risk associated with developing new technology through direct funding of R&D, the use of cost recovery type contracts and contract changes to accommodate changes in development programs, and other special contractual arrangements;
3. the need for public accountability, which leads to elaborate procedures for letting contracts, monitoring their progress and auditing final results;
4. the equally elaborate budget process within DOD and in Congress; and,
5. the close relationships and continual exchange of people and information between major defence contractors and the programme offices in the DOD.

These characteristics of the defence market have tended to favour a certain set of business firms, those that have learned to deal successfully in the specialized environment. Indeed one can trace a parallel evolution in the complexity of the defence marketplace and the adaptive mechanisms developed in response by the defence firms. Major defence firms and the defence-oriented divisions of larger diversified corporate conglomerates maintain specialized staffs to deal with complex DOD paperwork requirements for programme monitoring and financial reporting. Overhead categories and sources of finance are tailored to exploit the complexities of the acquisition regulations. For example, through the DOD's programme for independent research and development (IR&D), defence contractors can recover a share of the costs of their in-house R&D in overhead charges on their DOD contracts. Scientific and engineering manpower can be stockpiled to enhance the firm's ability to compete for new projects; since the firm is typically selling its technological capability rather than an existing product, the promise of future performance is important.

The military emphasis on new technology supports literally hundreds of programmes for new weapons developments. For the individual firm, a contract to develop a new weapon is the best insurance of winning the production contract later on, and there is intense rivalry for these development contracts. The incentives in the system tend to favour early optimism on technological risk and costs, continuous technical elaboration during the development cycle, and subsequent delays and cost increases over original estimates. More bluntly, these phenomena are known as buy-ins, gold-plating and cost overruns, and they are a persistent feature of new weapons development.

Marketing in the defence industry involves an extensive range of company contacts in DOD project offices, the Pentagon and Congress. The companies work closely with the military services, supplying much of the information needed to defend their programmes in Congress. The two-way flow of personnel between defence firms and the Pentagon facilitates these contacts; each year scores of retired military officers go to work for major defence contractors while employees of the defence firms are recruited to serve in the civilian offices of the Pentagon.[5] The interest of the firms and of their military customers coincide in promoting the weapons systems under development, since both stand to benefit from a continuation of on-going programmes. In addition, the defence firms, through various industry associations, lobby for a strong national defence and large defence budget, thus helping to create a favourable environment for their specific projects.

With successful adjustment to the peculiar demands of the defence market comes longevity. Turnover rates among the top 50 prime contractors for the Department of Defense have been low for at least twenty years.[6] New entry into the market is impeded by the need both for technological expertise to qualify as a bidder and for the ability to deal with government regulations. Exit from the market is also difficult, once a firm has adapted to the defence environment, because of different attitudes towards cost control in the civilian-oriented markets and the different marketing skills required in those markets. In recent years the major source of change in the ranked list of prime contractors has been the greater prominence of oil companies, owing to the post-1973 price increases for fuel, and regional shifts in the representation of construction and oil companies reflecting shifting patterns of US involvement abroad. Whereas in the 1960s the major oil suppliers in Southeast Asia were prominent on the list, along with a number of specialized charter airlines heavily involved in the US military effort in Vietnam such as Flying Tiger Line, in the 1980s Hellas Motor Oil Company and Saudi Maintenance Co., Ltd., are new entrants. Production of military hardware, however, has been performed by a stable set of prime contractors; such variability as has occurred has typically been through mergers, rather than new entry.

Thus the defence industry is not competitive in the economist's sense of the word; in particular, price is rarely the most important factor in determining which firm gets a contract. (Indeed, in FY 1980, 64 per cent of all prime contracts were let on a sole-source basis, with no competition at all.[7]) Technological rivalry and competition for contracts, where it occurs, are played out in an environment that compensates failure as well as success and seldom allows a major firm to lose a number of successive contracts.

To the public, the defence industry in the United States is virtually identified with the set of large aerospace firms like General Dynamics, McDonnell Douglas, United Technologies, Boeing, Lockheed and Grumman (as can be seen from Table 1.2). The top ten prime contractors received 30 per cent of all prime contracts in FY 1980. Eight of these companies are in the aerospace industry and primarily engaged in the manufacture of aircraft, missiles and aircraft engines, although several of the firms also have divisions operating in other defence-related fields. Lockheed, for example, has subsidiaries in electronics and shipbuilding, General Dynamics owns Electric Boat, and United Technologies owns Norden Systems, which specializes in communications, as well as Pratt and Whitney (jet engines) and Sikorsky (helicopters). As the technological focus has shifted from the aircraft itself to the avionics and missile systems mounted on it, the major defence firms have diversified into new technological areas, either through acquisition of other companies, as when Rockwell bought Collins Radio, or through developing new capabilities themselves. The move into space is a good example of this adaptive mechanism; nearly all the major NASA contractors are also major DOD contractors.

The defence contractors in Table 1.2 show widely varying levels of dependence on the defence market; many of these companies have followed a policy of diversifying into civilian markets (as well as new defence-related technologies) by acquiring subsidiaries. Raytheon, for example, is the parent firm of the publishers D.C. Heath and Co., and United Technologies has acquired Carrier Corporation, which makes air conditioners, and Otis Elevators. General Electric, of course, has always been predominantly a civilian-oriented firm. Diversification by acquisition avoids the difficulties of penetrating unfamiliar markets and leaves the defence-oriented divisions of the firm to continue to operate in their own special environment. It also provides a financial buffer against the instabilities of the defence market, which occur, for example, when a major programme is cancelled.

Table 1.2: Leading Department of Defense Prime Contractors, FY 1980

Company	Principal defence products	Rank on list of 100 prime contractors			DOD Prime Contract as percentage of Sales
		1980	1970	1958-60	1978[a]
General Dynamics	aircraft, missiles nuclear submarines	1	2	2	47.3
McDonnell Douglas	aircraft, missiles	2	5	8 Douglas 13 McDonnell	72.6
United Technologies	aircraft engines, helicopters	3	6	6	28.6
Boeing	AWACS, missiles	4	12	1	39.3
General Electric	nuclear reactors for submarines, aircraft engines	5	3	4	8.7
Lockheed	missiles, aircraft	6	1	3	46.6
Hughes Aircraft	missiles, radar	7	10	10	60.7
Raytheon	missiles, electronics	8	18	12	36.9
Tenneco (Newport News)	ships	9	27	28 (Newport News)	10.0
Grumman	aircraft	10	8	17	92.0

Note: a. Does not include subcontracts.

Sources: US Department of Defense, '100 Companies Receiving the Largest Dollar Volume of Prime Contract Awards', Washington, DC: Fiscal Years 1970 and 1980; Merton J. Peck and Frederic M. Scherer, *The Weapons Acquisition Process*, Boston: Harvard University Press, 1962, pp. 602-11; and US Department of Defense, 'Program Managers Newsletter', May-June 1978, p. 25.

Important Defence Industries

The aerospace (including missiles) and electronics industries are the most important industries in defence contracting in the US. Shipbuilding is a distant third. In financial year 1980, procurement awards for major hardware comprised 61 per cent of all DOD prime contract awards, and of this amount almost three-quarters went to the categories aircraft, missile and space systems, and electronics and communications. Shipbuilding accounted for another 9 per cent of the dollar value of major hardware contracts.[8] A similar pattern exists for RDT&E contract awards except that shipbuilding involves very little R&D, since most of the new technology on ships resides in the equipment, such as aircraft and communications equipment, for which the ship itself is merely the platform.[9]

This heavy concentration of DOD contracts in a few industries is the result of the importance given to advanced technology in performing military missions, and the absence, noted above, of an established arsenal tradition in aerospace and missiles, so that these systems have been almost wholly developed in private industry. Table 1.3 lists the top ten industries supplying the Department of Defense in 1978. The industries are defined at the four-digit level of the Standard Industrial Classification, a degree of refinement that reveals market dependencies which would tend to be lost at higher levels of aggregation. The data on shipments to DOD, however, give only partial coverage at the subcontract level and do not include shipments from DOD-owned facilities, so they are not a complete measure of the importance of the defence market. In addition, a set of industries supplying ordnance and ammunition (SIC 3482-84 and 3489) is heavily dependent on the DOD market, but its sales in peacetime are not large enough to merit inclusion in Table 1.3.

The data in Table 1.3 indicate that for the most part there is a dual dependence between industry and customer: in the industries that sell a lot to the DOD, the DOD is an important customer. The two exceptions on the list are petroleum refining and electronic computing equipment, both markets in which large DOD purchases are dwarfed by commercial sales. At a higher level of aggregation tank manufacturing would be swamped by the civilian automobile industry. For the main areas of defence spending in aerospace, missiles and ships, however, the DOD dominates the market.

For aerospace firms the defence market has meant a high level of spending on R&D, largely financed through defence RDT&E contracts

Table 1.3: Ten Industries with the Largest Value of Shipments to the US Department of Defense, 1978

Industry	SIC Code	Shipments to DOD ($ million)	DOD as percentage industry total
Aircraft	3721	7,879.2	46.6
Radio and TV communication equipment	3662	7,730.3	46.1
Guided missiles and space vehicles	3761	3,874.3	69.0
Aircraft engines and parts	3724	3,286.5	44.6
Shipbuilding and repair	3721	3,152.6	47.2
Aircraft equipment (n.e.c.)[a]	3728	2,125.4	36.6
Petroleum refining	2911	1,584.4	1.6
Tanks and tank components	3795	689.1	76.8
Electronic computing equipment	3573	635.2	3.1
Space propulsion units	3764	591.4	58.6

Note: a. n.e.c. – not elsewhere classified.
Source: US Bureau of the Census, *Shipments to Federal Government Agencies, 1978*, MA-175 (78)-1, Washington, DC: 1980.

and the IR&D programme, and a consequent rapid rate of technological obsolescence. Funds for R&D as a percentage of sales ran as high as 20 per cent in the 1960s, declining to 13 per cent in the 1970s, a figure that is still four times the average figure for all US industry.[10] Over three-quarters of this investment has been government-funded, and as indicated, even company funds for R&D contain a large component of government financing through the IR&D programme.[11]

Defence business has also brought access to government-furnished plant and equipment, and a tacit commitment by the government to maintain its major contractors in business.[12] With the decline in numbers of military aircraft procured each year from about 1,500 in 1961 to 734 (including foreign military sales) in 1979,[13] considerable excess capacity has appeared in the aircraft industry. The excess capacity has been sustained by swelling the overhead accounts of the remaining contracts: besides raising costs this also tends to discourage investment in new plant and equipment by the aircraft companies. It is not surprising, therefore, that foreign military sales and diversification into other product lines, both military and civilian, have become important strategies for the aerospace industry during the 1970s. The market for commercial aircraft is important to firms like Boeing, Lockheed and McDonnell Douglas but this market, too, is subject to cyclic demand, and these cycles have tended to mirror the fluctuations in military procurement of aircraft.[14] In any case, shifting of resources between the two sectors to offset market fluctuations is limited by the fact that, on the military side, much of the plant and equipment is owned by the government and is not available for production of civilian aircraft.

Shipbuilding, although equally dependent on the military market, has not experienced the rapid rate of technological change of the aerospace industry; it remains an industry dominated by traditional designs and building practices. Since the last Navy-built nuclear submarines were delivered in the early 1970s, the Navy has relied entirely on privately-owned yards for new construction, with the Navy yards that remain doing only repair and overhaul work. There are currently twelve private yards building ships for the Navy, and for specific types of ships the number is fewer. For example, Newport News is the only yard with the capacity to build large aircraft carriers, and only two yards – Electric Boat and Newport News – are currently qualified to build nuclear-powered submarines.

Government involvement in the shipbuilding market is not limited to the Navy programme. Almost all nations subsidize shipbuilding; in the US the government pays the difference between the cost of domestic production of a ship and the world price. Such subsidies are typically justified by reference to the strategic importance of shipbuilding, but they have created a world-wide over-capacity in shipbuilding, which has been strongly evident during the downturn in the demand for ships after 1973.

For the US shipbuilding industry, the naval programme could have served as a moderating influence, but instead, it has been marked by its own instability. A combination of budget constraints, disagreements over the role of large, nuclear-powered aircraft carriers, and, more broadly, the future role and missions of the Navy has led to repeated changes in the five-year shipbuilding plans. This instability has been mirrored by fluctuations in the shipyards themselves, with large swings in employment in individual yards, and a chronic loss of skilled manpower to civilian industries. Labour turnover rates run as high as 75 per cent per year, a state of affairs that has saddled the firms with high training costs and lowered labour productivity.[15] Moreover, except for Todd Shipyards, all of the private naval shipyards in the US have been acquired by diversified conglomerates such as Congoleum and Litton. The change in ownership has tended to disrupt the traditionally close ties between the shipyards and the Navy and has added to the problems caused by inflation and instability in long-range planning. So unsatisfactory have relations between the Navy and the industry become, that Secretary of the Navy, John Lehman, suggested in March 1981 that the Navy might seek to purchase some of its ships abroad.

Electronics and communications is the third major technological area in which the DOD spends large sums of money. Here, however, the DOD does not dominate the market except in specialized product lines; for the electrical and electronics industry as a whole shipments to DOD were only 12 per cent of total shipments in 1978.[16] In particular, the dramatic advances in technology in semiconductors and integrated circuits, although originally stimulated by military needs, have found ever-expanding commercial applications, so that in recent years the military share of the semiconductor market has been less than 10 per cent.[17] Ironically, the very vigour of the civilian market has placed the DOD at a disadvantage because the military market is too small to be of interest to the leading semiconductor firms. The DOD has responded by cultivating a small number of suppliers (including the capability of some of the large defence contractors like Hughes Aircraft) which are prepared to manufacture particular semiconductor components to DOD specifications. A current DOD initiative to fund technological development of very high speed integrated circuits (VHSIC) attracted the participation of these defence-oriented firms, but not of Intel and other leading firms in the civilian semiconductor industry.* It appears that the special requirements for defence contracting, such as adherence to military specifications for ruggedness and reliability, custom designs, and government paperwork requirements, discourage entry into the defence market even in an industry characterized by rapid growth and new firms.

Hidden behind the major aerospace firms and other prime contractors for DOD is a vast array of subcontractors and lower tier suppliers, spread over many industries. Roughly half the value of prime contracts is let out in subcontracts to other firms, and this amount is shared roughly half-and-half between divisions of other prime contractors and smaller firms. Some of these smaller firms sell only a small part of their output in the defence market; for them the prime contractor is just another customer. Others are specialized suppliers of parts for military systems and are extremely vulnerable to cycles in the defence market or to cancellation of major programmes. These smaller firms are

*The leading firms in the second phase of the VHISC program are Honeywell, IBM, Texas Instruments, Hughes Aircraft, TRW and Westinghouse, all firms in the top 50 of the 100 prime contractors list.

unable to shift the financial and technological risks that they face to the government, as prime contractors do, because they are typically locked into fixed price contracts and do not have government-furnished plant and equipment.

There is evidence that these supplier firms have been leaving the defence market as a consequence of spending cuts after Vietnam, leading to less competition and, in some cases, to shortages of critical subcomponents. Jacques Gansler, among others, has argued that the weakness of the lower tiers of the industry would prevent a surge in defence production if it were needed, despite the excess capacity present at the level of the prime contractors.[18] It is difficult, however, to assess the overall situation because of the absence of any comprehensive data base for subcontracting. Given changing technology and product mix and the static membership of the group of top prime contractors, contractor turnover in the lower tiers is a source of needed flexibility; that is, the observed exit of firms from the ranks of subcontractors could be considered a form of rationalization that does not occur at the prime contractor level because of the special relationships between government and firms. The increased defence spending proposed by the Reagan administration is likely to provide a test of the surge capacity of the industry, since it will require expansion of existing firms or new entry in the lower tiers of the defence market. The degree to which the spending increases are accommodated without serious bottlenecks developing will be a test of the responsiveness of the industrial base.

The High Technology Product

The resources allocated to military research and development currently equal about one-fourth of total US spending for R&D, and in past years the fraction was much higher. Throughout the 1950s and 1960s military-related R&D, including space, regularly accounted for 50 per cent of total national spending for R&D and 85 per cent or more of the federal government's share.[19] The output from this large scale commitment has been rapid advances in a vast array of technical fields, resulting in new weapons of great sophistication. The fascination with new technology and the emphasis placed on qualitative superiority in weapons has had important consequences for US force structure and military posture.

It has also had effects on the civilian technology base. Development of certain technologies has been accelerated by the military R&D programmes – the most salient examples are computers, light-water nuclear reactors, jet engines and integrated circuits. But military technology has become more and more specialized with the passage of time, diverging from the common technology base shared with civilian users. Most of the funds for military R&D are spent in the engineering development phase of the development cycles of specific weapons, and this work is not likely to have direct civilian applications. Thus the impact of military R&D programmes on the economy as a whole is probably more a consequence of the effect of military demands on the overall mix of R&D than a result of identifiable spillover from military to civilian technology. Emphasis on some areas of technology by the military is by implication de-emphasis of other areas with more purely civilian applications. In particular, the position of the aerospace industry as a leader in the performance of R&D is largely the result of the military investment in new aircraft and missiles. While the country has reaped the advantages of new technology for air transportation, the railroads have become virtually obsolete. This is an illustration of the opportunity cost to the economy of the bias towards military programmes in the national programme of R&D.

Patterns of Defence Contracting and Employment

DOD contracts are distributed geographically in a pattern that reflects the regional concentration of the favoured industries (see Table 1.4). The data for 1977 (the most recent year of the Census of Manufactures) show that the pattern of defence contracts does not follow the pattern of manufacturing activity in general. The big differences are the greater importance of defence contracting relative to general manufacturing in the New England and Pacific regions, and the opposite situation in the East North Central region (which covers the industrial states of the Midwest).

California, home of many large aerospace and electronics firms, leads the nation in military prime contracts, with 21 per cent of the total value in FY 1979. The next-ranked state, New York, had 9 per cent, and Texas and Connecticut each had around 7 per cent of the total.[20] The Pacific region, with 25 per cent of the total prime contract awards in FY 1979, had the largest share of awards in three of seven major weapons categories – missiles and space systems, weapons, and electronics and communications – and was second or third in three

Table 1.4: Regional Distribution of Manufacturing Employment, Value-added and Military Prime Contract Awards, 1977

Region[a]	Percentage Distribution by Region		
	Calendar Year 1977		Fiscal Year 1977
	Manuf. Employees	Manuf. Value-added	Prime Contract Awards[b]
New England	7.1	6.1	11.2
Middle Atlantic	18.5	17.6	15.7
East North Central	25.4	27.4	9.3
West North Central	6.6	7.0	8.3
South Atlantic	14.4	12.4	13.4
East South Central	6.8	6.2	4.1
West South Central	7.4	8.9	7.8
Mountain	2.4	2.3	3.4
Pacific	11.5	12.2	26.9
Total US	19,597,000	$585,096 million	$45,540 million

Note a. The States included in the various US Census regions are: New England – Maine, New Hampshire, Vermont, Massachusetts, Rhode Island, and Connecticut; Middle Atlantic – New York, New Jersey, Pennsylvania; East North Central – Ohio, Indiana, Illinois, Michigan, Wisconsin; West North Central – Minnesota, Iowa, Missouri, North Dakota, South Dakota, Nebraska, Kansas, South Atlantic – Delaware, Maryland, District of Columbia, Virginia, West Virginia, North Carolina, South Carolina, Georgia, Florida; East South Central – Kentucky, Tennessee, Alabama, Mississippi; West South Central – Arkansas, Louisiana, Oklahoma, Texas; Mountain – Montana, Idaho, Wyoming, Colorado, New Mexico, Arizona, Utah, Nevada; and, Pacific – Washington, Oregon, California, Alaska, Hawaii.
b. Total Prime Contracts less work outside the US, actions of $10,000 or less and work not assigned to a state.
Source: *Statistical Abstract of the United States, 1980*; US Department of Defense, Directorate for Information, *Military Prime Contract Awards by Region and State, FY 1977-79*, Washington, DC: n.d.

other categories.[21]

The underlying reasons for the regional clusters of defence firms, and hence, of prime contracts, range from the advantages conferred by the good climate of the sunbelt states, both in attracting workers and reducing production costs, to the hold of traditional locations in older technologies. The fast-growing aerospace and electronics industries have concentrated in the Western and Southern states, whereas tank production is largely co-located with its parent industry, automobiles, in Michigan, and shipbuilding, naturally enough, is distributed among the coastal states.

Defence employment echoes the pattern of regional and industrial concentrations of defence spending. For the economy as a whole, defence-related employment was 5.058 million in 1980, or 4.7 per cent of the labour force and 5.1 per cent of total employment. Of this number, DOD civil service employees and active-duty military were 3.036 million and employment in defence-related industry was estimated at 2.022 million, an estimate that includes indirect employment in supplier industries (subcontractors and lower tiers), but does not include the multiplier effects on the economy of DOD purchases of goods and services. Although defence-related employment in industry is not a large fraction of total employment in the United States, it is, of course, significant in certain industries and locations. To a first approximation these are the industries and regions showing high concentration of DOD spending in Tables 1.3 and 1.4, namely, the aerospace and shipbuilding industries and New England and Pacific regions.

Defence-related employment also falls unevenly across occupations. Table 1.5 gives data for 1970 on the occupational structure of defence-generated employment compared to the economy as a whole. The biggest difference in the occupational profiles is the low level of service workers in the defence economy. These workers lie largely outside the defence establishment in the surrounding economy. Blue collar workers, which include the machinists and sheetmetal workers needed on the production lines and the civilian mechanics and repairmen hired

Table 1.5: Occupational Profiles of Total US Employment and Defence-generated Employment, FY 1970

Occupational Group	Percentage Distribution of	
	Total US Employment	Defence-generated Employment in Private Industry
White collar workers	46.8	40.2
Professional & technical	13.6	14.4
Managers, officials & proprietors	10.0	7.1
Clerical workers	17.2	16.0
Sales workers	6.0	2.8
Blue collar workers	36.6	54.0
Craftsmen & foremen	13.0	19.6
Operatives	18.7	30.5
Nonfarm labourers	4.9	3.9
Service workers	12.4	3.6
Farm workers	4.2	2.2

Source: Richard Dempsey and Douglas Schmude, 'Occupational Impact of Defense Expenditures', *Monthly Labour Review* no. 94 (December 1971): 13.

directly by DOD to service equipment, are more heavily represented in defence employment than in the economy as a whole.

In some occupations, notably engineers and scientists, defence work is an important source of employment. Dempsey and Schmude estimate that in 1970 one out of five engineers was employed in defence-related work and nearly one out of four physicists.[22] These numbers are sensitive to cyclic influences; for example, the recession in the aerospace industry in the early 1970s caused a drop in employment of scientists and engineers doing R&D in the aerospace industry from 92,200 in 1970 to 70,800 in 1972.[23] Roughly, it can be estimated that employment of scientists and engineers in defence work will parallel the proportion of national R&D funds spent on defence, so that the current fraction of all scientists and engineers employed in defence work would be close to one-fourth. Other occupations for which defence work is a significant source of employment are aircraft mechanics – 40 per cent of these jobs were defence-related in 1970 – and machinists.

Capital Investment and Profit Rates

Capital investment in the defence industry is important for its contribution to productivity and as a component of the opportunity costs of defence spending. Unfortunately, it is not possible to make an accurate estimate of the capital employed in defence production for reasons that go beyond the well-known difficulties of measuring capital. The large diversified defence firms do not report defence-related capital assets separately, so there is no basis for estimating private capital investment in the defence industry. There are data for the government-owned plant and equipment supplied to defence contractors, but these data reflect acquisition costs, with no allowance for depreciation. Since most of the plant and much of the equipment is quite old, dating to the Korean War or earlier, its depreciated value is small; conversely, its replacement value in inflated dollars is much larger than the reported costs.

Table 1.6 gives some information on plant and equipment furnished to contractors by the government. Because of the limitations noted above, the table is more interesting for what it shows of the composition and distribution of the plant and equipment than for the dollar values. Plant buildings, included in the second column, include such facilities as the Air Force plant at Fort Worth that is used by General

Dynamics and Lockheed's 'Skunk Works' in California. In the equipment category, government-owned machine tools are especially important in the munitions industry, both in government-owned facilities and in contractor facilities. The total acquisition cost of industrial plant equipment owned by the DOD was $3.6 billion in 1979, of which approximately 45 per cent was in the custody of private contractors.[24] The remainder was in use in government arsenals or being held inactive in the Industrial Reserve. Equipment in the latter category is of questionable value; 75 per cent is more than twenty years old.[25]

Despite the lack of complete data, some generalizations are possible. First, it is likely that the defence-oriented firms invest less in plant and equipment than other comparable firms. The contractual forms used by the DOD discourage capital investment by covering all allowable costs on cost-plus contracts and by giving only a low weight to a firm's investment in new plant and equipment in negotiated profit rates. Contractors are able to pass through higher manufacturing costs attributable to ageing or obsolescent capital equipment and are not rewarded for making cost-saving investment. The availability of government-owned plant and equipment is a further disincentive to investment.

In a 1976 study of defence contracting, *Profit '76,* the DOD estimated that the amount invested per dollar of sales in 168 government profit centres (divisions of firms selling mainly to DOD) was only 35¢, compared to 63¢ per dollar of sales for a Federal Trade Commission (FTC) sample of approximately 5,000 durable goods producers in the commercial market. The greatest investment per dollar of sales by the DOD contractors was in shipbuilding, but even there it was substantially below that of commercial shipyards.[26] These investment data, however, include operating capital, and defence firms need less working capital because they collect regular progress payments over the life of their DOD contracts. But even when this asymmetry is allowed for, and only investment in facilities is measured, the disparity between government profit centres and commercial durable goods producers in rates of investment remains.[27]

These figures help explain the profit performance of defence contractors in the United States. The defence industry often complains of its low negotiated profit rates as a percentage of sales, which in the early 1970s averaged only 4.7 per cent. A more significant measure, however, is profits defined as the rate of return on investment, and this rate for the defence industry is comparable to that earned by firms in civilian markets. Table 1.7 gives the summary results of the DOD's *Profit '76* study. In addition to the 168 government profit centres (in

Table 1.6: Distribution by Region of Department of Defense Real and Personal Property in Custody of Contractors[a]

Region	Land		Other Real Property		Industrial Plant Equipment		Other Plant Equipment	
	A[b]	B[c]	A	B	A	B	A	B
New England	9	2.0	13	3.3	69	11.2	73	3.7
Mid Atlantic	6	0.3	13	2.0	128	10.5	112	6.2
East North Central	9	30.9	18	19.1	107	20.6	93	7.3
West North Central	12	27.4	12	18.4	35	15.0	30	5.4
South Atlantic	7	7.7	11	9.1	67	9.9	89	17.5
East South Central	4	15.2	6	8.8	22	3.8	22	5.5
West South Central	7	10.0	11	11.5	36	12.0	39	4.6
Mountain	7	0.6	14	15.0	44	4.1	51	27.3
Pacific	9	5.5	24	13.2	134	13.8	163	22.2
Total US	70	$167.3 million	122	$3,274.0 million	642	$1,579.2 million	672	$2,781.0 million

Note: a. Recorded cost of property under control of prime contractors. b. A — number of contractors. c. B — per cent recorded cost (i.e., acquisition cost).
Source: US Department of Defense, Directorate for Information Operations, 'Real and Personal Property of the Department of Defense as of 30 September 1979', Washington, DC: n.d.

64 firms), data were collected from a sample of commercial profit centres of these firms (column 2). The sample showed remarkably high profit rates, but was considered too limited to use as a primary source of commercial profitability. Instead, the FTC sample of 5,000 manufacturers of durable goods was used.

Table 1.7: Profit Before Taxes/Total Assets[a] by Product Group, 5 Year Average 1970-74 (percentages)

	Profit Earned by 64 Defence Contractors		Profit Earned by Producers of Durable Goods[b]
	Government Profit centres	Commercial Profit centres	
Aircraft	11.2	9.0	6.9
Electronics	15.3	9.5	10.0
Missiles	20.0	27.8	6.9
Ships	5.8	4.2	–
Other	11.5	18.9	–
Total	13.5	17.6	10.7

Note: a. Less progress and advance payments. b. FTC data.
Source: US Department of Defense, *Profit '76 Summary Report*, Washington, DC.: December 1976, pp. II-28.

The return on investment for the defence profit centres was 13.5 per cent, compared to 10.7 per cent for the durable goods producers. The discrepancy in profit rates between sectors of the defence industry is closely related to their differing levels of capital turnover. Shipbuilding with a large capital investment per dollar of sales as well as a low rate of return on sales, has the lowest rate of return on investment, whereas missile production is quite profitable. Although it is often argued that the relatively good return on investment in the defence industry is a result of excluding government-supplied capital from the asset base, the *Profit '76* study found otherwise. Including government-supplied capital in the asset base had virtually no effect on the rate of return when it was properly depreciated in accordance with standard accounting procedures because of the age of the plant and equipment. Instead of a 13.5 per cent return, the figure dropped to 13.0 per cent.[28]

From other studies it appears that the return on investment of subcontractors is only half that earned by the major defence contractors, largely because subcontractors do not have easy access to the arrangements that make it possible for the prime contractors to operate with low capital investment and face a more competitive market.[29] This low level of profits, coupled with the higher risks borne by small subcontractors, provides a strong reason for the decline in numbers of subcontractors.

The implication of these findings for the defence industry is that the average return on investment should be adequate to attract resources to the industry, at least at the prime contractor level. But there is very little incentive for companies to undertake capital investment, since they cannot translate such investment into further profits. Contract prices are based on certified costs, with only a small allowance for investment in new production facilities. The negotiations between contractors and the DOD over costs and government-furnished equipment are more significant than profit levels in determining new investment in the industry.[30]

Foreign Military Sales

Weapons sales by the United States to other countries became an important element in the defence market during the 1970s. Totalling $1,390 million in FY 1971, foreign sales agreements reached $15,277 million by FY 1980, and deliveries in that year were $7,698 million. (Deliveries lag behind orders by several years in the case of large complex weapons systems and construction projects.)[31] This large increase in foreign military sales was a result of several factors, including a policy of shifting weapons transfers from the category of military assistance to sales and the effect of inflation on the cost of items being exported. The biggest change, however, was the large increase in sales to the newly rich, oil-producing countries in the Middle East. Sales to Iran and Saudi Arabia, the two largest customers, went from a combined total of approximately $370 million in 1971 to nearly $9 billion in 1976. The cumulative total for these two countries from 1971 to 1980 was $126 billion.[32] (These figures include construction and service contracts executed as foreign military sales.) The fall of the Shah of Iran at the end of 1978 resulted in the cancellation of most of the outstanding military sales agreements with that country, so that actual deliveries for 1971-80 will be somewhat smaller than the figure for sales, but the volume of weapons transferred is, nevertheless, enormous.

The US has a strong export surplus in its weapons trade, exporting $6,700 million FY 1978, compared to imports of $120 million. These exports were 4.7 per cent of total US exports; this figure may be compared to the value of agricultural exports, $18,311 million, and exports of all manufactured goods (including major weapons), $81,912 million.[33]

Foreign military sales raise a number of substantial foreign policy issues that lie beyond the scope of this chapter. The policies of the Carter Administration tended to discourage sales to some countries, although total sales continued to rise after 1977, owing largely to increased exports to Europe and to Saudi Arabia. The Reagan Administration has announced changes in US policy towards foreign military sales. They are now to be regarded as 'a vital and constructive instrument of American foreign policy',[34] so that to the extent that official policy has inhibited sales, fresh increases may now be expected.

Whatever one may think of the new policy for foreign military sales on political grounds, from the perspective of the defence industry it appears highly desirable. Foreign sales yield higher profits than sales to DOD and are an outlet for excess capacity. The surge in US foreign military sales in the early 1970s coincided with the reduction in defence spending after the US withdrawal from Vietnam, and thus were a partial offset to those cutbacks. In FY 1976 foreign military sales agreements were $14,674 million, compared to $41,976 million in DOD prime contracts. This was the high point in the ratio of the two figures, since after 1976 foreign sales were somewhat constrained, and the US defence budget for new weapons began to rise. In FY 1980, foreign sales agreements were 20 per cent of prime contract awards.[35]

Table 1.8 shows the breakdown by weapons category of cumulative arms transfers from 1950 to 1980. These DOD data include construction and support services such as training, technical assistance, and supply operations that have been contracted for through the foreign military sales programme. Some of these programmes might be considered to be quasi-civilian (for example, the construction of roads and airfields), but they typically involve contracted services from the major defence contractors, most of which have set up special subsidiaries to supply these services abroad. Among the weapons categories, aircraft dominate. Missiles and vehicles are next in importance for foreign military sales, and ammunition for military assistance programmes.

In recent years there has been a tendency to supply the most advanced weapons to foreign customers, rather than the older equipment that was distributed through the military assistance programmes. An economic argument in favour of this policy is that it reduces the overhead costs charged to current US military contracts by increasing

Table 1.8: US Foreign Military Sales and Military Assistance by Weapons Category, 1950-80 ($ million)

	Foreign Military Sales		Military Assistance	
	Ordered	Delivered	Programmed	Delivered
Aircraft	32,003	19,170	9,579	9,430
Ships	3,904	1,087	2,301	2,286
Vehicles & weapons	9,672	6,010	9,848	9,757
Ammunition	4,574	3,243	11,241	11,212
Missiles	12,335	5,891	1,541	1,394
Communication equipment	2,589	1,482	2,813	2,738
Other equipment & supplies	2,983	2,188	5,281	5,234
Construction	18,605	5,843	1,027	1,027
Support services	20,991	10,608	12,347	12,291

Source: US Department of Defense, Security Assistance Agency, *Foreign Military Sales and Military Assistance Facts*, Washington, DC: December 1980, p. 69.

total production. This benefit is likely to be of only marginal importance, however, relative to the political factors entering into decisions to sell modern weapons to other nations.

Industry interest in foreign military sales as a way of expanding its markets is strong. For certain contractors, these sales have become their major source of business; Northrop, for example, which manufactures the F-5 export fighter and has sold 1,731 to other countries, averaged 50 per cent of its business in foreign sales from 1976-80.[36] Other companies with a high dependence on foreign sales are Raytheon and Grumman.

The foreign market is important, but it does not greatly alter the environment in which the defence firms operate because most of the foreign sales are routed through DOD and are covered by government procurement regulations. The area of behaviour which is most affected is marketing, since strategies for selling to foreign governments are different than for DOD. The widespread practice of paying bribes or kickbacks caused a number of scandals in the 1970s and led to reform in reporting requirements for US companies. Probably more significant for the industry is the need to negotiate elaborate production-sharing arrangements in order to clinch sales to other industrialized nations, especially those in NATO, which are anxious to protect their own industrial base and employment. Institutionalization of these arrangements with an increase in cross-national teaming for weapons are areas where future foreign sales may have a profound effect on the US defence industry.

The Defence Industry and Prospects for Conversion

The US defence industry as described above is a large industry, but not dominant in the context of the whole economy. Spending is high, but it is less than 6 per cent of GNP. The major defence contractors are large firms, but not the largest ones in the United States. The very large US companies such as General Motors, American Telephone and Telegraph and the major oil firms do appear on the list of 100 top DOD prime contractors, but their defence business is only a small fraction of their total sales. Furthermore, the United States enjoys the advantage of a fully articulated economy, in the sense of a complete set of industries connected by relatively free markets, and thus it possesses an inherent degree of flexibility that may be lacking in centrally controlled economies or less-developed countries. *A priori* it seems that changes in government spending for new weapons, either increases or decreases, could be accommodated by the economy with little difficulty.

What is true for the whole, however, is not necessarily true in detail. The defence industry is really a collection of segments of distinct industries and the effects of defence spending differ by sector. Furthermore, the effect of incremented changes in defence spending may differ from the average relationships that can be inferred from the data in Tables 1.1-8 and this difference between average and marginal effects will, itself, vary between sectors of the defence industry. Thus, increased spending for communications equipment would increase demand for the output of firms in the electronics industry and the electronics divisions of the aerospace firms. The electronics industry is not heavily dependent on the DOD market and a change in DOD demand would be a small perturbation of its total sales. But if skilled engineers are in short supply to the industry (and they are), and the DOD demand is for products requiring custom-designed chips (and it is), then it may be more difficult for the electronics industry to expand output of defence-related items than for an industry already heavily committed to the DOD market but with excess capacity.

Turning to the problems of economic conversion, the economic difficulties of adjusting to reduced military spending are often thought to constitute a powerful political argument against any move towards disarmament. Decreases in levels of defence spending will have considerable impact for particular industries, regions and occupations precisely because changes are rarely made evenly across the board, but rather are concentrated in specific programmes. Just as closing a military base may have a serious impact on its local economy, reductions in procurement programmes can affect strongly those companies that are involved in the specific programmes. The regional grouping of defence contractors, in turn, insures the uneven regional impact from changes in defence spending.

Similarly, the effects on employment will vary by occupation. The largest effects are likely to be found among the scientists and engineers, for whom defence jobs represent one quarter of total employment. Lloyd Dumas has argued that the long-run diversion of scientific resources into military-related research in the US has caused a deterioration in civilian technology, and that conversion of these resources (mostly scientists and engineers) from military to civilian-oriented work is inhibited by the degree of specialization that marks military R&D.[37] The straightforward identification of military-related employment with a burden on the civilian economy must be modified, however, by recognition of the role played by the defence-related demand for scientists and engineers in expanding the supply of scientifically-trained people. The availability of well-paying jobs in the defence industry has attracted people to engineering as a career, and federal programmes have helped to pay for their education. Thus, not all of the scientists and engineers employed in defence-related work can be regarded as having been taken away from the civilian economy.

The analysis is further complicated on the demand side. The level of alternative employment in civilian industry is conditioned by national practices in the use of engineering skills; for example, Japanese firms employ more engineers on the floor, that is, directly involved in the production process, than do US firms. Nevertheless, the US has 58.7 scientists and engineers per 10,000 labour force population and, even if this number is reduced by one-quarter to adjust for defence-related work, the figure for the US is exceeded only by that for Japan among the OECD nations.[38] It appears that a shift in resources from military to civilian-oriented R&D will require stimulating civilian demand as well as retraining specialized defence engineers to function in the civilian marketplace.

These considerations serve to strengthen Dumas's argument that conversion of defence resources to civilian use will require careful planning in order to minimize the cost to those industries and workers that are heavily involved in defence work. These potential costs are a barrier to disarmament policies to the extent that the narrow interests of the defence industry can be translated into effective political pressures for maintaining a large defence programme. Fears of lost jobs and income from cancelled projects are a potent political force in the localities affected. The pluralist character of the US society bestows power on coherent, well-organized groups, power that in the case of defence spending is not balanced by the diffused interests that are harmed by high levels of defence spending. The close ties between defence contractors and the Pentagon and Congress make the translation of economic interest into political decisions particularly easy. An additional factor is the general bias in the US economy towards growth, so that policies that would reduce rather than increase markets are suspect. Defence company executives are not likely to regard growth in other parts of the economy as compensation for negative growth rates in their own companies.

Nevertheless, this list of obstacles to conversion would not form a serious barrier to disarmament if political conditions were suitable. As argued above, the level of defence-related activity is small relative to the whole economy, and cutbacks in the scale likely to be caused by any arms control agreement could be easily absorbed through compensatory policies directed at the industries most affected. Unfortunately, political conditions have rarely been suitable in the post-World War II period. While economic considerations play a part in the opposition in the United States towards any policy of disarmament, the most serious obstacles lie elsewhere.

Notes

1. M.G. Theodore Antonelli, 'American Mobilization During World War I', *Defense Management Journal* 12 (July 1976): 40-7.
2. Merritt Roe Smith, 'Military Arsenals and Industry before World War I' in *War, Business, and American Society*, ed., Benjamin F. Cooling, Port Washington, NY: Kennikat Press, 1977, p. 37.
3. Elsbeth E. Freudenthal, 'The Aviation Business in the 1930s' in *The History of the American Aircraft Industry*, ed., G.R. Simonson, Cambridge, Mass.: The MIT Press, 1968, p. 100.
4. *Budget of the United States Government, FY 1982*, p. 106, and *Special Analyses, Budget of the United States Government, FY 1982*, p. 322.
5. A study of eight major defence contractors found that from 1970 to 1979 they hired a total of 1,672 former employees of the DOD and NASA. In turn, 270 of their employees were hired by DOD or NASA. Gordon Adams, *The Iron Triangle: The Politics of Defense Contracting*, New York: Council on Economic Priorities, 1981, p. 78.
6. William Baldwin, *The Structure of the Defense Market, 1957-64*, Durham, NC: Duke University Press, 1967, pp. 16-24. When Baldwin's method of calculating turnover is applied to later years, the rates remain low.
7. US Department of Defense, Directorate for Information, Operations and Reports, 'Department of Defense Prime Contract Awards, Fiscal Year 1980', Washington, DC: February 1981, p. 54.

8. Ibid., p. 23.

9. Ibid., p. 24.

10. National Science Foundation, *National Patterns of Science and Technology Resources, 1980*, NSF 80-308, Washington, DC: 1980, pp. 44-6.

11. In aerospace, IR&D payments comprise an estimated 20 percent of 'company-financed' research. See Judith Reppy, 'Defense Department Payments for "Company-Financed" R&D', *Research Policy*, vol. 6 (1977): 406.

12. See James Kurth, 'Aerospace Production Lines and American Defense Spending', in *Testing the Theory of the Military-Industrial Complex*, ed., Steven Rosen, Lexington, Mass., Lexington Books, 1973.

13. Aerospace Industries Association of America, Inc. (AIA), *Aerospace Facts and Figures 1980/81*, Washington, DC: 1980, p. 34.

14. See Table 7.5 in Jacques S. Gansler, *The Defense Industry*, Cambridge, Mass.: The MIT Press, 1980, p. 174.

15. Ibid., p. 191.

16. US Bureau of the Census, *Shipments to Federal Government Agencies, 1978*, MA-175(78)-1, Washington, DC: 1980, p. 8.

17. *Business Week*, 3 December 1979, p. 68.

18. See Gansler, *The Defense Industry*, Chapter 6.

19. National Science Foundation, 'National Patterns of R&D Resources 1953-71', NSF 70-46, Washington, DC: n.d., p. 6.

20. US Department of Defense, Directorate for Information, 'Military Prime Contract Awards by Regions and State: Fiscal Years 1977, 1978, 1979', Washington, DC: May 1980, pp. A-3,4.

21. Ibid., pp. xiii-xiv.

22. Richard Dempsey and Douglas Schmude, 'Occupational Impact of Defense Expenditures', *Monthly Labor Review* no. 94 (December 1977): 13.

23. AIA, *Aerospace Facts and Figures*, p. 132.

24. US Department of Defense, Directorate for Information Operations, 'Real and Personal Property of the Department of Defense as of 30 September, 1979, Washington, DC: n.d., pp. 72, 97.

25. Otto Huntz, *et al.*, 'Machine Tool Industry Study Final Report', US Army Industrial Base Engineering Activity, Rock Island, Ill.: 1978, p. 24.

26. US Department of Defense, 'Profit '76 Summary Report', Washington, DC: December 1976, pp. II-34.

27. Ibid., pp. II-35.

28. Ibid., pp. II-25.

29. Gansler, *The Defense Industry*, pp. 138-42.

30. A 1979 study by the General Accounting Office found that the DOD's effort to increase company investment by rewarding it through higher profit rates had not succeeded because the weight given to new investment in negotiating profit rates was too low. US Comptroller General, 'Recent Changes in the Defense Department's Profit Policy – Intended Results Not Achieved', PSAD 79-38, Washington, DC: General Accounting Office, 8 March 1979.

31. US Department of Defense, Defense Security Assistance Agency, 'Foreign Military Sales and Military Assistance Facts', Washington, DC: December 1980, pp. 1-2.

32. Ibid.

33. US Arms Control and Disarmament Agency, *World Military Expenditures and Arms Transfers, 1969-1978*, Washington, DC: 1980, p. 115, and *Statistical Abstract of the United States, 1980*. The ACDA data exclude construction, training and technical services from foreign military sales.

34. *Aviation Week and Space Technology*, no. 114 (1 June 1971): 21.

35. US Department of Defense, Defense Security Assistance Agency, 'Foreign

Military Sales', pp. 1-2, and US Department of Defense, Directorate of Information Operations, '100 Companies Receiving the Largest Dollar Volume of Prime Contract Awards', various years.

36. Ibid., p. 68, and Northrop Corporation, *Annual Report*, 1980, p. 30.

37. Lloyd J. Dumas, 'Disarmament and Economy in Advanced Industrialized Countries – The US and the USSR', *Bulletin of Peace Proposals*, 12:1 (1981): 5-6.

38. National Science Foundation, *National Patterns of Science and Technology Resources, 1980*, NSF 80-308, Table 18. Some countries outside the OECD report higher figures, for example, the Soviet Union and Israel.

2 THE SOVIET UNION

David Holloway*

The Soviet defence industry is one of the largest in the world; according to American estimates, it is the largest of all. But it is impossible to establish just how big the Soviet defence sector is, for the Soviet authorities have shrouded it in secrecy. They do not reveal how many men the Soviet Union has under arms, or what equipment their forces have. Each year the Soviet government publishes a single figure for the defence budget, but (although it is not clear what this figure covers) it is far too small to pay for the men and equipment that Western governments estimate the Soviet Armed Forces to have.[1]

Western governments, in particular that of the United States, publish estimates of Soviet defence spending and defence production. These estimates are one of the most helpful sources for studying the Soviet defence industry, but they present problems too. Western intelligence services face considerable difficulties in estimating the scale and cost of the Soviet defence effort. This was shown in a rather dramatic way in 1976, when the CIA revised its estimate of the proportion of GNP that the Soviet Union was spending on defence from 6-8 per cent to 11-13 per cent.[2] Besides, official Western estimates of the Soviet defence effort are part of the political process through which Western defence budgets are established, and they may be affected by political pressures in ways that are difficult to discern. Hence, they must be treated with caution.

Private researchers who study the Soviet defence sector, however, have to make use of official Western estimates and Soviet national income accounts. The difficulties that these sources present should be borne in mind, because they introduce a considerable element of uncertainty into any analysis of the Soviet defence industry. More specifically, the quality and range of the sources restrict severely the kind of questions to which answers can be attempted, and the reliability of the answers obtained. In spite of the difficulties, it is worth attempting an analysis, because the Soviet defence industry is such an

*I would like to express my thanks to Kazimierz Poznznski, Judith Reppy and George Staller for helpful comments on an earlier draft of this chapter. Responsibility for errors rests with me.

important part of the Soviet economy, and because it has such a large impact on international relations.

Most studies of the Soviet defence industry are concerned primarily to compare it with the American. Here, however, the Soviet defence industry will be examined in the context of the Soviet strategy of economic development. This chapter will look at the historical growth and present size of the Soviet defence industry, at the structure of the industry and its main features, at arms transfers and their economic significance, and finally at the problem of conversion. It does not attempt to provide a detailed analysis of the way in which the estimates quoted have been compiled, but does make reference to sources that explore the difficult methodological issues involved in estimating the size of Soviet military production and research and development (R&D).

The Defence Industry

Defence Production

Imperial Russia was a considerable military power, with extensive artillery and small arms production and a large naval shipbuilding industry. Before the Revolution pioneering work was done on aircraft development; in 1917 1,897 aircraft were built. The First World War gave impetus to the development of the war industry but, like the rest of the economy, military production was affected by the Revolution and Civil War.[3] In the period of the New Economic Policy (1921-1927) the Soviet authorities gave little priority to military production. But when the Soviet leaders embarked on their industrialization drive in the late 1920s, one of their main aims was to create a strong defence industry. In 1933 the People's Commissar (i.e. Minister) of Army and Navy, K.E. Voroshilov, declared that the basic task of the First Five Year Plan (1928-1932) in the sphere of defence had been

> . . . the technical re-equipment of the Red Army on the basis of the successful growth of our industry. Therefore the main point in our work was the technical reconstruction of the Red Army on the basis of the production of new arms by our factories.[4]

The development of the Soviet defence industry began in the years of the First Five-Year Plan.

The industrialization drive was inspired, in significant measure, by the decision to make the Soviet Union strong, so that it could defend itself against the capitalist states by which it was encircled; in this it was very different from industrialization in Britain and the United States. One of the main aims of Soviet policy was to create a powerful defence industry, and this had a major effect on the whole industrial structure. The Soviet Union invested heavily in the defence industry, and large proportions of the output of other sectors went to meet the needs of defence. Military production grew rapidly during the 1930s (as can be seen from Table 2.1), although from a very small base. In the late 1920s the Soviet Union had virtually no tank industry, little or no warship construction, and low artillery and ammunition production; the aircraft industry was the brightest spot in an otherwise gloomy picture. Moreover, most Soviet weapons of the late 1920s were of pre-revolutionary or foreign design. In the early 1930s foreign weapons were acquired and used as the basis of Soviet designs. By 1940 the Soviet Union had a strong defence industry and 'an indigenous military technology',[5] and had gone a long way towards attaining the goal of strategic self-sufficiency.[6]

Table 2.1: Annual Average Production of Basic Types of Armament Units

	1930-31[1]	1932-34[1]	1935-37[1]		1940[2]
Aircraft					
in all	860	2,595	3,578		10,565
bombers	100	252	568	)	all military
fighters	120	326	1,278	)	8,331
Tanks	740	3,371	3,139		2,794
Artillery					
in all	1,911	3,778	5,020		15,300
small calibre	1,040	2,196	3,609		
medium calibre	870	1,602	1,381		
Rifles ('000)	174	256	397		1,461
	1928-32[2]	1933-36[2]	1939[2]		
Naval shipbuilding (tons displacement)	12,000 total	84,000 total	30,460 total		

Sources: (1) M.V. Zakharov, 'Kommunisticheskaya partiya i tekhnicheskoe perevooruzhenie armii i flota v gody predvoennykh pyatiletok', *Voenno-istoricheskii zhurnal*, 1971, no. 2, p. 7.
(2) Julian Cooper, ***Defence Production and the Soviet Economy 1929-1941***, CREES Discussion Paper, Birmingham, UK: University of Birmingham, 1976, pp. 46-50, quoting Soviet sources.

In spite of the losses and dislocation caused by the German invasion of 22 June 1941, the Soviet defence industry withstood the test of war. By the end of 1941 more than 1,300 major industrial plants and ten million people had been evacuated to the East.[7] In the second quarter of 1942 tank production was more than four times as great as in the second quarter of 1941; aircraft production had grown by 70 per cent; the output of machine-guns had risen fourfold; and ammunition production had increased by 80 per cent. In 1942 military production constituted 63.9 per cent of industrial output, compared with 26 per cent in 1940. In 1943 the proportion fell to 58.3 per cent; in 1944 to 51.3 per cent; and in 1945 to 40.1 per cent.[8] Stalin claimed in 1946 that victory showed that his policies of collectivization and industrialization had been correct, and declared that the Soviet Union would build up its strength once again in order to be ready for all contingencies.

The transition from wartime to peacetime production proved difficult to accomplish. During the war little preparation had been made for conversion to civilian production. In many cases new production processes and new machinery were needed; supply shortages had to be eliminated and workers retrained. One of the main problems was that the composition of civilian output was more complex than that of war production; in 1946, for example, plants of the tank industry, which during the war had had seven or eight basic products, had to adjust to the production of over 40 basic types of locomotive, wagon and other transport machinery. Productivity fell as a consequence. In 1946 industrial output was 17 per cent lower than in 1945, as a result of the sharp drop in military production and the difficulties of conversion. Industrial growth began again in 1947.[9]

While the Soviet Union cut military production it also launched major programmes to develop nuclear weapons, long-range rockets, jet propulsion and radar. During the war it had become clear that science was affecting warfare in a new way and that the Soviet Union lagged behind advances that had been made in Britain, Germany and the United States. The cost of these development programmes is not known, though the then Minister of Finance later wrote that they required 'significant resources' and affected post-war reconstruction under the fourth Five-Year Plan (1946-1950).[10] These new technologies illustrated the increasing importance of R&D in the creation of new weapons and ushered in what Soviet writers refer to as the 'military-technical revolution'. But civilian industry was reconstructed largely on the basis of pre-war technology. Thus the relationship between military and civilian production changed: before the war the two had been intimately linked; after the war they grew apart, with the defence industry moving ahead rapidly in technology. It was not until 1955 that the Soviet leaders began to give urgent attention to the problem of technological innovation in civilian industry.[11]

The Soviet Union has published no figures for the production of arms and equipment since 1945. Data have to be taken from Western sources, and these are particularly poor for the years 1945-1960. Table 2.2 gives some figures for weapons production in this period, but these are very patchy and not at all reliable; even the lower figure for aircraft production in 1960 is probably too high by a factor of ten since missile production had been expanded, and more complex aircraft were being produced. Moreover, these figures give little indication of trends in production or of changes in the structure of the industry. There is some evidence from other sources of a drop in conventional arms production during the mid-1950s, when the Soviet Armed Forces began to acquire nuclear weapons and their means of delivery.[12] But an upturn in defence production may have taken place before the end of the decade: the US Defense Intelligence Agency stated in 1981 that the Soviet defence industry had 'grown steadily and consistently over the past 20-25 years'.[13] The Seven Year Plan, which began in 1959, seems to have included provision for increased military production, certainly of

Table 2.2: Production of Basic Types of Armament, 1950-60

	1950	1955	1960
Aircraft	7,200-24,000[1]		5,000[2]-(12-18,000)[1]
Tanks	6,000[3]	6,000[4]	
Armoured personnel carriers		10,000[4]	
Warships (tons displacement)	100,000[5]	100,000[5]	100,000[5]

Sources: (1) Asher Lee, *The Soviet Air Force*, New York: The John Day Company, 1962.
(2) R.E. Stockwell, 'Soviet Aircraft Production', p. 249, in *The Soviet Air and Rocket Forces*, ed. Asher Lee, London: Weidenfeld and Nicolson, 1959. Stockwell writes that this figure '. . . strikes a rather good average among the various estimates', although Lee's estimates range from 12,000 to 18,000.
(3) United States, NSC 100, *A Report to the NSC by the Chairman, National Security Resources Board and Recommended Policies and Actions in Light of the Grave World Situation*, Washington, DC: 11 January 1951, p. 9.
(4) N. Galai, *Bulletin of the Institute for the Study of the History and Culture of the USSR*, no. 7 (1954), pp. 6, 13. The figures are for 1953.
(5) S. Breyer, *Guide to the Soviet Navy*, Annapolis, Md.: US Naval Institute, 1970, p. 227.

strategic missiles, and perhaps also of conventional arms.[14] In December 1959 the Strategic Rocket Forces were established as a separate service.[15]

The period from 1959 to 1964 remains a confused one in Soviet military policy. Khrushchev appears to have wanted to cut back the production of conventional weapons, perhaps in an effort to keep defence spending down. But it is not clear from the available figures how far his statements about the obsolescence of surface warships and aircraft affected R&D and production, although he did say in 1960 that bomber production had been reduced, and would be stopped.[16]

The period 1955 to 1965 was one of upheaval in Soviet military affairs, caused by the transition to a new policy for the nuclear age. The same turmoil has not existed since 1965. The figures in Table 2.3 do not

Table 2.3: Production of Basic Types of Armament, 1965-80

	1966[1]	1970	1975	1976[4]	1980[4]
Combat Aircraft	900	900/1,000[2]	900/1,000[2]	1,230	1,340
Helicopters	350	500[1]	700[2]-1,100[3]	1,400	750
Tanks	3,100	3,800[1]	2,600[3]-3,000+[2]	2,500	3,000
Armoured Vehicles	2,800	4,000[1]	3,700[3]-4,000+[2]	4,500	5,500
Artillery	1,000+	1,600[1]	1,000+[2]-1,400[3]	3,300	.1,850
Infantry Weapons ('000)				250	400
Naval Shipbuilding (tons displacement)		1965-75: annual average – 150,000[5]			

Sources: (1) US, Department of Defense, *Annual Defense Department Report. FY 1978*, Washington, DC: 1976, p. 27, Chart V-2.
(2) 'Ministry of Defense Release', London: Ministry of Defense, 1976. The figures are for the rate of production for 'recent years'.
(3) Cecil Brownlow, 'DOD Request Represents $7.2 Billion in Real Growth', *Aviation Week and Space Technology*, 26 January 1976, p. 20. The figures are for average production for 1973-5.
(4) US Congress, Joint Economic Committee, *Statement of Major General Richard X. Larkin, Deputy Director and Edward M. Collins, Vice Director for Foreign Intelligence, Defense Intelligence Agency, before the Subcommittee on International Trade, Finance and Security Economics*, Washington, DC: 8 July 1981, pp. 86-87.
(5) 'US/USSR Naval Force Comparisons', *NATO Review*, December 1976, p. 11.

show any sharp year-to-year fluctuations in the production of conventional arms between 1965 and 1980. This does not, of course, mean constant levels of production for each type of equipment. Tank production, for example, rose in the late 1960s to a peak of about 4,250, and then fell to under 2,500 in 1975, only to rise again to 3,000 in 1980.[17]

According to DIA estimates, fighter production from 1965 to 1970 was between 800 and 900 a year. From 1970 to 1975 it rose to 1,200 a year, and in 1980 it was 1,300.[18] The overall picture to emerge from the data on arms production is one of expansion over the fifteen year period.[19]

The Size of the Defence Sector

The production of these arms and equipment requires a large defence industry. An incomplete list of defence plants in the 1930s identifies 54 major production plants; in 1937-8 most Soviet factories appear to have been devoting at least some part of their capacity to defence purposes.[20] The Defense Intelligence Agency claims that there are now 134 final assembly plants that are involved in the production of weapons as end products (see Table 2.4) and over '3,500 installations

Table 2.4: Defence Industry Plants in the Soviet Union

Final assembly plants involved in the production of weapons as end products	Number
Ground forces materiel	24 plants
Naval materiel	24 shipyards
Aircraft materiel	37 plants
Missile materiel	49 plants
Total	134 units

Note: According to *Military Production in the USSR*, London: Ministry of Defence, 1976, p. B-1, the Ministry of Aircraft Production has '30 to 40 production-related factories, though not all are currently engaged in series production. Many of them have been expanded in recent years.' Another source says that the Ministry has five plants producing fighters, nine producing bombers, transports and larger civil aircraft, ten manufacturing helicopters, light and utility aircraft, and ten producing aero-engines. Alexander Boyd, *The Soviet Air Force Since 1918*, London: MacDonalds and Janes, 1977, p. 277.
Source: US Congress, Joint Economic Committee, *Statement of Major General Richard X. Larkin, Deputy Director and Edward M. Collins, Vice Director for Foreign Intelligence, Defense Intelligence Agency, Before the Subcommittee on International Trade, Finance and Security Economics*, Washington, DC: 8 July 1981, p. 83.

which provide support to these assembly plants'.[21] Other sources have claimed that 60 per cent of all industrial plants in the Soviet Union are engaged in military production in one way or another.[22]

No figures exist for the numbers of people employed in the defence industry. Estimates have been derived from other data about the Soviet defence effort, but it is not possible to make an independent check of these. One study has estimated the size of the military-industrial work force as 4.8 million for 1979, in the following way:

> ... assuming (as in the US) a 1:1 relation between personnel in armed forces and supporting civilian industry, adjusted for USSR productivity in military industry assumed 75 per cent of US, then USSR military industry estimated as using 4.8 million employees.[23]

Another figure can be derived from the estimate that 20 per cent of total industrial output is devoted to defence (see Table 2.5). Assuming an equal level of productivity for civilian and military production, the total number of employees working on military production in 1975 will have been about seven million.[24] Making the same assumptions about the machinebuilding and metalworking industry, one third of the output of which is said to go to defence (see Table 2.5), one can estimate that the number of employees in that sector engaged in defence production in 1975 was about 4.5 million.[25] From these very rough estimates one can conclude that in the Soviet Union in the 1970s between four and seven million people were employed in military production – but this conclusion is no better than the doubtful assumptions on which it rests.

Table 2.5 gives estimates of the proportion of total industrial output, and of the output of the various industrial sectors, devoted to defence. These are derived from different sources, but most come from the CIA. There are some divergencies among the figures. In particular, Lee estimates that the proportion of Machinebuilding Final Demand devoted to defence rose from 33 per cent to 48 per cent between 1967 and 1975, while the CIA concludes that a constant proportion of machinebuilding and metalworking final output went to defence in the same period. The figures in Table 2.5 do point to the importance of defence in the output of major sectors of industry, and in particular of high-technology branches. For the purposes of comparison it may be noted that in 1940 26 per cent of total industrial output went to defence.[26]

Military Expenditure as a Proportion of GNP

Table 2.6 gives some estimates of the proportion of GNP that the Soviet

Table 2.5: Proportion of Industrial Output Devoted to Defence (percentage)

Sector	1967	1970	1976	1977
Total industrial output			20.0[4]	
Machinebuilding and metalworking	33.3[2]	33.3[2]	33.3[2]	33.3[2]
Machinebuilding	33.0[5]	40.0[5]	48.0[5,a]	
Metallurgical				20.0[3]
Chemical				16.6[3]
Energy				16.6[3]
Radioelectronics		70.0[1]		
Aircraft industry	66.6[2]	66.6[2]	66.6[2]	66.6[2]
Shipbuilding	66.6[2]	66.6[2]	66.6[2]	66.6[2]

Note: a. Data from 1975.

Sources: (1) Andrew Sheren, in US Congress, Joint Economic Committee, Compendium: *Economic Performance and the Military Burden of the Soviet Union*, Washington, DC: US Govt. Printing Office, 1970, pp. 129, 131.

(2) US, Central Intelligence Agency, National Foreign Assessment Center, *Estimated Soviet Defense Spending: Trends and Prospects*, SR 78-10121, Washington, DC: June 1978, p. 2: '. . . during the 1967-77 period, defense consumed approximately one-third of the final product of machinebuilding and metalworking, the branch of Soviet industry that produces civilian investment goods as well as military hardware. In ruble cost terms, about two-thirds of the aircraft and over two-thirds of the ships and boats produced in the Soviet Union went to the defense sector.'

(3) US Congress, Joint Economic Committee, Subcommittee on Priorities and Economy in Government, Hearings: *Allocation of Resources in the Soviet Union and China – 1977*, Part 3, Washington, DC: US Govt. Printing Office, 1977, p. 38, Statement of Admiral Stansfield Turner, Director of the CIA.

(4) US, Central Intelligence Agency, National Foreign Assessment Center, *Estimated Soviet Defense Spending in Rubles*, SR 76-101210, Washington, DC: May 1976, p. 16.

(5) William T. Lee, *The Estimation of Soviet Defense Expenditures, 1955-1975. An Unconventional Approach*, New York: Praeger, 1977, p. 102. The figure is for 'National Security Durables as a Proportion of Machinebuilding Final Demand'. For a critical discussion of Lee's approach, see Hanson, 'Estimating Soviet Defense Expenditure', footnote 2.

Union spends on defence. It should be noted that there is a major difference between the method used for the estimates in rows 1-4 and that used for the estimates in rows 5-7. The former are derived, by various methods, from Soviet statistics for the budgetary allocations to 'defence' and 'science', and for the output of the machinebuilding industry. There are considerable problems in using these statistics, because it is not clear what they cover, and because the Soviet authorities take pains to hide information about their defence production. (For example, the 'defence' heading does not seem to include weapons procurement, and some other items are probably excluded as well.)

Table 2.6: Western Estimates of Soviet Defence Expenditure as a Proportion of Gross National Product (percentages)

Source		1950	1955	1960	1965	1970	1975	1980
(1) Bergson (current ruble factor cost)		10.9	10.3					
(1950 ruble factor cost)		10.9	10.7					
(2) Cohn	(A)	11.6	12.3	9.7	11.7	12.8[a]		
(1955 rubles)	(B)	11.4	12.3	8.6	9.9	10.0		
(current	(A)	12.6	12.3	9.2	10.7	10.4		
rubles)	(B)	10.8	12.3	8.4	9.1	10.1		
(3) Becker			(1958-64: not more than 10 per cent)					
(4) Lee			11.5	9.0	10.0	11.5	14.0/15.0	18.0
(5) ACDA						13.0	12.6	
(6) CIA (1974)				10.0				
(7) CIA (1978, 1980)						11.0/13.0	11.0/13.0	12.0/14.0

Note: a. Data from 1969.
Sources: (1) A. Bergson, *The Real National Income of Soviet Russia since 1928*, Cambridge, Mass.: Harvard University Press, 1971, p. 245.
(2) S. Cohn in US Congress, Joint Economic Committee, Compendium: *Soviet Economic Prospects for the Seventies*, Washington, DC: US Govt. Printing Office, 1973, pp. 159-60. The alternative A and B time series are based on different assumptions about the composition of the 'defence' allocation in the Soviet budget and the proportion of 'science' allocation that goes to defence.
(3) Abraham Becker, *Soviet National Income, 1958-64*, Berkeley and Los Angeles: University of California Press, 1969, pp. 164-5.
(4) William T. Lee, *The Estimation of Soviet Defense Expenditures, 1955-1975. An Unconventional Approach*, New York: Praeger, 1977, p. 98. The figure for 1980 is taken from Lee's testimony in US Senate, Committee on Armed Services, Subcommittee on General Procurement, Hearings: *Soviet Defense Expenditures and Related Programs*, Washington, DC: US Govt. Printing Office, 1980, p. 9.
(5) US, Arms Control and Disarmament Agency, *World Military Expenditures and Arms Transfers, 1969-1978*, Publication 108, Washington, DC: 1980, p. 66.
(6) US Congress, Joint Economic Committee, Subcommittee on Priorities and Economy in Government, Hearings: *Allocation of Resources in the Soviet Union and China – 1974*, Washington, DC: US Govt. Printing Office, 1974, p. 25.
(7) US, Central Intelligence Agency, National Foreign Assessment Center, *Estimated Soviet Defense Spending: Trends and Prospects*, SR 78-10121, Washington, DC: June 1978, p. i. The 1980 figure is taken from US Congress, Joint Economic Committee, Subcommittee on Priorities and Economy in Government, Hearings: *Allocation of Resources in the Soviet Union and China – 1980*, Part 6, Washington, DC: US Govt. Printing Office, 1981, p. 136.

The estimates in rows 5-7 are based on the 'direct costing' method.[27] This starts from the physical components and the activities of the Soviet Armed Forces – weapons, personnel, buildings, supplies and so on. Where ruble prices are not available, these basic elements are priced in dollars, which are then converted into rubles. This approach is available only to the Western intelligence services which possess detailed information about the physical elements of the Soviet defence effort. It has the advantage of avoiding many of the difficulties associated with other methods, but it is open to criticism, particularly in the way in which ruble prices are assigned to Soviet activities. The very fact that the CIA had to revise its estimates so drastically in the mid-1970s show that this method is far from foolproof.

It is striking, however, that the estimates in Table 2.6, which are derived by different methods, show a fairly consistent pattern, except for 1975 and 1980. The proportion of GNP devoted to defence falls in the range of 9 to 12 per cent, except for 1960 when, all the estimates agree, defence spending was lower than for other years in the table. The estimates by the CIA and Lee show a rise in the proportion spent on defence in the 1970s, though Lee's estimate is much higher than the CIA's. This general pattern fits with what is known about the output of the defence sector in the postwar period. Of course, the CIA and ACDA figures are derived in part from the data about the defence production and so do not constitute independent evidence. But the other estimates are not directly linked to the data on defence output, and so may be taken as an independent check.

The Military R&D Effort

The development of advanced military technology requires a large R&D effort. This is perhaps the most difficult aspect of the Soviet defence industry to explore, and it is clear from Table 2.7 that there is a wide variation in the estimates of the size of military R&D expenditure. The first problem is to decide what total R&D outlays are; the second is to decide what proportion of these is devoted to defence. The figures in the table are not all comparable. Soviet 'science' expenditures cover research in the humanities and the social sciences, which would not be included in Western R&D figures.[28] Some of the Western estimates of Soviet military R&D are based on the argument that total R&D expenditure is much greater than Soviet 'science' outlays, and assume that some military R&D – in particular prototype development – is not covered by the 'science' outlays. The Western estimates are derived by making assumptions about the proportion of the estimated R&D expenditure devoted to defence. Unfortunately, there is no good

Table 2.7: Estimates of Soviet Military R&D Outlays (billions of rubles)

	1960	1965	1970	1975	1980
(1) Total 'science' outlays: current & capital	3.9	6.9	11.7	17.4	19.3[a]
(2) Nimitz[b]	1.6-2.0	2.6-3.3	3.3-4.8[c]		
(3) CIA (1976, 1978)[d]			8.0-10.0	10.0-12.0	10.6-12.6[e]
(4) Lee[b]	2.9	4.7	7.5	11.3	
	4.6	7.7	12.4	18.6	

Notes: a. Data from 1978; b. Current rubles; c. Data from 1968; d. 1970 rubles; e. Data from 1977.

Sources: (1) L. Nolting, *Sources of Financing the Stages of the Research, Development and Innovation Cycle in the USSR*, Foreign Economic Reports, no. 3, Washington, DC: US Department of Commerce, 1973, p. 10. The figure for 1975 is taken from *Narodnoe Khozyaistvo SSSR v 1975 g.*, Moscow: 1976, p. 744. The 1978 figure is taken from *Narodnoe Khozyaistvo SSSR v 1978 g.*, Moscow: 1979, p. 535.

(2) N. Nimitz, *The Structure of Soviet Outlays on R&D in 1960 and 1968*, R-1207-DDRE, Santa Monica, Calif.: Rand Corporation, June 1974, p. vii. The figures cover defence and space R&D; they do not cover capital investment.

(3) US, Central Intelligence Agency, National Foreign Assessment Center, *Estimated Soviet Defense Spending in Rubles, 1970-75*, SR 76-10121U, Washington, DC: May 1976, p. 13. The higher figure includes all space R&D. The 1977 figure is taken from US, Central Intelligence Agency, National Foreign Assessment Center, *Estimated Soviet Defense Spending: Trends and Prospects*, SR 78-10121, Washington, DC: June 1978, p. ii.

(4) W. Lee, *The Estimation of Soviet Defense Expenditures 1955-75. An Unconventional Approach*, New York: Praeger, 1977, p. 294. The higher estimates are derived by adding 30 per cent to estimated Soviet R&D expenditures to allow for suspected understatements in the figures. Lee's estimates cover military and space R&D. They also cover capital investments in R&D plant.

evidence to indicate what that proportion is; assumptions have varied from 40 to 80 per cent.[29] The latter figure appears too high in view of the scale of Soviet civilian R&D. Fifty per cent seems more likely to be the correct proportion, though that has probably changed over time. If one assumes that 50 per cent of the Soviet R&D effort is devoted to defence, and that the same proportion applies to R&D personnel, then the number of scientists and engineers employed in Soviet military R&D was in the order of 400,000 in the late 1970s.[30]

Summary

A number of general conclusions can be drawn from this survey of the growth and size of the defence industry. First, the Soviet Union sustains a very large military-industrial and military-technological effort, and has done so since the First Five-Year Plan. Only at the end of the war with Germany and during the mid- to late 1950s does the military effort appear to have slackened; expenditure either dropped or held steady, and the proportion of GNP devoted to defence fell as a consequence.

Second, although there have been major shifts in the kinds of weapons produced over the last twenty-five years, the available figures for total weapons production do not show sharp percentage fluctuations from year to year, while the changeover from one weapon to another appears to be relatively orderly (except perhaps in the period 1955-65). This may result from the way in which the figures are derived (with actual production estimated in part from production capacity), but it may also be a consequence of the fact that defence production is centrally planned.

Third, the Soviet defence industry is large when compared in terms of output with the defence industries of other countries. Its size is to be understood in the context of Soviet policies. The Soviet leaders have put great stress on building up the power of the Soviet state. The creation of military power was one of the main aims of Soviet industrialization policy, and Soviet industry was planned accordingly. After the war the high technology branches were built up to meet the demands of the arms race with the United States. The Soviet GNP is approximately 60 per cent of the American, and the defence industry is inevitably more important in the Soviet than the American economy, given the Soviet goal of matching American military power.[31] The United States' allies in NATO are economically and militarily more powerful than the Soviet Union's allies in the Warsaw Pact, and this places an extra burden on the Soviet Union. Moreover, since the mid-1960s between 10 per cent (in ruble terms) and 20 per cent (in dollar terms) of Soviet defence expenditure has gone to sustain forces facing China.[32]

Main Features of the Defence Industry

The last section looked at the output of the defence industry, and at the resources devoted to defence. This section will examine some of the main features of the defence industry's organization and performance. Here too the specific character of the Soviet industrialization drive had a marked effect on the defence industry. The organization of the defence sector reflects the high priority the Soviet leaders have given to defence.

In the late 1920s and early 1930s, at the start of the Soviet industrialization drive, there existed a conflict between immediate investment in military production and the creation of a strong civilian industry which would eventually provide the basis for military power. The political leadership tried to minimize the conflict between military and civilian purposes by ensuring that civilian industry could be shifted to military production if the need arose. Three main principles of military-economic policy were elaborated:

(1) to maintain a high level of armaments production;
(2) to ensure the flexibility of the economy (for example, in shifting from civilian to military output, raising the rate of arms production, or introducing new weapons);
(3) to secure the viability of the economy in wartime.

The war with Germany showed how important these principles were, and they are still taken as the most important indicators of a state's 'economic potential', of its ability to provide for the material needs of society while producing everything necessary for war.[33]

The organization of the defence sector is similar to that of Soviet industry as a whole. The production plants and R&D establishments are subordinate to ministries, whose activities are in turn planned and coordinated by higher agencies. This basic organizational form was established in the 1930s. At first most defence production was carried out in enterprises of the People's Commissariat (Ministry) of Heavy Industry. In 1936 a People's Commissariat of the Defence Industry was established, and in 1939 this in turn was divided into four separate commissariats, for ammunition, armaments, ships and aircraft. The main feature of the defence industry organization in the postwar period has been the creation of new ministries to take charge of the new science-based branches of defence production: the Ministry of Medium Machinebuilding was set up in 1953, with responsibility for the nuclear weapons programme. The Ministry of the Radio Industry was set up in 1954, and from it two further ministries have been created: for the Electronics Industry (1961) and for the Communications Equipment Industry (1974). The Ministry of General Machinebuilding, which is responsible for the development and production of strategic missiles, was established in 1965.[34]

There are now nine ministries with primary responsibility for military production (see Table 2.8). The enterprises of these ministries also produce goods for civilian use, and military goods are produced in plants of ministries not in this group; consequently, the defence industry, as defined by output, is not coextensive with the ministries in the defence industry group. In 1971 Mr. Brezhnev stated that 42 per cent of the output of the defence industry was for civilian purposes; but it is not clear whether he was referring to the Ministry of the Defence Industry (which is responsible for the production of conventional army materiel such as tanks, armoured vehicles, artillery pieces and so on) or to the industry as a whole.[35] According to CIA estimates, one third of shipbuilding and aircraft production, and about the same proportion of electronics output, are civilian. Little information is available, however, about the military output of ministries outside the defence industry group, although the CIA does estimate that one sixth of the Chemical Industry's output goes to defence (see Table 2.5).

Table 2.8: Ministries in the Defence Industry Group

Ministry	Output
Aviation Industry	Aircraft and aircraft parts
Defence Industry	Conventional army materiel
Shipbuilding Industry	Ships
Electronics Industry	Electronic components and equipment
Radio Industry	Electronic components and equipment
Means of Communication	Electronic components and equipment
Medium Machinebuilding	Nuclear weapons
General Machinebuilding	Strategic missiles
Machinebuilding	Ammunition

Sources: Sheren, *Economic Performance*, pp. 123-32, and M. Agursky, *The Research Institute of Machine-Building Technology*, Soviet Institutions Series Paper, no. 8, Jerusalem: The Soviet and East European Research Center, The Hebrew University of Jerusalem, September 1976, p. 6.

Although the defence sector is organized in much the same way as the rest of Soviet industry, it does have several distinguishing features. The most important of these is the high priority that it has enjoyed for over 50 years. The Soviet leaders have been determined to build up Soviet military power, and they have been willing to devote resources to this end. They have also created institutional arrangements to ensure that the defence industry works as well as possible. The Military-Industrial Commission, which is headed by Deputy Premier L.V. Smirnov, appears to have responsibility for coordinating large-scale military R&D and production programmes. The defence industry has traditionally had first priority in the allocation of scarce resources: skilled manpower, qualified scientists and engineers, and scarce materials and supplies. In the Soviet economy, which is beset by shortages and supply bottlenecks, this must be of considerable value in helping to ensure the smooth running of the defence industry, though it must make bottlenecks in the rest of the economy even worse. Thus high priority is not merely a matter of the political leaders' attitude; it is also built into the arrangements for the day-to-day management of production. Many of those in charge of the defence industry have held their positions for a long time – Smirnov, for example, has been in his post since 1963 – and this too will help to perpetuate the existing mode of operations. It is worth noting, however, that these arrangements are not good enough to eliminate all supply problems. From the scanty evidence available, it appears that the ministries in the defence industry group try to minimize their dependence on supplies from other ministries by building up their own supply industries; in other words, these industrial branches are characterized by a high degree of vertical integration.[36] This would presumably not be the case if these minstries were satisfied that they could obtain the materials and components they needed, in good time and of the right quality, from other ministries.

The powerful position of the customer in the weapons acquisition process is another important feature of the defence industry. The Ministry of Defence prepares plans for weapons development and production, chooses among competing designs, concludes agreements with the design bureaus (which develop the systems), supervises the development work, conducts trials, concludes production agreements, supervises production, conducts acceptance trials and assimilates the equipment.[37] This gives the Ministry considerable control over the defence industry and a degree of consumer power unusual in the Soviet Union. The Ministry had the necessary institutional arrangements and technical resources to carry out this role. One important instrument of Ministry supervision is the system of military representatives (*voennye predstaviteli*) in the design bureaus and production plants. These have three major functions: to prevent production bottlenecks by speeding up the supply of materials and parts; to police the pricing of military products; and to ensure that military production meets all quality standards.[38]

Studies of Soviet R&D suggest that the defence sector is more successful at innovation than civilian industry, and the high priority of the defence sector and the powerful role of the Ministry of Defence do contribute to the industry's success in developing and producing weapons of high quality. At the same time, however, comparisons of

Soviet and US military technology indicate that American military R&D is more innovative than Soviet military R&D.[39] In 1977 Admiral Turner, Director of the CIA, testified that

> . . . while virtually all of the Soviet inventory of weapons falls within US production technology, the Soviets simply do not have the technology required to produce many of the US weapons nor could they produce close substitutes.[40]

In the same hearings Admiral Turner implied that as much as 30 per cent of American military technology was beyond the technological capacity of the Soviet Union to produce.[41]

Soviet design philosophy gives great importance to simplicity, commonality and evolutionary development in weapons, and these features are encouraged by the organizational arrangements of the weapons acquisition process.[42] These characteristics will make for cheaper designs than would complexity and the striving after revolutionary change. But there is little in the organization of the defence industry to indicate that these qualities are obtained in an especially efficient or economical way. The stress in planning has been on obtaining the desired performance rather than on reducing costs. One of the reasons why the CIA revised its estimate of Soviet military expenditure in the mid-1970s was that it concluded that defence production was much less efficient than had previously been thought.[43]

Moreover, it should be stressed that simplicity and evolutionary development are relative, not absolute, features of Soviet weapons design, and that the Soviet defence industry, like its counterparts in other countries, faces the problem of rising intergenerational costs. One Soviet textbook lists the reasons for this as follows:

> . . . first, the use of critical, costly raw materials, advanced expensive equipment, and a large amount of increasingly expensive electrical energy and electronic equipment; second, the high relative share of expenditures for scientific research and experimental design work, which entails the hiring of a large number of skilled workers and the onetime production of the equipment necessary for these projects; third, the production of military products in small series in peacetime; and fourth, the necessity of putting out the needed type of produce in a very short time.[44]

A steady level of military production will become increasingly costly as more complex and expensive systems are produced.

Arms Transfers

Foreign sales now absorb a substantial part of the output of the Soviet defence industry. These sales go to two major groups of countries: to other members of the Warsaw Pact and close allies such as Vietnam, North Korea, Mongolia and Cuba, and to non-Communist less developed countries (LDCs) such as Libya, Algeria and Iraq. The Soviet Union has always been the major supplier of arms to the Warsaw Pact and in the 1970s Soviet arms sales to LDCs grew rapidly as the world arms market expanded. The US Arms Control and Disarmament Agency estimates that from 1974 to 1978 the Soviet Union transferred arms valued at $27 billion of which $8.5 billion went to close Soviet allies, and $18 billion to non-Communist LDCs.[45]

The Soviet Union publishes no data about its arms exports, so that here again foreign estimates have to be used. These estimates are better for Soviet arms transfers to the LDCs than they are for Soviet transfers to other Warsaw Pact countries. The most striking feature of Soviet arms sales has been the expansion of sales to LDCs in the 1970s. According to CIA estimates, the value of Soviet military agreements with non-Communist LDCs amounted between 1955-9 to $690 million; in 1960-6 to $3,830 million; in 1967-73 to $8,665 million; and in 1974-9 to $34,155 million.[46] Three major developments have contributed to this rapid growth: the 1967 and 1973 Middle East Wars, which made it necessary for the Arab countries to re-equip; Israel's deep penetration raids into Egypt in 1970, which prompted the Soviet Union to start supplying modern weapons to Egypt, and later to other customers; and the rise in oil prices in 1973-4, which led the Soviet Union increasingly to seek commercial and financial returns from its arms exports.[47]

From 1974 to 1978 Soviet arms sales (to all customers) amounted to between 12 and 15 per cent of Soviet exports, and were therefore an important element in Soviet foreign trade. To indicate the degree of importance of Soviet arms exports as an earner of hard currency, Checinski has calculated that Soviet arms exports between 1976 and 1979 of approximately $27.8 billion were 2.36 times higher than the total value of manufactured goods exported to the West in the same period, and equalled about 60 per cent of the costs of manufactured goods that the USSR imported from the West in the same period.[48]

The significance of arms exports can also be evaluated in terms of their importance for the defence industry. American estimates of Soviet defence spending and arms transfers suggest that in 1974-7 the value of arms exports (measured in domestic rubles) was equivalent to 16.6-19.9 per cent of Soviet outlays for the procurement of weapons for the Soviet Armed Forces. About two-thirds of arms exports went to non-Communist LDCs, and one-third to Warsaw Pact and other Communist countries. Perhaps one-sixth of total Soviet weapons production in the period was for export or for replacement of equipment that had been exported.

The estimate of the proportion of defence industry output going for export is derived in the following way. The CIA estimates that in 1975 about 36 per cent of Soviet defence spending, when measured in rubles, went to procure new weapons and equipment for the Soviet Armed Forces.[49] It is reasonable to suppose that this proportion held good for 1974-7, given the apparently stable pattern of Soviet defence spending.[50]

The CIA also estimates that arms flows to non-Communist LDCs in 1974-7 when measured in domestic rubles were equivalent to 5-6 per cent of estimated Soviet defence expenditure.[51] Not all arms flows are weapons. According to the CIA, 60 per cent of Soviet arms sales in this period comprised weapon systems, 33 per cent support, and 7 per cent services.[52] But support includes important items of defence production (communication systems, radar, spare parts, ammunition and so on).[53] Because this estimate is in US dollars, it may underestimate the equipment component of the arms flows. A ruble valuation would probably assign a lower price to services. It seems plausible to assume that military products constitute about 80 per cent of Soviet arms flows to non-Communist LDCs. Consequently, the value of these arms transfers was equivalent in 1974-7 to between 11.1 and 13.3 per cent of the outlays on the procurement of weapons and equipment for the Soviet Armed Forces.

According to the US Arms Control and Disarmament Agency, sales to Warsaw Pact and other Communist countries amounted to nearly half the value of sales to non-Communist LDCs in 1974-8 when measured in dollars.[54] (The dollar valuation is not significant here for our purposes since it is proportions that are at issue and no serious index number problem is apparent.) Sales to the Warsaw Pact and other Communist countries can therefore be estimated as equivalent to 5.5-6.6 per cent of the outlays on the procurement of weapons for the Soviet Armed Forces.

If it is assumed that exports are drawn from current production, then the proportion of total Soviet weapons production (here calculated as weapons production for domestic use plus sales to non-Communist LDCs plus sales to Warsaw Pact and other Communist countries) going for export was between 14 and 17 per cent (figures rounded). The assumption that exports come from current production is probably not wholly justified, even though the Soviet Union now exports mainly modern equipment.[55] But unless stocks were being drawn down (and there is no evidence of this), between 14 and 17 per cent of military production in 1974-7 went for export or for replacement of exported equipment. It must be stressed, however, that this estimate should be treated as provisional, given the number of steps and assumptions involved.

Some branches of the Soviet defence industry export more than others; the Soviet Union does not, for example, sell nuclear weapons or strategic missiles. Tables 2.9, 2.10 and 2.11 provide figures for some of the types of equipment exported between 1967 and 1980. It is clear from these tables that exports to non-Communist LDCs have been absorbing an increasingly high proportion of the output of the Ministry of the Defence Industry, which produces tanks, armoured vehicles and artillery, and of the Ministry of the Aviation Industry; a substantial part of the output of the Ministry of the Shipbuilding Industry is also sold abroad. In the period 1977-80 the Soviet Union produced 46 major surface combatants, and exported 25; 208 minor surface combatants and exported 107; 48 submarines and exported five.[56] Between 1967 and 1980 the number of combat aircraft exported to non-Communist LDCs was equal to one-third of the number produced by the Soviet Union. The same is true of tanks. The figure for artillery is much higher, that for armoured vehicles lower, and that for helicopters much lower. The tables should not, of course, be taken to indicate that exports are drawn from current production. Unfortunately the available data do not make it possible to say what types of weapon are exported in what numbers. Western studies show that in the 1950s and 1960s the Soviet Union exported older models, but that in the 1970s it sold its most modern equipment. In some cases non-Communist LDCs received the latest Soviet models before the Warsaw Pact forces did.[57]

In the 1950s and 1960s the Soviet Union offered its customers favourable terms: large discounts off list-prices, repayment periods of ten years at 2 per cent interest, and acceptance of local commodities in repayment.[58] Although Soviet equipment appears still to be comparatively cheap, the Soviet Union has come increasingly to insist on pay-

Table 2.9: Exports of Basic Types of Armament to Non-Communist LDCs, as a Proportion of Soviet Production, 1967-73

Units	Production (1)	Exports (2)	(2) as percentage of (1)
Combat aircraft	6,750	1,855	27
Helicopters	5,550	516	9
Tanks	27,750	6,164	24
Armoured vehicles	28,700	3,574	12
Artillery	13,200	10,069	76

Note: The categories of weapons are not comparable in every case. For example, the export figure for tanks includes self-propelled guns, while the production figure may not. Hence, the proportions given should be treated as approximate.
Sources: Column 1: The production figures for combat aircraft and helicopters are calculated from US Congress, Joint Economic Committee, Subcommittee on Priorities and Economy in Government, Hearings: *Allocation of Resources in the Soviet Union and China – 1978*, Part 4, Washington, DC: US Govt. Printing Office, 1978, p. 227. The figures for tank production are calculated from ibid., p. 225. Although the numbers are deleted on these charts, they can be calculated by inserting figures from the CIA testimony given in US Congress, Joint Economic Committee, *Statement of Major General Richard X. Larkin, Deputy Director and Edward M. Collins, Vice Director for Foreign Intelligence, Defense Intelligence Agency. Allocation of Resources in the Soviet Union and China – 1981*, Washington, DC: 8 July 1981, pp. 86-7. The figures for armoured vehicles and artillery production are taken from US Department of Defense, *Annual Defense Department Report. FY 1978*, Washington, DC: n.d., p. 27, Chart V-2. Column 2: The export figures are calculated from US Arms Control and Disarmament Agency, *World Military Expenditures and Arms Transfers, 1967-1976*, Publication 98, Washington,DC: July 1978, p. 161, and US, Central Intelligence Agency, National Foreign Assessment Center, *Arms Flows to LDCs: US-Soviet Comparisons, 1974-1977*, ER 78-10494OU, Washington, DC: November 1978, p. 8.

ment in hard currency.[59] This has been partly in response to the increased revenues of the oil-producing states, some of which – Iraq, Libya and Algeria – have been among the largest Soviet customers. But even Ethiopia, it seems, was asked to pay for its Soviet arms in hard currency.[60] The net result of this policy has been that 'by 1970 and in every year until 1978, arms exports kept the USSR's trade with LDCs out of the red'.[61] In 1975 the Soviet Union is estimated to have earned $800 million in hard currency from arms exports; in 1976 and 1977 the figure rose to $1.5 billion. It has been estimated that between 1971 and 1980 65 per cent of Soviet arms sales to LDCs were for foreign currency and brought in $21 billion in hard currency.[62] Hard currency earnings may continue to grow, given the big jump in the volume of agreements concluded between 1974 and 1979.

Table 2.10: Exports of Basic Types of Armament to Non-Communist LDCs, as a Proportion of Soviet Production, 1974-78

Units	Production (1)	Exports (2)	(2) as percentage of (1)
Combat aircraft	6,335	1,850	29
Helicopters	5,500	460	8
Tanks	12,425	5,750	46
Armoured personnel carriers	22,200	6,595	30
Anti-aircraft artillery	3,250	2,335	71
Field artillery	8,600	3,620	41
Surface-to-air missiles	227,000	15,050	7

Sources: Column 1: The production figures for 1976-8 are from US Congress, Joint Economic Committee, ***Statement of Major General Richard X. Larkin, Deputy Director and Edward M. Collins, Vice Director for Foreign Intelligence, Defense Intelligence Agency***, Washington, DC: 8 July 1981, pp. 86-7. The 1974 and 1975 figures for combat aircraft, helicopters, and tanks are from US Congress, Joint Economic Committee, Subcommittee on Priorities and Economy in Government, Hearings: ***Allocation of Resources in the Soviet Union and China – 1978***, Part 4, Washington, DC: US Govt. Printing Office, 1978, pp. 225, 227. The 1974 and 1975 figures for armoured personnel carriers and field artillery are calculated by averaging the figures for 1975 in Table 2.3 above. The 1974 and 1975 figures for anti-aircraft artillery and surface-to-air missile production are derived by taking the average production for 1976-8. This may lead to some errors, but is unlikely to affect the proportions in a substantial way.
Column 2: The figures for export are taken from US, Arms Control and Disarmament Agency, ***World Military Expenditures and Arms Transfers, 1969-1978***, Publication 108, Washington, DC: December 1980, p. 163.

Table 2.11: Exports of Basic Types of Armament to Non-Communist LDCs, as a Proportion of Soviet Production, 1977-80

Units	Production (1)	Exports (2)	(2) as percentage of (1)
Combat aircraft	5,100	2,100	41
Tanks and self-propelled guns	13,800	5,750	43
Armoured personnel carriers and armoured vehicles	21,000	7,600	36
Artillery pieces	12,100	10,500	87
Surface-to-air missiles	200,000	13,600	7

Sources: Column 1: US Congress, Joint Economic Committee, ***Statement of Major General Richard X. Larkin, Deputy Director and Edward M. Collins, Vice Director for Foreign Intelligence, Defense Intelligence Agency***, Washington, DC: 8 July 1981, pp. 86-7.
Column 2: Judith Miller, 'US Weapon Sales Expected to Dip', ***New York Times***, 4 October 1981.

While Soviet arms transfers may be primarily political in purpose, they do have two economic advantages and one military one for the Soviet Union. First, they reduce the unit cost of some types of equipment by making possible longer production runs, and they keep productive capacity employed. Second, they have become an increasingly important source of hard currency, at a time when hard currency earnings from other sources, for example from oil exports, may be about to decline (though only in 1973 has more than 10 per cent of hard currency earnings come from arms sales). Third, equipment that is exported may be used in battle and thus provide information about its performance in combat conditions. It is not clear how far the defence industry has been adapted to meet the demand for arms sales, though overall production plans must take account of the export market as well as of the needs of the Soviet Armed Forces. Recent expansion of tank and aircraft production facilities may be partly a response to the increase in arms sales.[63]

Conclusion

No one who studies the Soviet defence industry can fail to be impressed by the secrecy that surrounds it, and by the unsatisfactory nature of the estimates that have to be used in analyzing it. Because the sources are unreliable, any analysis must be provisional. The danger always exists that figures, once quoted, will assume a life of their own and that the attendant qualifications will be forgotten. Nevertheless, some general points can be made about the Soviet defence industry.

First, the Soviet Union has given high priority to the creation of military power, has established special mechanisms for extracting resources from the society and channelling them into the defence industry, and has tried to shield defence production from the shortcomings of the rest of the economy. As a result, the Soviet defence industry is in some important respects more effective than civilian industry. This does not mean, however, that the two sectors are wholly isolated from each other. On the one hand, the absence of a powerful and dynamic civilian technological base has certainly imposed constraints on the Soviet military-technological effort. On the other, the development and production of weapons has placed considerable demands on the economy as a whole. The defence sector is drawing resources from an economy that has been profoundly conditioned by the requirements of Soviet security. It is the demands of the state rather than of the individual consumer that have determined what is produced; hence the consumer goods market is underemphasized – as is indicated, for example, by the high proportion of radioelectronics production going to defence.

Second, foreign technology has played an important role in the Soviet defence industry, particularly in three periods: in the years of the first three Five-Year Plans (and especially of the first) when the defence industry was being established; during the war with Germany, when equipment was obtained from the Soviet Union's allies; at the end of that war, when the acquisition of German technology was of great importance for the post-war weapons development programmes. Since then, acquisition of foreign technology has played a smaller role in Soviet weapons technology, although there have been reports that some recent advances, for example in missile accuracy, have resulted from imports of foreign machinery. It is difficult, without detailed study, to assess the importance of foreign technology for the Soviet defence industry, but Soviet efforts to acquire technology abroad suggest that it has not been negligible. Its importance was perhaps greatest in the 1930s and in the mid-1940s. Since the late 1940s the Western powers have tried to restrict the export of strategic technology and this has forced the Soviet Union to rely more on its own R&D.[64]

Third, the system of economic planning and management that was formed in the Soviet Union in the 1930s has been called a 'war economy', not because of the priority given to defence, but because the Soviet leaders pursued their objectives through a high level of investment and the centralized allocation of resources.[65] This economic system was successful in pursuing the goals of military power and economic growth. But now a contradiction has emerged between the two goals. The present economic system (which is not fundamentally different from that of the 1930s) is successful in protecting the defence sector and ensuring that it performs more effectively than the rest of the economy. But it has now become a brake on civilian industry, in particular because it does not encourage the efficient use of resources and because it hampers technological innovation. The planning system worked well when large increments of capital and labour were available for industry each year; but it does not work well when economic growth has to come from more 'intensive' use of existing factors of production.[66]

In the 1970s the Soviet leaders tried to make the defence industry contribute more to the civilian economy by transferring some of its organizational and managerial advantages to civilian industry, particularly in the sphere of R&D. The effect of these measures do not appear to have been great, in part because not all the conditions that account for the relative success of the defence sector (for example the special role of the customer) can be reproduced in civilian industry.[67] It may be that better economic performance will come only through reform of the system of planning and management, and such a reform might affect the protection and privileges that the defence sector now enjoys.

All the available evidence suggests that the defence industry is continuing to expand. The DIA reported in 1980 that 'the growth in total floorspace has averaged nearly 3 per cent per year in the defence industry in the past five years'.[68] This expansion may be designed to accommodate civilian production, to meet the increased demand generated by foreign sales, or to expand production for Soviet forces. In spite of a steadily declining rate of economic growth (GNP grew at 5.8 per cent a year from 1959 to 1960, and at 2.8 per cent a year from 1976 to 1980) the defence industry has continued to receive high priority.[69]

No special consideration has been given in the Soviet Union to the problems of conversion from military to civilian production.[70] It is assumed that when the time comes, the planning system will be able to cope. Moreover, the Soviet economy is a 'deficit' economy: there exists a high level of pent-up demands for consumer goods and services, a shortage of labour in the industrial areas of the European part of the Soviet Union, and a shortage of funds for investing in the exploitation of the natural resources of Siberia. At the macro level, therefore, it seems clear that the resources now devoted to defence could be absorbed by the economy.

It is much more difficult to make any judgment about conversion at the micro level, because the data needed for specific analysis are not available. Soviet planners have made arrangements for conversion from civilian to military production, and if dual-purpose plants are now being used for military production, reconversion should not present great difficulties. It may be, indeed, that this grey area between purely military and purely civilian production is considerable, and this might ease the process of conversion. Conversion might also be eased by the fact that Soviet industry, including the defence industry, is concentrated in large industrial centres. Nevertheless, conversion at the plant level would certainly entail detailed planning and programmes for retraining the workforce,[71] and it might well lead to disruptions in production.

The defence industry enjoys a privileged position in the Soviet economy at present, with higher pay and better conditions of work than the civilian sector. The fear of losing these might lead to opposition to conversion from within the defence industry. Certainly the loss of these privileged conditions would require new patterns of behaviour from managers, engineers and workers.

It is only to be expected that, since military products constitute one-fifth of the final output of Soviet industry, conversion to civilian production would cause problems. It might even be necessary, if economic policy were to emphasize the production of consumer goods, to reform the system of economic planning and management to make production more responsive to consumer demand. But one can only conclude by saying (as most discussions of conversion do) that the difficulties of conversion, although considerable, are not insuperable. They could be overcome if the appropriate political conditions prevailed, and the political will existed to surmount them.

Notes

1. On the problem of estimating Soviet defence spending see, for example, F.D. Holzman, *Financial Checks on Soviet Defense Expenditures*, Lexington, Mass., Toronto and London: Lexington Books, 1975; Arthur J. Alexander *et al.*, *The Significance of Divergent US-USSR Military Expenditure*, N-1000-AF, Santa Monica, Calif.: Rand Corporation, February 1979, Abraham S. Becker, *The Burden of Soviet Defense. A Political-Economic Essay*, R-2752-AF, Santa Monica, Calif.: Rand Corporation, 1981; Franklyn D. Holzman, 'Soviet Military Spending: Assessing the Numbers Game', *International Security* 6 (Spring 1982): 78-101. Beyond the problem of estimation lies the problem of definition.The civilian and defence sectors of the Soviet economy are not wholly distinct, many products have both civilian and military uses, and many plants produce both civilian and military goods. In this paper 'defence industry' and 'defence sector' are used interchangeably to denote that part of the economy that produces military goods for the Soviet Armed Forces and for arms transfers to other countries. The 'defence industry group' refers to the ministries most closely engaged in military production; on this, see pages 62-67.

2. On the background see, for example, Philip Hanson, ' Estimating Soviet Defense Expenditure', *Soviet Studies* 30 (July 1978): 403-10.

3. L.G. Beskrovnyi, *Russkaya armiya i flot v XIX veke*, Moscow: Nauka, 1973, pp. 269-97 and 501-32; D.A. Kovalenko, *Oboronnaya promyshlennost' Sovetskoi Rossii v 1918-1920gg*, Moscow: Nauka, 1970, pp. 17-64 on the pre-revolutionary defence industry; Norman Stone, *The Eastern Front 1914-1917*, New York: Charles Scribner's Sons, 1975, pp. 194-211 on the political war-economy. The figure for aircraft production is taken from B.V. Shavrov, *Istoriya konstruktsii samoletov v SSSR do 1938 goda*, Moscow: Machinostroenie, 1969, p. 213.

4. K.E. Voroshilov, 'Strengthening the Defence of the USSR', in *From the First to the Second Five Year Plan*, Moscow and Leningrad: Cooperative Publishing Society of Foreign Workers in the USSR, 1933, p. 350.

5. A.C. Sutton, *Western Technology and Soviet Economic Development 1930 to 1945*, Stanford, Calif.: Hoover Institution Press, 1971, p. 248.

6. Julian Cooper, *Defence Production and the Soviet Economy 1929-41*, CREES Discussion Paper, Birmingham, UK: University of Birmingham, 1976, p. 29.

7. *Istoriya Velikoi Otechestvennoi Voiny Sovetskogo Soyuza*, vol. 2, Moscow: Voenizdat, 1961, p. 148.

8. G.S. Kravchenko, *Ekonomika SSSR v gody Velikoi Otechestvennoi Voiny*, Moscow: Ekonomika, 1970, pp. 202, 230, 351. In 1941 industrial production was 98 per cent of the 1940 level; in 1942, 77 per cent, in 1943, 90 per cent, in 1944, 104 per cent, and in 1945, 92 per cent. Ibid., p. 358.

9. E. Yu. Lokshin, *Promyshlennost' SSSR 1940-63*, Moscow: Politizdat, 1964, pp. 108-13, 122.

10. A.G. Zverev, *Zapiski ministra*, Moscow: Politizdat, 1973, p. 227.

11. For an analysis of the evolution of Soviet science policy in the post-war period see E. Zaleski *et al.*, *Science Policy in the USSR*, Paris: OECD, 1969.

12. See, for example, W.T. Lee, *The Estimation of Soviet Defense Expenditures, 1955-75. An Unconventional Approach*, New York and London: Praeger Publishers, 1977, p. 54. According to Khrushchev, the Soviet Armed Forces dropped from 5,763,000 men in 1955 to 3,623,000 in 1958. This would certainly be compatible with a drop in the production of conventional arms. See Khrushchev's speech in *Pravda*, 15 January 1960.

13. US Congress, Joint Economic Committee, *Statement of Major General Richard X. Larkin, Deputy Director and Edward M. Collins, Vice Director for Foreign Intelligence, Defense Intelligence Agency. Allocation of Resources in the Soviet Union and China – 1981*, Washington, DC: 8 July 1981, p. 83. [Hereafter cited as *JEC-1981*.]

14. Deployment of the SS-4 MRBM began in 1959; by 1965 750 SS-4 and SS-SS-5 M/IRBMs had been deployed. Deployment of the ICBM force got underway as a substantial effort in 1962, although four SS-6 ICBMs had been deployed before then. The important point to note is that the production of strategic missiles (by Soviet definition, these are missiles with a range of 1,000km or more, and thus include M/IRBMs) expanded under the Seven-Year Plan. There had been some production of missiles before that (for example, of the SS-3 MRBM, deployment of which began in 1955). According to *Military Production in the USSR*, London: Ministry of Defence, 1976, p. A-1, ICBM production reached about 400 a year in the late 1960s. According to the same source, about 1,000 SLBMs had been produced by 1979. M/IRBM production was at a virtual standstill between the mid-1960s and 1976.

15. It is difficult to obtain estimates of strategic missile production. Good figures for deployment exist, but deployment and production figures are not the same, as the following data show:

		1976	1977	1978	1979	1980
ICBMs:	production	300	300	200	200	200
	deployment	140	150	170	170	120
SLBMs:	production	150	175	225	175	175
	deployment	96	80	32	32	80
IRBMs:	production	50	100	100	100	100
	deployment		20	80	20	40

(Sources: The production figures are taken from *JEC-1981*, p. 87; the deployment figures for ICBMs and SLBMs are taken from Stockholm International Peace

Research Institute, *World Armaments and Disarmament, SIPRI Yearbook 1980*, London: Taylor and Francis, Ltd., 1980, pp. XLII-XLIII; the figures for IRBM deployment are taken from *The Military Balance*, London, International Institute for Strategic Studies, various years.)

Even allowing for a lag of one to two years between production and deployment (and assuming the figures to be accurate), it is clear that there is a considerable difference between the figures for production and those for deployment. This may perhaps be accounted for by the need to produce missiles for testing and also by stockpiling. According to American sources, the silos for the SS-17 and SS-18 ICBMs (both of which use cold-launch techniques) could be reused to fire other missiles (see US, Department of Defense, *Soviet Military Power*, Washington, DC.: 1981, p. 56). It is not clear whether the production totals for ICBMs and IRBMs include launchers for the space programme; the Soviet Union conducted about 75 launches a year in the earlier years. See Milan Kocourek, 'Rocket Technology', p. 511, in *The Technological Level of Soviet Industry*, eds. R. Amann, J. Cooper and R.W. Davies, New Haven, Conn.: Yale University Press, 1977.

16. In January 1960 Khrushchev stated that 'Our state possesses powerful rocket equipment. Military aviation and the Navy have lost their former significance with the contemporary development of military technology. This type of weapon is not being reduced, but replaced. Military aviation is almost all being replaced by rocket equipment. We have now sharply reduced and, apparently, will proceed to reduce further and even to stop the production of bombers and other obsolete equipment. In the Navy the submarine fleet is acquiring great significance, but surface ships can no longer play the role they played in the past'. *Pravda*, 15 January 1960.

17. US Congress, Joint Economic Committee, Subcommittee on Priorities and Economy in Government, Hearings: *Allocation of Resources in the Soviet Union and China – 1980*, Part 4, Washington, DC: US Govt. Printing Office, 1978, p. 225, [hereafter *JEC-1978*]; and *JEC-1981*, p. 86. The numbers in the chart in *JEC-1978* have been deleted, but can be estimated from the information in *JEC-1981*, pp. 86-7, and in the US, Department of Defense, *Annual Defense Department Report. FY 1978*, Washington, DC: 1975, p. 27.

18. *JEC-1978*, p. 227; *JEC-1981*, p. 87.

19. *JEC-1978*, p. 224.

20. Cooper, *Defence Production*, pp. 22-23. A.C. Sutton, *Western Technology*, p. 344, writes that a 'search of the OKW (*Oberkommando der Wehrmacht*) files fails to reveal a single plant in 1937-38 that was not devoting part of its capacity to war purposes. The German intelligence lists of plants producing war equipment were at the same time, in fact, comprehensive lists of all Soviet plants'.

21. *JEC-1981*, p. 83.

22. *Peking Review*, no. 45, 1975, p. 20. There were more than 40,000 enterprises in the Soviet Union in 1976 (cf. *Narodnoe Khozyaistvo SSSR za 60 let*, Moscow: Statistika, 1977, p. 170).

23. Seymour Melman, *Barriers to Conversion from Military to Civilian Industry*, paper prepared for the United Nations Centre for Disarmament, *ad hoc* Group of Governmental Experts on the Relationship between Disarmament and Development, New York: Columbia University, April 1980, mimeo, p. 5.

24. Stephen Rapawy, 'Regional Employment Trends in the USSR', p. 604, in Compendium: *Soviet Economy in a Time of Change*, vol. I, US Congress, Joint Economic Committee, Washington, DC: US Govt. Printing Office, 1979. A very high estimate of Soviet defence-industrial employment of 14 million was made by Checinski. See Michael Checinski, *The Military-Industrial Complex in the USSR: Its Influence on R&D and Industrial Planning and on International Trade*,

SWP-AZ2302, Fo.P1.III.2a/81, Ebenhausen, FRG: Stiftung Wissenschaft und Politik, September 1981, pp. 41-5.

25. Rapawy, 'Regional Employment Trends in the USSR', p. 609. These figures may be underestimates if, as in the United States, the defence sector is more labour-intensive than civilian industry. Labour-intensity in the US results from the need to produce small numbers of many high-technology items of equipment.

26. Cooper, *Defence Production*, p. 51.

27. For a critical analysis of the methodology of the CIA estimates, see Franklyn D. Holzman, 'Are the Soviets Really Outspending the US on Defense?', *International Security* 4 (Spring 1980): 86-104. See also Stockholm International Peace Research Institute, *World Armaments and Disarmament, SIPRI Yearbook 1979*. London: Taylor and Francis, Ltd, 1979, pp. 28-32. The problem of estimating Soviet defence spending is also discussed in Holzman, *Financial Checks on Soviet Defense Expenditures*, Alexander *et al.*, *The Signifiance of Divergent US-USSR Military Expenditure*; Becker, *The Burden of Soviet Defense*; and Holzman, 'Soviet Military Spending: Assessing the Numbers Game'.

28. For a survey of Western estimates of Soviet military R&D expenditure see David Holloway, 'Innovation in the Defense Sector', appendix, in *Innovation in Soviet Industry*, eds. R. Amann and J. Cooper, New Haven and London: Yale University Press, 1982.

29. Ibid., for further discussion.

30. Louvan E. Nolting and Murray Feshbach, 'R&D Employment in the USSR – Definitions, Statistics and Comparisons' in US Congress, Joint Economic Committee, *Soviet Economy in a Time of Change*, p. 746. Nolting and Feshbach give a figure of 828,100 scientists and engineers employed in R&D in the Soviet Union in 1978. According to the FY 1981 Department of Defense Program for Research, Development and Acquisition, 'over half' of Soviet 'scientific workers' (estimated at over 1,000,000) were engaged in military R&D. US Congress, House Armed Services Committee, Hearings: *Military Posture, Research and Development, Title II*, Washington, DC: US Govt. Printing Office, 1980, p. 79. What these figures actually signify is another matter, since the Soviet Union is less innovative than the United States (see Amann, Cooper and Davies eds., 'Rocket Technology') and the productivity of Soviet R&D manpower is therefore considerably lower than that of American R&D manpower.

31. Soviet security objectives are of course broader than merely matching American power. One Soviet source defines the external function of the Soviet Armed Forces as follows: 'defence of [the] state against attack by an aggressor; defence in community with the other fraternal armies of the whole socialist system and of each state of that system; help to peoples struggling for their freedom and independence; defence of peace in the whole world'. D.A. Volkogonov *et al.*, eds., *Voina i armiya*, Moscow: Voenizdat, 1977, p. 355.

32. US Congress, House Permanent Select Committee on Intelligence, Hearings: *CIA Estimates of Soviet Defense Spending*, Washington, DC: US Govt. Printing Office, 1980, p. 83.

33. *Marksizm-Leninizm o voine i armii*, 5th edn., Moscow: Voenizdat, 1968, pp. 258-9.

34. For further discussion of the institutional structure of the defence industry see Holloway, 'Innovation in The Defense Sector'.

35. *Materialy XXIV s'yezda KPSS*, Moscow: 1971, p. 46.

36. For further discussion see Holloway, 'Innovation in The Defense Sector'.

37. See also David Holloway, 'The Soviet Style of Military R&D' in *The Genesis of New Weapons' Decisionmaking for Military R&D*, eds. F.A. Long and Judith Reppy, New York and Oxford: Pergamon Press, 1980, pp. 139-58.

38. US Congress, Joint Economic Committee, Subcommittee on Economy and Priorities in Government, Hearings: *Allocation of Resources in the Soviet Union and China – 1977*, Part 3, Washington, DC: US Govt. Printing Office, 1977, p. 40 [hereafter cited as *JEC-1977*].
39. See, for example, the case studies in Amann, Cooper and Davies (eds.), *The Technological Level of Soviet Industry*.
40. *JEC-1977*, p. 40.
41. Ibid., p. 87.
42. See Holloway, 'The Soviet Style of Military R&D'; also Arthur J. Alexander, *Decisionmaking in Soviet Weapons Procurement*, Adelphi Papers Nos. 147/8, London: International Institute of Strategic Studies, Winter 1978/9.
43. US, Central Intelligence Agency, *Estimated Soviet Defense Spending in Rubles*, SR 76-10121U, Washington, DC: May 1976, p. 1.
44. P.V. Sokolov, *Political Economy*, Moscow: Voenizdat, 1974; quoted in *JEC-1981*, p. 40.
45. US, Arms Control and Disarmament Agency, *World Military Expenditures and Arms Transfers, 1969-78*, Publication 108, Washington, DC: 1980, pp. 159-62. In addition, the Soviet Union exported $525m. worth of arms to Yugoslavia, and $50m. to Finland.
46. US, Central Intelligence Agency, National Foreign Assessment Center, *Communist Aid Activities in Non-Communist Less Developed Countries, 1979 and 1954-79*, ER 80-10318U, Washington, DC: October 1980, p. 5. Kampuchea, Laos and Vietnam are included for the years before 1975. The figures in the table refer to estimated Soviet export prices. See US, Central Intelligence Agency, National Foreign Assessment Center, *Arms Flows to LDCs: US-Soviet Comparisons, 1974-7*, ER 78-10494, Washington, DC: November 1978, pp. 3-4.
47. CIA, *Communist Aid Activities in Non-Communist Less Developed Countries, 1979 and 1954-79*, p. 4.
48. Information on Soviet arms exports as a percentage of total Soviet exports from, US ACDA, *World Military Expenditures and Arms Transfers, 1969-1978*, p. 150. Checinski's estimates are found in, Checinski, *The Military-Industrial Complex in the USSR*.
49. US, Central Intelligence Agency, *Estimated Soviet Defense Spending in Rubles*, p. 13.
50. US, Central Intelligence Agency, National Foreign Assessment Center, *Estimated Soviet Defense Spending: Trends and Prospects*, SR 78-10121, Washington, DC: June 1978, p. 5.
51. US, Central Intelligence Agency, National Foreign Assessment Center, *Arms Flows to LDCs: US-Soviet Comparisons, 1974-77*, p. 6.
52. Ibid., p. ii.
53. Ibid., p. 1.
54. US, ACDA, *World Military Expenditures and Arms Transfers, 1969-1978*, pp. 158-62.
55. US, CIA, *Communist Aid Activities in Non-Communist Less Developed Countries, 1979 and 1954-79*, p. 1, and Roger F. Pajak, 'Soviet Arms Transfers as an Instrument of Influence', *Survival* 23 (July-August 1981): 168.
56. Production figures are from *JEC-1981*, p. 86. Export figures are from US Congress, House Committee on Foreign Affairs, Subcommittee on International Security and Scientific Affairs, Report: *Changing Perspectives on US Arms Transfer Policy*, Washington, DC: US Govt. Printing Office, 25 September 1981, p. 16.
57. US, CIA, *Communist Aid Activities in Non-Communist Less Developed Countries, 1979, and 1954-79*, p. 1, and Pajak, 'Soviet Arms Transfers', p. 168.
58. Pajak, 'Soviet Arms Transfers', p. 167.
59. CIA, *Communist Aid Activities in Non-Communist Less Developed*

Countries, 1979 and 1954-79, p. 5.

60. Ibid.

61. Ibid.

62. US, Central Intelligence Agency, National Foreign Assessment Center, *The Soviet Economy in 1976-77 and Outlook for 1978*, Washington, DC: August 1978, pp. 11-12; Wharton Econometric Forecasting Associates, *Centrally Planned Economies. Current Analysis*, Washington, DC: 22 January 1982, p. 2.

63. The CIA has reported increases in capacity for submarine, aircraft and tank production over the last ten years. *JEC-1981*, p. 84.

64. For some further discussion see David Holloway, 'Innovation in the Defense Sector'.

65. Oskar Lange, *Papers in Economics and Sociology*, Oxford: Pergamon Press, and Warsaw: Panstwowe Wydawnictwo Naukowe, 1970, p. 102.

66. For a discussion of these problems see Joseph S. Beiliner, *The Innovation Decision in Soviet Industry*, Cambridge: MIT Press, 1976.

67. See Julian Cooper, 'Innovation for Innovation', In R. Amann and J. Cooper, eds., *Innovation in Soviet Industry*.

68. *JEC-1981*, p. 83.

69. Abram Bergson, 'Soviet Economic Slowdown and the 1981-85 Plan', *Problems of Communism*, 30 (May-June 1981): 24-36.

70. Melman, *Barriers to Conversion from Military to Civilian Industry*, p. 29.

71. Ibid., pp. 33-49.

3 FRANCE

Edward A. Kolodziej

Historical Background

The development of France as a modern nation-state has been closely associated with the growth of an armaments industry. Almost simultaneously with the introduction of firearms in France in the fourteenth century, the crown assumed monopolistic control over the production of powder. Conferring the right to make or buy powder marked the first entry of the state and crown into the fabrication and sale of arms and munitions. The production of hand weapons and firearms similarly fell under progressive state control.[1] The wars of Louis XIV required greater national oversight and assured supplies other than those then afforded by private outlets. Beginning with naval armament, Colbert created arsenals at Toulon and Rochefort, upgraded the foundries at Lyon and Strasbourg, and organized another at Douai in 1669. The standardization of equipment developed slowly. By the end of the eighteenth century production of heavy armaments was increasingly uniform throughout France.[2]

After the fall of the crown, revolutionary leaders were no less preoccupied with arms manufacture than the kings who had preceded them. The Committee on Public Safety under Lazare Carnot's impulsion created hundreds of arms manufacturers and brought them under state direction to protect the revolution from internal subversion. By 1794 Paris was producing 750 muskets a day, an output greater than the rest of Europe.[3]

The Industrial Revolution and the growth of a capitalist economy prompted a major transformation of France's armaments industry. The superiority of Prussian arms and the use of railroads to speed troops to battle spurred these changes. In 1885, the legislature of the newly formed Third Republic laid the foundation for modern arms production. Private arms manufacture was given priority over government arsenals. Legislators expected that private manufacturers, motivated by profits and patriotism, would produce more and better arms. There was also the hope that foreign sales would boost French commerce. Fears that the bourgeois-dominated Third Republic might be overturned -- a fear prompted by the Commune scare – also encouraged the new accent on private production.[4] The success of these reforms may be measured by the quantity and quality of France's arms production in World War I. France assumed the lead in aircraft production and supplied not only its own needs but also most of those of the American expeditionary army. The development of the 75 mm. light artillery piece, used in profusion with devastating effect on German troops in World War I, illustrated the high quality of French arms.

The French arms industry, which had grown to historic high levels in World War I, declined after 1918 as the nation returned to peacetime. The defensive strategy of the French High Command also relaxed demands on the arms production system. German rearmament in the 1930s forced a re-examination of industrial policy at approximately the point that the Leftist Popular Front gained ground. Its election in 1936 paved the way for the nationalizations of selected, but key, private firms and installations engaged in producing arms.[5]

The French defeat and German occupation in World War II decimated much of France's arms industry. It took much of the life of the Fourth Republic from 1948 to 1958 to reconstitute and renovate the French arms production system. Successive governments were faced with the conflicting tasks of re-building the civilian economy while fighting a series of colonial wars. Meeting these immediate problems left few resources for the rapid regeneration of French arms production capacity. Government arsenals and shipyards were, however, gradually rebuilt. The aircraft industry was reorganized in 1949. As early as 1947, work began on several military models. The Ouragon 450, produced by Dassault, was the first military jet aircraft sold to the French air force; Israel and India subsequently purchased an additional 164 copies.[6] This success was soon followed by the development of the Mystère fighter, the Fouga Magister trainer, the Noratlas transport, Djinn and Alouette helicopters, and a series of tactical missiles. All of these systems were sold to French and foreign armed forces.[7] By 1949 work began on the MX-series of armoured vehicles that became the mainstay of the French army and a product in steady demand by ground armies around the world. An Atomic Energy Commission was also established as early as 1946. Fuelled heavily by defence funds, it organized France's explosion of an atomic device in 1959. A half decade later, France created a nuclear strikingforce,[8] initially composed of Mirage IV aircraft, armed with kiloton-yield atomic weapons.

American economic and military aid indirectly assisted the renovation of the French arms industry and underwrote France's colonial wars. France received more than $4 billion in military aid, twice as much as any other NATO partner.[9] 'Off-shore' sales contracts in the 1950s stimulated French military production and broadened its capacity. In 1954, the US placed an order for 224 Mystère IV aircraft that were to be delivered to the French air force. Copies were subsequently sold to Israel (60) and to India (110).

French arms production plans under the Fifth Republic developed through a series of *loi-programmes* or programme-laws. These basic laws, first introduced into legislation in 1960, relate French military doctrine, weapon systems, and force levels to France's industrial capacity and economic resources over a five-year period. Four have been developed covering 1961-5, 1966-70, 1971-5, and 1977-82.[10] These documents establish arms production goals and detail the financial arrangements to support targeted levels of production. Each year the production schedule and appropriations are updated to take account of a variety of factors, including economic conditions, price changes, availability of raw materials, employment problems, and technological and scientific developments.[11]

Organization of the Arms Industry[12]

At the hub of the French arms industry is the General Direction for Armament (Délégation Générale pour l'Armement or DGA). It is a technical service within the Ministry of Defence which has the overall mission of coordinating the complex sprawl of manufacturing, research, and development centres concerned with arms design, testing, and production. Created in 1961, it centralized the separate technical arms production agencies that were attached to the armed services. DGA is the interface between the armed forces and arms industry. The military requirements of the military services are translated into procurement, research, and development programmes that are under DGA direction. Over the years there has developed a very close relation between DGA personnel and the leadership and managers of the arms industry. DGA, composed largely of military engineers, has therefore tended to reflect more of an industrial than an armed forces orientation.

DGA not only oversees the arms industry but is also heavily engaged in arms production itself. Historically, French arms production has been composed of three distinct organizational structures. The first is an elaborate arsenal and shipbuilding complex under the direction of the DGA. It has centres throughout France, extending along the

Atlantic and Mediterranean coasts from Cherbourg to Toulon and along a north-south line between Paris and Toulouse. Installations, however, have not been placed along the land approaches in the northwest and northeast of France, because these areas have been the scene of repeated invasions over the centuries.

Semi-public firms form the second tier of the French arms production system. Most of these firms produce both civilian and military goods. DGA is responsible for the direction of the military part of output in cooperation with these separately organized and funded public companies. Until the nationalization of major armament sectors in 1981, there were four major groupings. The first and most important is the Société Nationale des Industries Aérospatiale (SNIAS). Its creation marked an evolution of more than 40 years of gradual consolidation of the aircraft industry into a predominantly nationalized public sector. Aérospatiale controls approximately 40 per cent of France's aerospace industry. Its helicopter, ballistic missile and space, and tactical missile divisions are key production components of France's nuclear striking force and conventional arms.

SNECMA (Société Nationale d'Etude et de Construction de Moteurs d'Aviation) is France's principal aircraft engine manufacturer and in Europe second only in size to Britain's Rolls-Royce. Formed shortly after World War II, it produces most of the engines for Aérospatiale and Dassault. A smaller, private firm, Hispano-Suiza, was absorbed by SNECMA in 1968. SNECMA's initial success can be attributed to its development of the Atar engine. It is presently perfecting the M53 engine for France's F1 and Mirage 2000 and 4000. It is also collaborating with General Electric (GE) in the production of the CF6-50 engine, the power unit for the Airbus A300 and an option for airlines flying Boeing 747s and DC-10s. With GE, it is also developing the CFM-56 engine for airline needs in the 1990s. Its Olympus 593 engine, built with Rolls-Royce, is the Concorde's power plant.

The National Society of Powder and Explosives (Société Nationale des Poudres et Explosifs, SNPE) is another important semi-public firm. Its activities are divided into four programmes: self-propulsion systems, powder and explosives, engineering processes, and chemical development. Its nine installations are spread evenly across France along two north-south axes between Brest and Toulouse in the west and Le Bouchet in the north and St Chamas in the east.[13]

The Renault group may also be included in this unique French creation of the semi-public firm. Reformed after World War II, it was essentially nationalized as the Regie Nationale des Usines Renault for industrial and commercial purposes 'endowed with a civil personality and financial autonomy'. Five ministries have representatives on the Administrative Council, including defence, to assure governmental control and direction.[14] Renault also purchased controlling shares in Berliet and Saviem in 1975, and these corporations were subsequently grouped into a single unit in 1978. With sales in 160 countries Renault is the second largest group in France and the sixteenth largest European firm in terms of annual business receipts.[15]

The third and, until recently, the fastest growing segment of the French arms industry has been the private sector. The election of a Socialist government in the spring of 1981, has, however, significantly reduced the private sector's share of arms production. Its nationalization programme comprised nine industrial groups, including Thomson-Brandt, Dassault-Bréguet and Matra, which are major arms producers.[16] The previous balance between private and public firms in electronics, airframe, and missile development has now been tipped in favour of the public sector. With Aérospatiale, Dassault-Bréguet dominates the French airframe industry. Matra and Aérospatiale share the missile market equally. SNECMA is France's principal engine manufacturer; Turboméca, a private firm, is a distant second. The nationalization of Thomson-Brandt assures government direction of the armaments sector of the electronics industry. The smaller Dassault Electronique, a subsidiary of Dassault-Bréguet, has also fallen under governmental control with the takeover of the parent corporation.

Most major naval military construction is undertaken in government-run shipyards. Private firms, comprising six main companies, build small craft and provide repair and maintenance services. With upgrading, firms like Chantier d'Atlantique and Chantier Naval de la Ciotat are capable of constructing large surface warships. Dubigeon Normandie at Nantes constructs the Daphné class diesel submarine and Constructions Méchanique de Normandie (CMN) at Cherbourg produces mine-sweepers, fast patrol boats, and missile-firing gunships.[17]

Defence Budget: Impact on Governmental Spending, GNP, and Military Procurement

Defence spending tends to be viewed from two different and often conflicting perspectives in France. The armed services, as consumers of the budget, approach defence spending from the perspective of the material support that it will furnish French forces. Neither the requirements of maintaining a strong and competitive economy nor the imperatives of preserving an economically viable defence industry are the principal priorities of this group. While an indigenous arms production system is valued, the quality, quantity, and accessibility of the weapons to be acquired is of prime concern. A focus on military requirements and performance has led to proposals for weapons systems, like the Future Combat Airplane in the 1970s, that were beyond the financial possibilities of the government. It has also heightened pressures for more and better arms. Conventional weapons have been a particularly sensitive area of concern since defence spending has consistently fallen below projected levels of weapons acquisition.

Another and growing sector of the security community, including DGA personnel and their allies in industry, is more attuned to the economic implications of defence spending, the constraints and opportunities afforded by overall and targeted spending for economic development, and the requirements of global competition not only in arms but in industrial areas, like electronics, computers, and nuclear energy, associated with military production. While the development of an independent nuclear force was primarily motivated by strategic considerations, it was also viewed by governmental proponents as a vehicle for the modernization of the French economy.[18]

The May 1968 uprising, which almost topped the de Gaulle government and threatened the Fifth Republic, brought to the surface widespread dissatisfaction with increased defence expenditures at the expense of social programmes. Fifth Republic leaders could ignore these constraints on the level and distribution of defence spending only at their peril. A second shock was delivered by the oil crisis of the early 1970s. Two-thirds of France's energy needs are produced by oil, and over 90 per cent of France's oil needs must be imported. The oil crisis increased inflationary pressures within the economy. The result was a decreased competitive position in world trade and increased unemployment. In the context of these deteriorating economic conditions, undiminished by domestic pressures for social expenditures, the government pursued several courses of action that had important implications for defence spending, arms procurement, and the development and support structure for the arms industry.

Table 3.1 provides an initial point of departure to analyze trends in governmental decision-making. It reveals a decline between 1960 and 1976 in defence spending relative to overall governmental expenditures and to gross national product. Also of interest is the decade-long

Table 3.1: French Defence Budget and Selected Components as a Percentage of Total Governmental Budget and Gross National Product ($ billion)

Year	Defence[a] Budget (initial)	Defence Procurement Costs		Defence Personnel Costs		Total Government Budget (initial)	GNP	Defence Expenditure as percentage of	
								Total Budget[b]	GNP
	$	$	%	$	%	$	$		
1960	3.35	1.20	35.7	2.17	64.3	11.8	61.1	28.4	5.5
1961	3.41	1.16	34.1	2.25	65.9	12.7	66.5	26.8	5.1
1962	3.50	1.13	32.4	2.37	67.6	14.2	74.4	24.6	4.7
1963	3.76	1.59	42.2	2.17	57.8	15.7	83.4	23.9	4.5
1964	4.02	1.85	45.9	2.17	54.1	17.5	92.5	23.0	4.3
1965	4.21	2.10	49.9	2.11	50.1	18.7	99.2	22.5	4.2
1966	4.46	2.28	51.2	2.18	48.8	20.5	107.9	21.8	4.1
1967	4.77	2.47	51.8	2.30	48.2	23.1	116.4	20.6	4.1
1968	5.06	2.63	51.9	2.43	48.1	25.2	127.6	20.1	4.0
1969	5.08	2.52	49.6	2.56	50.4	28.4	139.3	17.9	3.6
1970	4.90	2.36	48.1	2.54	51.9	27.8	140.9	17.6	3.5
1971	5.21	2.44	46.9	2.77	53.1	29.2	157.5	17.8	3.3
1972	6.17	2.88	46.6	3.29	53.4	35.0	194.5	17.6	3.2
1973	7.81	3.70	47.4	4.11	52.6	44.1	250.2	17.7	3.1
1974	7.95	3.70	46.6	4.25	53.4	45.7	264.4	17.4	3.0
1975	10.22	4.45	43.5	5.77	56.5	60.5	335.3	17.0	3.1
1976	10.46	4.38	41.9	6.08	58.1	61.2	346.8	17.1	3.0
1977	11.89	4.87	41.0	7.02	59.0	68.3	384.1	17.4	3.1
1978	15.00	6.32	42.1	8.67	57.9	88.7	474.2	16.9	3.2
1979	18.12	7.83	43.2	10.27	56.7	107.8	574.6	16.8	3.2
1980	20.97	9.44	45.0	11.53	55.0	124.1		16.9	

Notes: a. 'Defence Procurement Costs' plus 'Defence Personnel Costs' will not always equal the figure given for 'Defence Budget (initial)' due to rounding while changing francs into dollars.
b. Note that oscillations in percentage increases in budget are due partly to rate of inflation and shifting exchange rate, expressed in dollars. For example, between 1967 and 1970, defence spending (*crédit de paiements*) increased from 26.4 to 27.19 billion francs. However, the rate of the franc declined relative to the dollar and, therefore, the dollar value of defence spending is shown to have fallen.
Source: France, Assemblée Nationale, Commission de la Défense Nationale et des Forces Armées, ***Avis sur le projet de loi de finances pour 1978, Défense: Dépenses en Capital***, October 11, 1977, pp. 13-17. Rate of exchange for the francs is drawn from International Monetary Fund, ***International Financial Statistics: 1977***, XXI, no. 5 (May 1977), pp. 166-7 and *Ibid.*, May 1981, p. 152. Note slight discrepancies in GNP registered by IMF and the parliamentary report. Differences are due partly to calculation but are not substantial. Years 1960-77 are drawn from the parliamentary report noted above; 1977-9, from the IMF report, 1980. Figures for years 1977 through 1980 are drawn from France, Assemblée Nationale, Commission des Finances de l'Economie Générale et du Plan (CFEGP), ***Rapport sur le projet de loi de finances pour 1980, Défense Considérations Générales***, no. 1292, 2 October 1979, pp. 27, 81, 108-10.

decline in the proportion of defence funds devoted to arms procurement after 1968. The defence budget is divided into two major titles or categories, covering spending on personnel (Title III) and procurement (Title V). The latter is a key indicator of support for the defence industry, since governmental purchases comprise a majority of the business receipts for the arms industry. It is precisely this segment, however, that experienced the largest decline after 1968.

The percentage of governmental spending devoted to defence has progressively fallen since 1960. With the end of the Algerian War in 1962, spending for defence declined from 24.6 per cent of the government's budget to a low of 17 per cent in 1975. A drop in percentage can be noted for each successive year until 1975. Sharpest declines are registered between 1960 and 1962 as the Algerian War wound to a close; between 1968 and 1969 in the wake of the May uprising; and between 1973 and 1975 during the period of adjustment to higher oil prices and inflation. The slide downward was to be arrested after 1976 with the adoption of the fourth five-year plan for defence. A heightened accent on conventional forces with continued development and modernization of tactical and strategic nuclear forces was to reverse the downward trend.[19] This upward trend was expected to continue until the end of the current defence plan in 1982, levelling off defence expenditures at approximately 20 per cent of the government's budget; however that has not happened.

The same declining trend in defence spending relative to GNP can also be noted. The decline is steady and smooth. At the height of the Algerian War in 1960, over 5 per cent of GNP was earmarked for defence. By 1965, defence spending had dropped more than one percentage point. The slide continued gradually until 1968. Defence spending stagnated between 1968 and 1972 as the government responded to pressures for non-defence spending. Meanwhile, GNP continued an upward progression after the upheavals of 1968 were weathered. By 1974 approximately 3 per cent of GNP was attributable to defence spending. A slightly greater proportion of GNP has been scheduled for defence needs under the fourth programme-law. Annual increases in defence spending since 1977 suggest that a higher spending plateau will be reached by the end of the five-year cycle in 1982.[20]

These same downward trends may be measured in constant francs which may provide a better indicator of the reduced role played by defence spending in governmental expenditures on GNP. Table 3.2 indexes the growth in GNP, governmental expenditures, and defence

Table 3.2: Indexed Growth of Gross National Product, State Expenditures and Defence Spending for Selected Years, 1959-80 (1959 = 100)

Year	GNP	State Expenditures	Defence Spending
1959	100	100	100
1962	135	125	110
1965	181	165	132
1968	230	223	159
1971	326	289	183
1974	478	394	243
1977	702	600	371
1980	1,008	939	562

Source: France, Assemblée Nationale, Commission des Finances de l'Economie Générale et du Plan, *Rapport*, no. 1292, 2 October 1979, p. 83.

spending from 1959 to 1980 in terms of 1959 prices. GNP and governmental spending have been roughly in step with each other. GNP has increased ten times while governmental spending has grown a little more than nine times. Defence spending, however, has progressed by about five and one-half times during this period.

Of equal interest is the proportion of the defence budget devoted to personnel and to procurement. During the Algerian war, it was understandable that most of the defence budget would be spent on personnel. This division is reflected in the years between 1960 and 1962 when approximately two-thirds of defence expenditures were devoted to personnel. This pattern changed abruptly in 1963 when procurement expenditures jumped more than ten percentage points. Procurement was increasingly favoured during this period, reaching a height of almost 52 per cent in 1968. The shift to personnel expenditures for much of the succeeding decade is related not only to the uprising of 1968 and the need to hold the loyalty of the armed forces but also to long overdue reforms in personnel services, facilities, training, salary, social benefits, and pensions that had been subordinated to the drive to develop a nuclear force in the 1960s. After 1968 the government was under pressure to spend more on social programmes nationally and on socially oriented programmes within the military.

Within the procurement envelope, strategic and tactical nuclear weapons programmes, which in number of personnel absorb only a small fraction of French military forces, have been favoured since the 1960s. As Table 3.3 notes, spending for nuclear forces jumped from

Table 3.3: Defence Spending for Research, Development and Procurement for Nuclear and Conventional Weapons for Selected Years, 1963-78 ($ billion)

Force Type	1963		1966		1969		1972		1975		1978	
	Amount	%	Amount	%	Amount	%	Amount	%	Amount	%	Amount	%
Nuclear	.327	20.6	1.129	49.5	1.033	41.1	1.008	35.0	1.495	33.6	2.055	32.6
Conventional	1.259	79.4	1.153	50.5	1.483	58.9	1.869	65.0	2.950	66.4	4.256	67.4
Total	1.586	100.0	2.282	100.0	2.516	100.0	2.877	100.0	4.445	100.0	6.311	100.0

Source: France, Assemblée Nationale, CDNFA, *Avis*, No. 1350, 11 October 1977, pp. 22-3.

one-fifth to one-half of the procurement budget between 1963 and 1966. Defence spending speeded construction of the nuclear reactor at Marcoule from which the plutonium required to build the first French atomic device was extracted and the enrichment plant at Pierrelatte. These facilities also became foundation stones in the development of the civilian atomic energy programme. The latter is important for French plans to decrease reliance on imported oil and to develop an export capability to partially offset the initially high capitalization cost of nuclear energy development. After the heavy start-up spending of the 1960s, the military nuclear programme now absorbs approximately one-third of the procurement, research, and development budget of the defence ministry.

Economic Growth, Arms Production and Sales

An Export-driven Arms Industry

The competing imperatives of military preparedness and economic growth dictated two general strategies to relax the tension between these goals. The first, and dominant, strategy centred on an increasingly export-driven arms industry. The second, which is discussed later, focused on the encouragement of joint weapons development programmes with other states. The implications of the first strategy went beyond the expected objective of arms producers to cut unit costs through increased series production, to finance additional research and development, and to facilitate the purchase of a larger number of weapons by indigenous armed forces than could be supported by the defence budget alone. By the end of the 1970s, French arms exports had become a key component of France's trade structure and an important part of its overall strategy to maintain its balance of payments position and politically tolerable levels of domestic economic growth and employment. Increasingly, weapons were treated like any other commodity for sale.[21]

For several reasons, the French arms industry achieved a powerful global position. American preoccupation with the Vietnam war during the 1960s opened new markets for French arms orders as American arms production was geared to the war and to the needs of selected allies. French arms merchants, encouraged, supported, and in some cases led by DGA, were quick to fill growing world demand. Other potential competitors, like Great Britain, West Germany, and Japan, were either unable or unwilling to fill the void. The British government under Labour Party rule in the 1960s hesitated to increase foreign arms sales and relinquished Britain's place as the third largest arms seller to France by the end of the 1960s. West Germany and Japan had developed a profitable export industry based on civilian goods. As defeated powers in World War II, expanding arms production would have opened their governments to charges of militarization from domestic and foreign critics. The success of French arms, particularly Mirage aircraft in the 1967 Arab-Israeli war, further promoted sales. On the other hand, the weakness of French exports in non-military sectors, relative to its OECD competitors, encouraged increased reliance on military sales. Of particular interest to the French was the sale of capital goods, like heavy complex armaments, whose added domestic value was high relative to consumer products for semi-processed goods. Also the employment opportunities afforded by arms sales for highly skilled, technical personnel, were considerable. These varied economic and political factors gave France a comparative advantage in selling arms abroad. Moreover, the size of the defence industry, measured by personnel, essentially remained stable from the late 1960s into the 1980s despite a decreasing demand for procurement by French military forces and continued social and potentially disruptive demands for increased social expenditures, economic growth and full employment. For French planners, arms sales were progressively viewed as a support for output that could not otherwise be sustained by domestic demand and as a complement to sagging civilian sales abroad.[22] For French planners, producing arms was an instrument of economic and social welfare. Arms could also act as product leaders to expand trade abroad and to open markets for civilian goods and services.

French dependence on arms transfers may be better understood by a brief glimpse at overall French economic dependence on increased trade. Table 3.4 relates French imports and exports to GNP for 1955, 1965, and 1975. Trade as a component of GNP rose from 20 per cent in the 1950s to 32 per cent in the 1970s. During this period, France experienced the highest economic growth rate in Western Europe, averaging 5.8 per cent a year between 1960 and 1972. However, decreased productivity and increased pressures for social spending narrowed France's competitive position by the middle 1970s. Sharp increases in oil prices increased the costs of production and upset its balance of payments position. The value of oil imports was 9.4 per cent of French imports in 1973; it doubled to 18 per cent in 1976 although the monthly volume of oils imports increased by only 12.7 per cent as a consequence of conservation efforts initiated by the French government and industry.

To finance increased oil imports, France embarked upon an active

Table 3.4: Ratio of French Imports and Exports to Gross National Product, 1955, 1965 and 1975 (billion francs)

	1955	1965	1975
(a) Exports	17.36	50.24	227.20
(b) Imports	16.74	51.27	231.18
(c) GNP	172.20	489.80	1,442.40
Ratio[a] in %:			
(a) + (b) / (c)	20	21	32

Note: a. Rounded to nearest 1 per cent.
Source: International Monetary Fund, *International Financial Statistics: 1977, Supplement, Annual Data: 1952-76*, XXI, no. 5 (May 1977), p. 132.

trade expansion programme in which increased arms sales played a key role. Arms sales increased both in absolute amount and as a percentage of overall trade. Table 3.5 summarizes the value of French arms deliveries from 1972 through 1977. These increased from $800 million in 1972 to almost $3 billion in 1977. Sales of aircraft, including missiles, were the largest component although the percentage share of this sector to total arms deliveries fell from 75 per cent in 1972 to a little more than 60 per cent in 1977. Sales of ground, naval, and electronics equipment increased at a faster rate. Meanwhile, French exports increased in this period from $26.43 billion to $64.97 billion for an increase of approximately 150 per cent. Arms sales grew at an even faster rate over these five years, reaching 300 per cent by 1977. The rate of increase in arms exports was, therefore, almost twice that of the growth in total exports.

Table 3.6 presents several revealing measures of the importance of French arms sales to the French economy. Arms transfers have grown from 3 per cent of overall exports in 1972 to 4.6 per cent in 1977. This is a substantial and growing proportion of French trade. Arms are not merely a supplement to the defence budget, off-setting the high cost of weapons development and production; they are integral components of France's competitive position in international markets.

Viewed from the perspectives of oil imports and balance of payments, arms exports are also key elements of French prosperity. The cost to France of oil more than quadrupled between 1972 and 1977, rising from $2.7 billion to $11.9 billion in 1977. Arms sales covered approximately 30 per cent of French oil imports in 1972; this percentage fell to a low of 14 per cent in 1973; and rose again to 25 per cent in 1977. The jump in oil prices, however, required a rapid

Table 3.5: French Arms Deliveries, 1972-77 ($ millions)

	1972	1973	1974	1975	1976	1977
Arms Deliveries[a]	800	1,175	1,386	1,944	2,435	2,992
(a) Air	607	867	871	1,166	1,695	1,830
(b) Ground	103	174	269	316	514	688
(c) Naval	23	64	117	196	37	174
(d) Electronics	68	70	130	266	190	299

Note: a. 'Arms Deliveries' totals may not equal the sum of 'Air', 'Ground', 'Naval', and 'Electronics' deliveries due to rounding when converting francs into dollars.

Source: France, Sénat, Commission des Finances, *Rapport général sur le projet de loi de finances pour 1975*, no. 99, III, Annexe 40, *Défense, Dépenses en Capital*, 1974; France, Assemblée Nationale, Commission de la Défense Nationale et des Forces Armées, *Avis sur le projet de loi de finances pour 1978, Défense: Dépenses en Capital*, 11 October 1977; and France, Assemblée Nationale, Commission des Finances de l'Economie Générale et du Plan (CFEGP), *Rapport sur le projet de loi de finances pour 1980. Défense: Considérations Générales*, no. 1292, 2 October 1979. Exchange rate from International Monetary Fund, *International Financial Statistics: 1977*, XXI, no. 5 (May 1977), pp. 166-7, and (May 1981), p. 152.

Table 3.6: Arms Transfers Related to Exports, Oil Imports and Commercial Balances ($ billion)

	1972	1973	1974	1975	1976	1977
EXPORTS	26.430	36.480	46.160	53.010	57.160	64.970
Arms deliveries/Exports	3.000	3.200	3.000	3.700	4.300	4.600
OIL IMPORTS	2.700	3.500	9.800	9.700	11.500	11.900
Arms deliveries/Oil imports	29.600	33.600	14.100	20.000	21.200	25.100
IMPORTS	27.000	37.550	52.840	53.940	64.460	70.500
Balance: Exports and Imports	-.570	-1.070	-6.680	-.930	-7.300	-5.530
Arms Sales	.800	1.175	1.386	1.944	2.435	2.992
Deficit without Arms Sales	-1.370	-2.245	-8.066	-2.874	-9.735	-8.522

Source: France, Assemblée Nationale, Commission de la Défense Nationale et des Forces Armées, ***Avis sur le projet de loi de finances pour 1978, Défense: Dépenses en Capital***, 11 October 1977, and France, Assemblée Nationale, Commission des Finances de l'Economie Générale et du Plan (CFEGP), ***Rapport sur le projet de loi de finances pour 1980. Défense: Considérations Générales***, no. 1292, 2 October 1979. Exchange rates from International Monetary Fund, *International Financial Statistics: 1977*, XXI, no. 5 (May 1977), pp. 166-7, and (May 1981), p. 152.

expansion in French arms sales merely to maintain the ratio of arms deliveries to oil imports of the pre-oil crisis period. While civilian goods and services certainly contributed their share to conserving oil imports, arms sales comprised a larger proportion of the increase. Similarly, arms deliveries account for a notable share of France's efforts to maintain an equilibrium in trade. As Table 3.6 suggests, France's trade deficits would have been sizeable if it had not been able to step up its arms deliveries. In the absence of arms deliveries, deficits in 1972 might conceivably have risen from $570 million to $1.37 billion; in 1977 the deficits might have increased from $5.53 billion to $8.52 billion.

Two other features of French arms transfers also bear examination. These refer to the contribution of arms transfers to total capital exports and to the geographic distribution of French deliveries. During the 1970s French arms sales provided significant impetus to the expansion of France's capital goods exports. Overall capital exports are estimated to have increased by 48 per cent between 1974 and 1976 while arms deliveries expanded by 75 per cent.[24] These percentages suggest the weaker competitive position of France's civilian capital goods industry relative to its principal competitors within the OECD. Arms deliveries, moreover, account for a significant share of France's trade with the developing world. In 1970, less than 50 per cent of French arms deliveries were outside the Atlantic community and Europe. By 1976, the proportions had significantly shifted to 15 per cent to the developed world and 85 per cent to developing countries.[25] Within this latter sector, it was not surprising to discover that some of the largest gains in trade were with oil producers. The first major breakthrough occurred with Libya in 1970 with the sale of 110 Mirage aircraft to the Gaddafi regime. This spectacular sale was followed by others to Saudi Arabia and Iraq throughout the 1970s and into the 1980s. This circumstance puts not only French arms deliveries but also overall French trade in a precarious position. The success of both are partly dependent on the vicissitudes of regional politics, ranging from internal upheavals, like Iran, to conflicts between regional states, like Iran and Iraq and the Arab states and Israel.

Size of the Arms Industry

If domestic spending for arms is combined with arms deliveries, some notion can be gained of the overall importance of arms production to the French economy. These totals are likely to underestimate the economic impact of arms production since precise data are not publicly available with respect to the contribution of the arms industry to each

Table 3.7: Estimated Business Turnover for French Armament Industry, 1972-77 ($ billion)

	1972	1973	1974	1975	1976	1977
(1) Domestic Procurement	2.880	3.700	3.700	4.450	4.380	4.870
(2) Delivery of arms to other states	.800	1.175	1.386	1.944	2.435	2.992
Total (1) and (2)	3.680	4.880	5.090	6.390	6.820	7.860
Arms sales as percentage of business turnover	21.7	24.1	27.2	30.4	35.7	38.1
Arms sales as percentage of domestic procurement	27.8	31.8	37.5	43.7	55.6	61.4

Source: France, Assemblée Nationale, Commission de la Défense Nationale et des Forces Armées, ***Avis sur le projet de loi de finances pour 1978, Défense: Dépenses en Capital***, 11 October 1977, and France, Assemblée Nationale, Commission des Finances de l'Economie Générale et du Plan (CFEGP), ***Rapport sur le projet de loi de finances pour 1980, Défense: Considérations Générales***, no. 1292, 2 October 1979. Exchange rates from International Monetary Fund, ***International Financial Statistics, 1977***, XXI, no. 5 (May 1977), pp. 166-7, and (May 1981), p. 152.

sector of the French economy or to the economy as a whole. Table 3.7 combines domestic and foreign demand for French arms and relates arms deliveries to business turnover. By these measures of business turnover in arms production, the value of arms produced for sales to French and foreign armed forces doubled from $3.68 billion in 1972 to $7.86 billion in 1977, a faster rate of expansion than that experienced by the index of industrial prices during the same period. Also revealing is the increasing proportion of French arms exports in business turnover. In 1972, this ratio stood at 21.7 and increased steadily by approximately three percentage points a year to 38.1 per cent in 1977.

The gradual replacement of domestic demand by foreign sales may also be measured by the ratio of arms deliveries to the procurement portion of the defence budget. This ratio more than doubled from 1972 to 1977. In 1972, the ratio stood at 27.8 per cent. Five years later the ratio of arms sales to domestic demand had grown to 61.4 per cent. The implications of these measures are clear enough. While the size of the arms industry has remained relatively static over the past twenty years, a point elaborated below in the discussion of personnel engaged in arms production, the proportion of the industry's resources devoted to export has progressively grown each year.

Table 3.8 relates the business turnover figures calculated in Table 3.7 to GNP. Arms production represented approximately 1.8 per cent of GNP in 1972; it steadily rose to 2 per cent by 1977. As these figures suggest the rate of increase in value for arms production was greater than the growth in GNP. During this five year period, the value of arms production (in current prices) increased by 113 per cent; GNP rose only 92 per cent. Not only is France's trade position significantly dependent on arms export, but domestic production and employment are tied more tightly than ever to arms production and productivity.

Sectoral Dependency and Arms Production

The increasing dependence of the French economy and exports on arms is, not surprisingly, mirrored in the dependence of key industrial sectors on arms contracts from the French state and abroad. The bulk of domestic arms spending is earmarked for French industry whose mixed composition has already been sketched above. Funds going to purely governmental agencies for the production of arms and research and development (R&D) is considerably less. For example, in 1979, only 28 per cent of the state's appropriations for military hardware and R&D were earmarked for the arsenals and establishments directly under DGA control and the Atomic Energy Commission. The remaining 72

Table 3.8: Business Turnover for Arms as a Percentage of GNP ($ billion)

	1972	1973	1974	1975	1976	1977
Business Turnover	3.68	4.88	5.09	6.39	6.82	7.86
GNP	199.65	256.85	275.43	336.51	352.20	384.10
Percentage	1.8	1.9	1.8	1.9	1.9	2.0

Source: France, Assemblée Nationale, Commission de la Défense Nationale et des Forces Armées, *Avis sur le projet de loi de finances pour 1978, Défense: Dépenses en Capital*, 11 October 1977, and France, Assemblée Nationale, Commission des Finances de l'Economie Générale et du Plan (CFEGP), *Rapport sur le projet de loi de finances pour 1980. Défense: Considérations Générales*, no. 1292, 2 October 1979. Exchange rates from International Monetary Fund, *International Financial Statistics, 1977*, XXI, no. 5 (May 1977), pp. 166-7, and (May 1981), p. 152.

per cent was divided among industrial suppliers. The distribution of contracts to industry across industrial sectors is outlined in Table 3.9 according to the distribution of contracts for arms production and R&D for 1979. Spending for electronics, including avionics, missiles and munitions, and nuclear development leads the list of major expenditures in both categories. These three areas comprise 75 per cent of the R&D contracts let principally by the Ministry of Defence through DGA and about 50 per cent of arms production. Naval aeronautics accounts for an additional 25 per cent of arms production.

How important these contracts are for the survival of selected but key industrial sectors can be better appreciated by closely examining Table 3.10 which summarizes business turnover for key industries in 1979, including aerospace, electronics, mechanics, automobiles, and armour. Currently the aerospace and electronics industries are heavily dependent on arms sales and R&D contracts for their survival. Efforts to diversify the French aerospace industry through the Airbus programme and through the sale of SNECMA engines, developed in cooperation with General Electric, have made some headway in world markets, but not nearly enough to protect the industry from any slackening of world demand for advanced fighter aircraft, helicopters, and tactical missiles. In the aerospace industry, the three main firms – Aérospatiale, Dassault-Bréguet and SNECMA – derived 50 per cent or more of their business receipts from arms. The percentages of turnover in arms production to total business receipts in 1979 were as follows: Aérospatiale, 50 per cent; Dassault-Bréguet, 94 per cent; and SNECMA, 75 per cent. For the three years between 1977 and 1979,

Table 3.9: Distribution of State Contracts for Arms in Key Industrial Sectors, 1979 (percentages)

Sector	Research and Development	Production
Electronics	27.2	22.2
Naval Aeronautics	14.9	26.2
Combat Vehicles	2.0	5.8
Engineering Materials and Vehicles	.4	4.8
Armaments	1.9	4.3
Munitions and Missiles	27.5	15.2
Naval Construction	2.5	12.4
Nuclear	20.1	6.7
Diverse	3.5	2.4
Total	100.0	100.0

Source: France, Assemblée Nationale, Commission des Finances, *Rapport sur le projet de loi de finances pour 1981, Dépenses en Capital*, no. 1976, 9 October 1980, pp. 272-3.

Dassault-Bréguet exported, respectively, 67, 69, and 66 per cent of their production in military equipment and aircraft.

The two major electronics firms of Thomson-CSF and Electronique Marcel Dassault (EMD) were similarly tied closely to arms sales. In 1979, 61 per cent of the business receipts of Thomson-CSF and 75 per cent of those of EMD were garnered from military contracts.[26] Matra, a major producer of tactical missiles and advanced electronic equipment, received almost 60 per cent of its orders from arms in 1978. Its dependency on arms has grown despite efforts by management to diversify its activities to include urban transport systems and even racing cars. Military contracts rose from 46 per cent of business turnover in 1974 to the 60 per cent level in 1978.[27]

Table 3.10: Industrial Sector Dependency on Domestic and Foreign Arms Contracts, 1979 (percentages)

Sector	Civil	Military
Aerospace	27	73
Electronics	36	64
Mechanics, Automobiles and Armour[a]	83	17

Note: a. Includes six major firms: Thomson-Brandt, Manurhin, Luchaire, Panhard, RVI, and SNPE.
Source: France, Assemblée Nationale, Commission des Finances, *Rapport sur le projet de loi de finances pour 1981, Dépenses en Capital*, no. 1976, 9 October 1980, pp. 176, 187, 189.

Research and Development

The amount and proportion of private and public spending for research and development has also become a public issue in France.[28] There has been considerable concern about whether France's scientific and engineering base is keeping pace with competitors elsewhere. A lag in R&D spending threatens the French position in several ways. First, the quality of French arms is placed in question and this affects the pursuit of an independent military posture. Choices are imposed on the arms industry that of necessity foreclose promising options. In the late 1970s several programmes had to be abandoned for lack of research and developmental funds. The Future Combat Aircraft gave way to the Mirage 2000, based on a different technology, because of skyrocketing costs. Similarly, in 1977, the army had to cutback spending for the design and development of a new generation of helicopters. In 1978, three other programmes – the cruise missile, observation satellites, and low altitude radar – were retarded. The Navy was also prevented from starting a number of proposed studies and of slowing the rate of work on several others. A Ministry of Defence supported study concluded that 'such a situation was all the more harmful since, alone, a sustained policy of research had permitted the directorate for Naval Construction to achieve the development of weapon systems competitive on an international level and capable of assuring arsenals of export outlets'.[29]

The economic competitiveness of the French arms industry, as the study just quoted recognizes, is also tied to adequate levels of R&D spending. Since the 1960s, however, France has been a net importer of technology. In 1973 the technological deficits stood at approximately $71 million and subsequently jumped to $250 million a year later. The deficits in 1975 and 1976 were also high, reaching almost $200 and $250 million, respectively. The French have also been unable to cover their technological purchases abroad. Even in 1975, one of France's best years, the rate of coverage of technological imports over exports reached only 51.2 per cent. The trend in technological exports over imports does not appear encouraging. If one uses an index of imports and exports of technological transfers based on 1963 prices, spending on foreign technology progressed from 100 in 1963 to 324 in 1974 while receipts received by French firms progressed from 100 to 289. France's relative position with respect to the value of technological imports during this period did not improve and may be said to have even slipped somewhat.[30]

Despite France's weak competitive position in high technology, there has been a steady rise in national spending for all forms of research and development since the 1960s. Expenditures rose from $1.3 billion in 1963 to $3.6 billion a decade later in 1972 and subsequently to $8.4 billion in 1978. This increase may also be measured in constant francs using 1959 as a base. R&D spending stood at 6.4 billion francs in 1963, at 18.3 billion francs in 1972 and at 37.7 billion francs in 1978.

Spending on military research and development, however, did not fare as well, prompting concern within the security community that France was losing out in the struggle to remain technologically competitive.[31] Between 1970 and 1974, during the years of the Pompidou administration, spending for military research and development averaged 5.92 per cent of the defence budget. For the first three years of the Giscard d'Estaing regime the average percentage dropped to 4.87 per cent, falling from 5.5 per cent in 1975 to 4.5 per cent in 1977. In constant francs, however, military R&D hardly changed from 1972 through 1977, oscillating between 2.2 and 2.4 billion francs during this period. An upward surge in R&D spending can be detected in the implementation of the fourth programme-law in the late 1970s. Between 1978 and 1981, spending for military research and development gradually rose from 2.5 billion francs to 3.2 billion francs.[32]

France's position remains clouded by continued dependency on foreign sources for key advanced technology. During the 1970s, 91 per cent of the total of France's payments for foreign technology was attributed to five states: the United States (53 per cent), Switzerland (17 per cent), Great Britain (8 per cent), West Germany (7 per cent), and the Netherlands (6 per cent). Principal imports concerned optics, chemicals, electronics, mechanical equipment, textiles, foundry materials and pumps. In military hardware, French industry lags in electronics and computers, semi-conductors, integrated circuits, radar, infra-red, diesel motors, and jet aircraft engines.[33] Offsetting this shaky position, however, is success in nuclear technology, turbine engines, small anti-tank missiles, helicopters, and combustible materials.

Foreign Bilateral and Multilateral Programmes

The scale and composition of French arms production has been increasingly internationalized. Despite repeated assertions of national independence, successive French Republic governments have signed more arms agreements with other European states than any other European power or any other state, including the United States, within the Atlantic Alliance. As suggested above, the incentives for cooperation are compelling: economies of scale to lower the rate of growth of expenditures and cost of new weapons, soaring expense of developing modern weapons, the need for access to new technologies beyond national means, strict national limits on available resources for defence spending, the desire to equip French forces with adequate numbers of advanced weapons to meet security requirements, and assured markets for weapons sales through the guaranteed purchases of the cooperating states. These factors have proven sufficiently strong to prompt France's membership in the Independent European Programmes Group (IEPG) within the Atlantic Alliance in order to promote joint weapons development, although it still refuses to return to the NATO framework or even to join the Eurogroup, whose function is to articulate a European perspective within the alliance.

Most of France's arms production relations, however, have been developed along more pragmatic, country-to-country lines to promote specific projects of interest to French armed forces. Table 3.11 lists the principal projects that have been launched under the Fifth Republic with other states, major French and foreign contractors, and armed service utilization of these arms or weapons systems. The list is largely confined to joint design, development, and production projects. These are distinguished from production programmes licensed by the French government or technical assistance provided by French corporations to foreign purchasers. The focus is on arms which the government has decided for a variety of reasons to develop or to produce jointly with one or more other states. Production licensing arrangements, shared production opportunities in executing foreign arms orders, and technology transfers are increasingly key parts of foreign arms sales. These arrangements are categorized, however, under external commercial relations of French arms suppliers rather than as genuine joint development and production programmes.[34]

The organizational forms created to execute an international accord assume a variety of patterns. In some instances a separate corporation has been established. Euromissile, for example, is a creature of Aérospatiale and Messerschmidt-Bölkow-Blohm (MBB). Created in 1972, it produces the Milan, Hot, and Roland missile systems. Aérospatiale and Turboméca have also developed a set of lucrative helicopter and tactical missile development and production arrangements with Westland, BAC, Hawker-Siddeley, and Rolls Royce. These have produced the Lynx, Puma, and Gazelle helicopters and the highly successful Exocet missile.

Table 3.11: Selected Joint Military Development and Production Programmes between France and Other European States

Programme	Cooperating Nations[a]	Major Contractors	Contracted or Start of Study[b]	Service
Transoll	G	Aérospatiale	1959	Army/Air Force
Atlantique marine patrol aircraft	G, GB, B, N	Bréguet, Sud-Aviation Dornier (G), Rolls-Royce (GB), ABAP (B), Fokker (N)	1960	Navy/Air Force
Hawk SAM missile	G, I, B, N	Consortium under SETEL grouped around Thomson-Houston	1960	Army
Martel ASM anti-radar (AS.37) TV-guided version (AJ168)	GB	Matra and Hawker-Siddeley	1963	Air Force/Navy
Milan anti-tank missile	G	Euromissile (composed of Aérospatiale and MBB)	1964	Army
Hot	G	Euromissile	1964	Army
Roland I (clear weather) and II (all weather) SAM	G	Euromissile	1964	Army
Jaguar dual-purpose training and attack aircraft (various models)	GB	SEPECAT, grouping British Aircraft Corporation (BAC) and Dassault-Bréguet; motors by Rolls-Royce (RR) and Turboméca	1964	Air Force
Helicopters Puma[c] Gazelle Lynx	GB	Puma/Gazelle, Aérospatiale and Turboméca; Lynx, Westland and RR	1967	Air Force/Army
Exocet[c] AM 38 and AM 39, air-to-surface (ASM) SSM naval missile	GB	Aérospatiale, BAC Hawker-Siddeley	1967	Air Force (AM 39) Navy (AM 38, AM39)

Alpha-Jet	G	Dassault-Bréguet and Dornier (airframe); SNECMA, Turboméca, MTU and KHD (motors)	1969	Air Force
Otomat (SSM, ASM) (Several successive versions)	I	Matra, Thomson-CSF Thomson-Brandt, Turboméca SNPE, Oto Melara	1969	Navy
NATO-ASSM	G	Euromissile	1977[d]	
PAH-2/HAC helicopter	G	Euromissile	1978[d]	
AS-21, ASM	G	Euromissile	1978[d]	
Leopard/AMX 30 Tank	G	GIAT, Krauss-Maffei & Krupp	1980[d]	

Notes: a. West Germany (G), Great Britain (GB), Belgium (B), Netherlands (N), Italy (I).
b. The sources are not always clear on these two points, the official accord between governments and the start of study and design of the project by one or more states.
c. Design is French; production essentially licensed by France to Great Britain.
d. These projects are in the development stage.
Sources: Various sources have been consulted. Most important are the annual review of world armaments issued by the Stockholm International Peace Research Institute (SIPRI), *World Armaments and Disarmament: The SIPRI Yearbook*, Stockholm: Almqvist & Wiksell, 1968-77, and London: Taylor and Francis, 1978-80; *Jane's All the World's Aircraft*, London: MacDonald, 1959-75; Defense Marketing Systems, *Foreign Military Markets, France*, Greenwich, Conn.: various dates; and, France, Assemblée Nationale, Commission de la Défense Nationale et des Forces Armées, *Avis sur le projet de loi de finances pour 1975, Défense: Dépenses en Capital*, no. 1233, pp. 93-6.

Dassault cooperated with Israel to develop the Jericho surface-to-surface missile. SEPECAT (Société Européenne de Production de l'Avion Ecole de Combat et Appui Tactique) was formed by Louis Bréguet and the British Aircraft Corporation to produce the Jaguar fighter-attack aircraft. In addition, Dassault-Bréguet built the Alpha Jet in cooperation with Dornier, while Matra has produced tactical missiles with British and Italian firms. Nor are the French adverse to joining international consortia. SETEL (Sociéte Européenne de Téléguidage) was formed around the Thomson-Houston group to produce Hawk SAMs under licence from US manufacturers. Moreover, in 1960, several French firms signed an accord with firms from West Germany, Belgium, the Netherlands, and Great Britain to build the French-inspired Atlantique Marine patrol aircraft.

These military production agreements have been complemented by civilian projects, like Concorde, Airbus A300 and Mercure civil aircraft, Olympus 593, M45H, CF6-50 and CFM-56 motors, and the Ariane rocket to launch European-sponsored satellites into space. The contract between SNECMA and General Electric has led to a partial break-through in the market for commercial jet engines since the engine jointly produced by both firms can be used as a substitute in recon-ditioning Boeing civil aircraft. The Ariane also promises to earn foreign exchange for France since it permits an alternative source of booster capability to the United States to launch foreign satellites.

These international civilian and military programmes entangle France in a set of industrial and commercial relations that cannot be easily broken even though a project, like the Concorde, is a financial disaster. Because national agreements are at stake, an international project, once commenced, tends to have greater stability and life expectancy than simply indigenous efforts that can be abandoned without provoking a foreign policy incident. The fabric of transnational and multilateral ties that is woven is not necessarily viewed as a constraint on national discretion and manoeuvrability. Administrators and technicians value the insulation that these programmes have from domestic pressures to cut or eliminate them. A market for the weapons which are produced is usually found among the cooperating states. The producer states also enjoy together a greater network of access to foreign markets whose exploitation can prove mutually profitable. It should also be noted that a major explanation for the range and com-plexity of France's international involvement in weapons development is attributable to the strength of its arms industry. Its vigour and sophistication act as magnets for joint projects. While the French government has been pursuing an active cooperation strategy, it also has something to offer other states bent on drawing on France's experience as a successful arms producer and seller.

The Labour Force of the Arms Industry

The arms industry also employs a large work force, many of whose members are highly skilled technicians, engineers, and professional personnel. Official estimates put employment in the arms industry at 280,000 or approximately 4.5 per cent of France's industrial force. The DGA has 75,000 people working on producing arms (50,000) or organizing and coordinating ground, naval, and air arms production (25,000). The latter oversee a total of 255,000 persons in industrial work. Of these, 105,000 work in the Paris area; the remaining 175,000 are distributed throughout France. Besides the DGA, 25,000 find themselves in nuclear arms production, 105,000 in the aeronautics industry, 40,000 in electronics, and 35,000 in associated industries. The government also estimates that 190,000 work for domestic consumption and 90,000 for exports.The 90,000 figure appears low in the light of increasing importance of arms sales as a percentage of overall arms production. A figure close to 130,000-140,000, or approximately half the work force, would seem closer to the mark.

While personnel are widely distributed throughout France, the Paris area has the largest concentration of workers, technicians, and central offices. This is especially true for private firms which prefer easy access to government agencies and foreign embassies, facilitating both governmental support and foreign sales. Finding a pool of skilled workers and technicians outside a metropolitan area or locating them in a region has proved difficult. Nevertheless, approximately 55,000 personnel are found in the west of France, largely devoted to arsenal work for the army and navy. Another 100,000 or more are located in the southeast of France and in the southwest near Toulouse where air and naval production and repair facilities are located.

Were it not for France's arsenal system, it is conceivable that arms production would be even more concentrated than it is. Brest, Tulle, and Tarbes, for example, are heavily dependent on arms production for what industry they have. These areas are subject to fluctuations in defence spending and to international competition. Their inability to diversify their economy, makes them vulnerable to swings in the economy and unstable levels of employment.

Conclusions

The French arms industry is the oldest national system for producing arms in the Western world. Despite the setbacks of World War II and the immediate postwar period, the French government and French firms have built so strong and sophisticated a production system that France now commands undisputed title to third place, behind the superpowers, of arms producers around the world. The economic benefits associated with producing arms, inextricably tied to a tenacious policy of national independence in defence, make it difficult for any regime in France, whether of the right or left, to change the existing pattern of dependency of the French economy and France's export position on arms production and sales. While the Socialist government of François Mitterand was quick to nationalize important segments of the arms industry in 1981, it was no less swift to assure foreign contractors that French arms would continue to flow without interruption. Meanwhile, the defence budget has not been cut and, indeed, a seventh nuclear submarine has been ordered and new weapons developments in satellites, tactical missiles, electronic warfare, nuclear weaponry, and aircraft have gone forward. If history is a guide, there would almost certainly be a French arms industry even if the economic burden were great. Thanks to French enterprise and ingenuity that economic burden has been somewhat lightened by the successful commercialization of French arms and know-how on world markets. For the foreseeable future, and for compelling strategic, economic and political reasons, the French arms industry is here to stay.

Notes

1. A. Bigant, *La Loi de Nationalisation des Usines de Guerre*, Paris: Editions Domat-Montchrestien, 1939, provides a brief survey of the development of the French armaments industry and the exercise of state control over its operations. Longer but of less use is H.L. Marquand, *La Question des Arsenaux: Guerre et Marine*, Paris: Plon, 1923.

2. Ibid., *La Loi de Nationalisation*, pp. 14-17, and Ernest Picard, *L'Artillerie Française au XVIIe*, Paris: Berger-Levrault, 1906.

3. Bernard and Fawn Brodie, *From Crossbow to H-Bomb*, rev. edn., Bloomington: University of Indiana, 1973, pp. 106-7.

4. Bigant, *La Loi de Nationalisation*, pp. 18-19.

5. Maurice Tardy, 'La Nationalisation et le Contrôle des Industries de Guerre', *Revue de France*, 1 September 1938, pp. 84-108. The best single source on military arms production in the interwar period is Robert Jacomet, *L'Armement de la France: 1936-1939*, Paris: Editions Lajeunesse, 1945.

6. *Jane's All the World's Aircraft*, London: Macdonald, 1954, p. 135.

7. Ibid., 1954-60 editions, *passim*.
8. See Lawrence Scheinman, *Atomic Energy Policy in France under the Fourth Republic*, Princeton, NJ: Princeton University Press, 1965.
9. United States, Department of Defense, *Foreign Military Sales and Military Assistance Facts*, Washington, DC: December 1978, p. 17.
10. The lapse for 1976 derived from delays arising from a major reformulation of planning and budgetary procedures within the budget decisional process of the Ministry of Defence. The fourth and latest military *loi-programme* is reviewed in the following parliamentary document: France, Assemblée Nationale, Commission de la Défense Nationale et des Forces Armées, *Rapport sur la programmation militaire pour les armées, 1977-1982*, no. 2292, 2 vols.
11. France, Assemblée Nationale, Commission des Finances de l'Economie Générale et du Plan, *Rapport sur le projet de loi de finances pour 1978*, no. 3131, Annexe no. 50, *Défense: Dépenses en Capital*, p. 3.
12. General discussions of the French arms industry may be found in Jean-François Dubos, *Ventes d'Armes: Une Politique*, Paris: Gallimard, 1974, pp. 65-87; Centre Local d'Information et de Coordination pour l'Action Non-Violente (CLICAN), *Les Trafics d'Armes de la France*, Paris: François Maspero, 1977, pp. 41-57; Françoise Sirjacques, *Determinanten der Französischen Rüstungspolitik*, Frankfurt: Peter Lang, 1977; and David Greenwood, *The Organisation of Defence Procurement and Production in France*, Aberdeen Studies in Defence Economics, Aberdeen: Centre for Defence Studies, 1980.
13. Yearly report of SNPE, 1974.
14. DAFSA, *Report on Renault*, December 1977. DAFSA is a private investment analysis group.
15. *Le Nouvel Economiste: Classements des Premières Sociétés Françaises et Européenes*, 1977, pp. 22, 180.
16. *Le Monde*, Bilan économique et social, 1981, pp. 28-35.
17. Greenwood, *The Organization of Defence Procurement*.
18. Robert Gilpin, *France in the Age of the Scientific State*, Princeton, NJ: Princeton University Press, 1968.
19. For an analysis of French strategic policy as a principal factor explaining increased defence spending, see Edward Kolodziej, 'French Security Policy: Decisions and Dilemmas', *Armed Forces and Society* 8 (Winter 1982): 185-222.
20. Please note that these percentages are lower than those reported by NATO, US sources, or the International Institute for Strategic Studies. NATO figures for 1975-80 appear in France, Assemblée Nationale, Commission de la Défense Nationale et des Forces Armées, *Avis sur le projet de loi de finances pour 1982, Défense: Politique de Défense de la France*, no. 473, 16 October 1981, p. 30. The range of *defence* spending is between 3.8 and 4.0 per cent of GNP in these sources.
21. See Edward Kolodziej, 'Determinants of French Arms Sales: Security Implications,' pp. 137-76, in *Threats, Weapons and Foreign Policy*, eds. Pat McGowan and Charles W. Kegley, Jr., Beverly Hills, Calif.: Sage, 1980, for further discussion and citations.
22. See France, Ministère de Défense, *Livre Blanc sur la Défense Nationale*, Tome I, 1972, for a statement of official policy in favour of active exports, especially pp. 54-5.
23. France, Institut National de la Statistique et des Etudes Economiques (INSEE), *Tableaux de l'Economie Française*, Paris: 1976, p. 6.
24. Kolodziej, in McGowan and Kegley, Jr., 'French Security Policy', pp. 156-7.
25. Ibid., p. 146.
26. France, Assemblée Nationale, Commission des Finances, *Rapport sur le*

projet de loi de finances pour 1981. Dépenses en Capital, no. 1976, 9 October 1980, p. 188.

27. Michel Béhar, 'L'Industrie d'Armement en France', *Cahiers Français* no. 201 (May-June 1981): 13.

28. See Gilpin, *France in the Age of the Scientific State*, and, more recently, *Recherche et Développement à Fins Militaires*, Paris: Les Cahiers de la Fondation pour les Etudes de Défense Nationale, 1978.

29. Ibid., p. 113.

30. Ibid., p. 110.

31. Ibid., and see discussion of the Mayer Report, commissioned but later embargoed for publication by the Ministry of Defence. Apparently one of the principal criticisms of the report concerned the level of spending for military R&D. *Le Monde*, 29 September 1976, p. 13.

32. France, Assemblée Nationale, Commission des Finances, *Rapport sur le projet de loi de finances pour 1981, Dépenses en Capital*, pp. 161-2.

33. *Recherche et Développement à Fins Militaires*, pp. 112-3.

34. This distinction underlies the categorizations of cooperation between France and other states which are sketched in the useful article of Marc Defourneaux, 'France and a European Armaments Policy', *NATO Review* no. 5 (1979): 19-25. See also his 'Indépendence Nationale et Coopération Internationale en Matière d'Armements', *Défense Nationale* 35 (February 1979): 35-48. For a probing review, see Elliott R. Goodman, 'France and Arms for the Atlantic Alliance: The Standardization-Interoperability Problem', *Orbis* 24 (Fall 1980): 541-71.

4 THE FEDERAL REPUBLIC OF GERMANY

Michael Brzoska

Four Phases of Arms Production in the Federal Republic of Germany

After World War II, the Allied forces treated German arms producing firms and their management and owners as they did the military, though not the party leadership. The better-known and more prominent leaders were jailed and arms production facilities were destroyed. Under Allied occupation laws, arms production was prohibited until 1951.[1] German arms suppliers for World War II, such as Krupp, promised never to produce arms again.

In 1950, however, it became clear that Germany would rearm. The West German government had opted for 'the West' in the developing 'cold war' confrontation. The way was paved towards greater German sovereignty and the employment of the resources available in West Germany in the interest of 'the West'.[2] The return to arms production was legalized in 1951 when the Allied prohibition of arms production was changed into an Allied supervision of arms production and arms exports. Some former arms producing firms that had not been successful in conversion into civilian fields started to produce military-related items for the West European armed forces. Among these were tents, communications equipment, optical instruments and special machinery for the armed forces, even small warships. In 1954, warships were exported to Ecuador and Indonesia.[3]

Thus, arms production did not have to start from scratch in 1955 when orders for the new Bundeswehr were placed by the federal government. It was made even easier by the fact that various designers, especially aircraft designers, had continued to operate in foreign countries, for example, in Spain (Dornier, Messerschmidt) and in Argentina (Tank). Even so, 1955 can be seen as the end of the first phase of – very modest – arms production in the Federal Republic of Germany (FRG), mainly for export. In 1955, control over arms production passed from the Allies to the West German government although the government agreed to Western European Union (WEU) prohibitions on German production of atomic, biological and chemical weapons as well as on mid- and long-range missiles and warships above a certain size. While the size limitations for warships were gradually changed until

Table 4.1: Arms Production and Employment in the FRG

	Arms Production Demand, 1,000 million DM	Domestic Part percentages	Domestic Arms Production Demand 1,000 million DM	Arms Exports million DM	Arms Production Value-added, 1,000 million DM	Arms Production Value-added as Part of Manufacturing Value-added, percentages	Number of People Working in Arms Production (1000s)
1956	2.1	40	0.8				
1957	3.0	40	1.2				
1958	3.6	40	1.4				
1959	4.2	41	1.7				
1960	3.7	50	1.8				
1961	5.3	66	3.5	56	3.6	1.98	253
1962	7.7	64	4.9	280	5.2	2.65	342
1963	9.1	55	5.0	283	5.3	2.57	332
1964	8.4	44	3.7	712	4.7	2.11	275
1965	7.2	60	4.3	407	4.7	1.88	249
1966	6.1	74	4.5	272	4.8	1.83	270
1967	7.7	75	5.8	231	6.0	2.18	217
1968	6.8	68	4.6	391	5.0	1.74	224
1969	7.6	70	5.3	396	5.7	1.75	224
1970	6.8	81	5.5	689	6.2	1.72	218
1971	7.2	85	6.1	453	6.6	1.68	240
1972	8.4	81	6.8	1,050	7.9	1.87	219
1973	9.2	83	7.6	371	8.0	1.71	203
1974	9.8	78	7.6	516	8.1	1.64	213
1975	10.2	80	8.1	967	9.1	1.83	218

1976	11.7	74	8.6	1,651	10.3	1.92	229
1977	12.0	81	9.7	1,925	11.6	2.02	229
1978	12.8	82	10.5	2,500[a]	13.0	2.09	229
1979	13.7	82	11.2	2,500[a]	13.7	2.01	231

Note: a. Estimate.

Sources: Columns 1 to 3: *Wehrtechnik*, no. 7 (1977): 101-3: *Wehrtechnik*, no. 12 (1978): 46-50; Federal Ministry of Defence, *White Paper 1979, The Security of the Federal Republic of Germany and the Development of the Federal Armed Forces*, Bonn: 1979, p. 36; C. Bielfeldt, *Rüstungsausgaben und Staatsinterventionismus*, Frankfurt: Campus, 1977, p. 88; Bundesminister der Verteidigung, *Haushaltsentwürfe*, no. 14, various years.

Column 4: US Arms Control and Disarmament Agency, *World Military Expenditures and Arms Transfers*, Washington, DC: various years.

Column 5: Columns 3 and 4.

Column 6: Column 5 and *Statistisches Jahrbuch der Bundesrepublik Deutschland*, Wiesbaden: various years, Table 'Produktionswert, Vorleistungen und Wertschöpfung nach zusammengefassten Wirtschaftsbereichen' and Table 'Entstehung des Auslandsproduktes und der Einkommen nach zusammangesfassten Wirtschaftsbereichen'.

Column 7: Column 6 and *Statistisches Jahrbuch der Bundesrepublik Deutschland*, Wiesbaden: various years, Tabelle 'Erwerbstätige nach Wirtschaftsbereichen'.

they were completely eliminated in 1981, there is still no West German arms production in the other four categories. Theoretically, all West German arms production continues to be controlled by the WEU, but this regulation has never had any practical consequences.

One important characteristic of the second phase was the sharp increase in the volume of orders: the first substantial procurement payments to West German arms producers reached $200 million in 1956. Also important was the fact that, at least for the governing parties, arms production had become politically acceptable once again. The opposition to rearmament and arms production was strongest in labour union circles, although there was also some opposition from industrialists who feared that arms production would draw too much labour and capital into non-productive areas.[4]

The second phse of arms production in the FRG was further distinguished by its technological limitations. The initial weapon systems in Bundeswehr arsenals were primarily foreign-bought and license-produced weapon systems; indigenous efforts were the exception. Among the major weapons produced in 1959, for example, only one type of light airplane (Do 27), the fast patrol boats (HAM-, Luerssen-, Jaguar-class), and one type of missile (Cobra) were designed and produced in the FRG. The limited technological capabilities in almost all areas of major weapons production were reflected by the fact that imports accounted for about 60 per cent of the procurement expenditure of the FRG in the late 1950s (see Table 4.1). This high import share implies that the extensive production of arms under licence that occurred in the late 1950s (transport aircraft Transall, light fighter-aircraft Fiat G 91, light tank HS-30) employed a limited amount of domestic inputs.

Changes became manifest in the early 1960s and this period marks the beginning of the third phase of FRG arms production. In 1959 the first destroyer of the Hamburg-class had been laid down with three more following over the next three years. In 1960, licence production of one of the most advanced fighter-aircraft of the time, the Lockheed F-104, began with the explicit intention of building an independent German aerospace industry.[5] In 1962, the design for the Bundeswehr's future main battle tank (the Leopard) was selected. The first Leopards rolled out of the factories in 1965. On the political front, the period of the late 1950s and early 1960s also brought decisive changes. The main opposition party (Social Democrats) and the unions accepted rearmament as it had been conceived by the conservative (Christian Democrat) government. Within industry there no longer was any objection to arms production and even Krupp started some limited arms production.

Throughout this third phase, arms production in the FRG expanded, though unevenly. Arms procurement is irregular by nature and the export of newly built equipment (in contrast to old equipment) throughout the 1960s and early 1970s was limited. Also, the technological breakthroughs in the rapidly changing field of military technology were limited – despite efforts in military research and development (R&D) (see Table 4.2 below). A second series of destroyers, this time equipped with missiles, was not built in the FRG but ordered from an American shipyard (Adams-class). The decision to import was made partly because the experiences with the Hamburg-class destroyers were rather disappointing. (The last one was commissioned six years after it had been laid down, while the Adams-class ships were each completed in three years.[6]) In addition, West German naval officials suspected that their own industry was not capable of integrating modern missile systems into a big ship. The aerospace industry also did not get off the ground. Several of its military and civilian projects (VFW 614, V/STOL aircraft) had to be cancelled before or after prototypes had been built. Nonetheless, federal funding was maintained and gradually the technological capabilities of the aerospace industry increased.

Most successful were the tank producers who exported Leopard I to a number of other NATO countries (Norway, Netherlands, Italy, Turkey, Belgium, Denmark, Canada) and to Australia, as well as the small arms producers, especially with the export of small arms production facilities.[7] These successes were, however, only relative. They arose mainly out of advances in civilian technology which could be applied to military products (engine, transmission, tracks and so on in the case of tanks, cold metal drawing technology in the case of small arms production) rather than advances in what are considered 'military technologies', that is, military electronics, airframe design, jet engines, or large guns.

Some, though not all, of these deficiencies were overcome in the second half of the 1970s. The second exclusively West German designed and constructed main battle tank, the Leopard II, carried a West German designed and built gun. (The Leopard I had had a British Vickers gun.) The first of six missile-armed frigates was laid down in 1978, the year the production of the Leopard II began. The multi-role combat aircraft (MRCA) Tornado, a 'swing-wing' aircraft of British/German/Italian design, was test-flown for the first time in

August 1974 and the first production aircraft reached the Air Force in 1980. Anti-tank and anti-aircraft missile systems (Milan, Hot, Roland) were mainly procured in the late 1970s by the French-German company Euromissile. The anti-tank helicopter for the Army was designed by Messerschmidt-Bölkow-Blohm (Bo 105).

Despite this impressive list, two facts have to be kept in mind. First, German military technology has some severe limitations, especially with regard to electronics, aircraft engine design and production, special metal forgings, and special metal production. Second, some of the programmes have succeeded only because technology was borrowed from partners in co-production programmes: the British in Tornado airframe and engine design and production, the French in Alpha-Jet airframe design production and in missile propulsion, the Americans in avionics and general electronics. With respect to technology, then, a fourth phase of arms production can be identified which began in the early 1970s for some weapon systems and in the late 1970s for others. This phase is characterized by independent production technology in most, though not all, major areas.

The fourth phase also has new political characteristics.One is the resurgence of a more aggressive political profile of the arms industry. While it had been fashionable throughout FRG history – that is, since 1949 – not to discuss one's arms production activities openly, from about 1976 this situation changed considerably. Military-related journals sprang up, with advertisements by arms producing firms occupying most pages. Firms that had once tried to disguise their arms production now boasted of it.[8] The second political characteristic has been the expansion of arms exports. Although the Federal government claims to have the most restrictive law on arms exports, a large increase in arms exports was observable in the second half of the 1970s. Closer analysis reveals that both phenomena can be explained by the economic crisis which began in 1973 and lasted the remainder of the decade. This crisis has mainly been an unemployment crisis.[9] By stressing the employment effect of arms manufactures, producers were successful in political circles in pushing through domestic programmes (for example, the six frigates, Type 122). They were so successful in obtaining numerous 'exceptions' for the export of ships (fast patrol boats, submarines, frigates, corvettes), aircraft (Alpha-Jet, Do 28, Transall), and tanks (TAM) that there are hardly any restrictions on exports left.[10]

West German Arms Production Today: the Dimension of the Industry

Is There an 'Arms Industry'?

What gives a finite number of firms the economist's stamp of a sector or industry? Generally, economic theory tells the analyst to consider competition among firms and characteristics of input, production and output. It is argued here that the criterion of somewhat similar political lobby interests is also important.

Competition among arms producers in the FRG is very limited. Only 2-4 per cent of the orders processed by the procurement agency, the Bundesamt für Wehrtechnik und Beschaffung (BWB) (Federal Office of Military Technology and Procurement) are open for tenders. The large majority, about 88 per cent of the order volume, is contracted with firms without any tenders at all.[11] The remaining 8-10 per cent is contracted after very limited competition between firms who are invited to make bids by the BWB. Competition is limited to the fringes of arms production, for example trucks, cars and support ships, some design of weapon systems and – to a limited extent – exports.

The common characteristic that could be used to distinguish arms production is that arms are produced. But what are 'arms'? Neither in theory nor in practice is it possible to find a good definition. Because of data limitations, this discussion will use a definition for procurement derived from official government publications and from what is called 'arms exports' in the US Arms Control and Disarmament Agency's publication, *World Military Expenditures and Arms Transfers.*[12] This leads to the unsatisfactory situation in which some 'civilian' goods for the armed forces, like ordinary cars or inflatable boats, are counted under the heading of arms production while light tanks for paramilitary units, like the Bundesgrenzschutz (Border Security Forces) are not included. Contrary to the situation existing in other countries, there is no arms producers' association or the like in the FRG. This does not mean that there are no lobbying groups,[13] but it makes the identification of an arms production sector somewhat more difficult.

Altogether there seems to be no good criterion for defining and isolating arms production. The mixture of 'near civilian' with 'hard-core military' industries therefore partially limits the political conclusions that can be drawn from the data that are presented here. This is further complicated by the official West German statistics where no separate data for any portion of arms production can be found.

The Size of the 'Market'

The limited definition of arms production employed here starts with Financial Plan 14 ('Defence') of the federal budget of the FRG. Financial Plan 14 covers only part of what is included in NATO's definition of military expenditure. Such expenditure categories as 'civil defence', 'pensions', and 'border troops' are excluded. The difference between military expenditure computed according to the FRG definition and according to the NATO definition is quantitatively substantial (for 1979, 36.7 million DM versus 45.4 million DM or 24 per cent). It is not possible to extend this study beyond Financial Plan 14, however, since the necessary breakdowns of data are only available for this financial plan. Included in this plan are the 'defence investment expenditures'. They contain the subtitles 'procurement', 'research and development', 'military infrastructure' and 'other investment' (mainly fuels). Infrastructure will be excluded from the following discussion since the 'military nature' of producing barracks, hospitals or roads is not obvious to the author. Similar considerations pertain to parts of the procurement expenditure (for example, cars) and 'other investment' (fuels). The disaggregation of the available data would, however, be so time-consuming that the available aggregate figures will be employed. The expenditure for 'maintenance' will be included. It is necessary for the efficiency of the existing weapon arsenals and it is also – at least partly – not distinguished by the nature of production from the construction of weapon systems. Very often production and maintenance occur in the same factories.

We thus have a workable definition of arms production: 'defence investment expenditure' minus 'military infrastructure' plus 'maintenance'. To measure the size of the German 'market' (as pointed out above, because of lack of competition, it is not really a 'market' in classical economic terms), the expenditures going to firms outside the FRG must be subtracted and arms exports added. This is done in Table 4.1. In 1978, for example, 'domestic demand for arms production' as defined above amounted to 12.8 million DM. Eighty-two per cent of this money went to firms located in the FRG. If payments to firms outside the FRG are excluded and an estimated 2.5 million DM in arms exports are added, there was a 'total demand for arms production' of roughly 13 million DM in 1978. This marked an increase of 10 per cent over 1977. This figure will be called 'arms production value added' since it refers to final demand only. It can be compared with the final demand data for manufacturing or industrial manufacturing. Arms production as defined here is totally included in the West German category of 'manufacturing' though not in the one of 'industry'.

From both Table 4.1 and Figure 4.1, it can be seen that 'arms production value-added' has not grown continuously and smoothly. Peaks can be observed in the early 1960s, the mid-1960s, and the early 1970s. The steep increase in the second half of the 1970s is at least partially due to higher inflation, but also to the large increase in arms exports. Arms production as part of manufacturing also did not develop in a regular way. On average, it has been around 2 per cent since the middle of the 1960s with a trough in 1971. This ratio is strongly influenced by fluctuations in manufacturing output. Arms production in 1979 was almost exactly 2 per cent of manufacturing only.[14] Measured by turnover, the FRG has the fifth largest arms industry in all industrial countries. Its size is about that of shipbuilding in the FRG and much larger than aerospace.

Figure 4.1: Arms Production, Export and Employment in the FRG

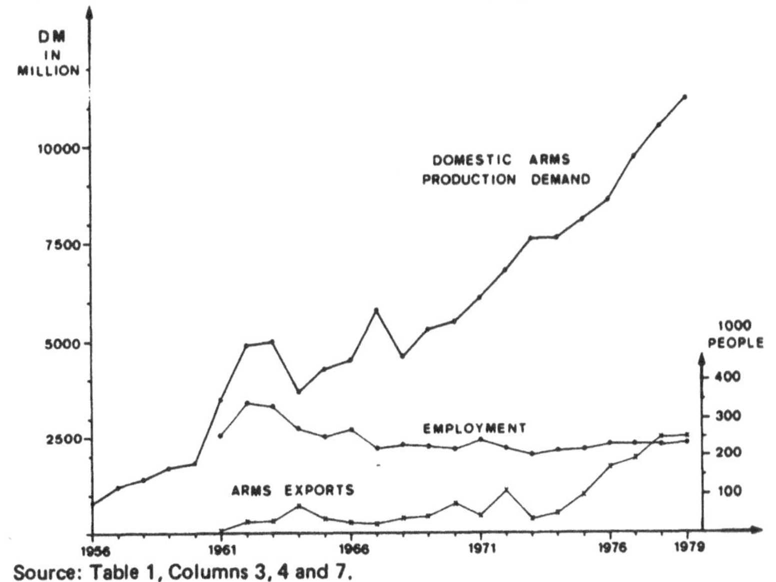

Source: Table 1, Columns 3, 4 and 7.

Employment in the Arms Industry

To estimate employment in the arms industry, a number of methods have been used. This has led to divergent projections. The Ministry of Defence gives a figure of 'a work force of about 200,000'[15]; the metalworkers union (to which most of the workers in arms production belong) has published a figure of 150,000.[16] Using input-output analysis, one of the leading economic research institutes in the FRG, the DIW (Deutsches Institut für Wirtschaftsforschung, Berlin) has calculated that in 1976, 173,000 people were directly employed in arms production and 119,000 were indirectly employed (for a combined total of 292,000).[17] A similar exercise produced the figure of 407,000 total employment for 1972.[18] Both these estimates are based on a more extensive definition of arms production than has been used in this analysis, that is, they include construction. Even broader defintions lead to even higher estimates.[19]

For the purposes of this discussion, it is assumed that the ratio of employees to value-added is the same for the arms industry as it is for the manufacturing sector as a whole. This is obviously a crude estimate but there is no better information available. With this restriction in mind, the number of people working in arms production can be arrived at by multiplying the percentage of arms production value-added in total manufacturing value-added by the total number of people working in manufacturing. The results of this calculation are presented in Table 4.1 and Figure 4.1.

The number of people working in the arms industry has decreased – though unevenly – between the early 1960s, when it reached its peak, and the second half of the 1970s. While almost 350,000 people are estimated to have worked either directly or indirectly in the arms industry in the FRG in 1962, it is calculated that the workforce declined to just above 200,000 in 1974. It has since increased by about 15 per cent and is estimated to have reached 231,000 in 1979.[20] It has to be remembered that these figures are only estimates. They are, again, heavily influenced by fluctuations in economic activity which change the ratio of people employed in a sector to that sector's value-added. There are some indications that the share of highly qualified personnel is greater in arms producing firms than in other sectors, but the information is scanty. Value-added per employee, for example, was found to be greater in arms production than in industry in general, indicating a higher pay-rate and, theoretically, qualifications.[21]

Military Research and Development

One part of 'defence investment expenditure' deserves special attention: military R&D. It tells us something about the technological capacity of a nation's industry with respect to actual and future arms production and also about the way in which the especially scarce resource of people working in research and development is divided between civilian and military uses.

Table 4.2: Military R&D, Selected Years

Year	Military R&D million DM	Military R&D as part of federal R&D, percentage	Military R&D as part of total R&D, percentage
1963	546	69.6	10.3
1968	982	36.9	9.4
1973	1,372	24.5	6.7
1978	1,707	17.3	5.7
1980	1,666[a]		

Note: a. provisional figure.
Source: G. Bräunling and D.-M. Harmsen, *Die Förderungsprinzipien und Instrumente der Forschungs- und Technologiepolitik*, Göttingen: Schwarz, 1975, Arbeitstabelle 1, SA 1; Bundesminister für Forschung und Technologie, *Bundesforschungsbericht VI*, Bonn: 1979, p. 137 ff., Tables 1 and 7.

In 1980, military R&D in the FRG amounted to about $910 million. This is equal to 4.3 per cent of the total 1980 military budget. Compared to total federal R&D, the growth of militry R&D during the last two decades has been limited. This is mainly due to the fact that, while federal authorities financed only a small portion of the total R&D carried out in the early 1960s, they now play a much more important role. Military R&D also declined as part of total R&D, reflecting the changing role of federal financing in the FRG. With just over 5 per cent of total R&D going to the military, Germany is not among those countries which devote a very substantial part of their R&D resources to military ends. The US, the USSR, France and Great Britain all direct much larger percentages of their R&D resources towards military ends. Even so, the FRG is still ranked high among all nations.[22]

About 10-12 per cent of total R&D expenditures have been classified as 'research' in recent years, the rest being referred to as 'development' (the distinction probably being made between 'basic' and 'applied' research). In the last few years, about 74 per cent of R&D funds have been spent on 'development and testing' of 'ready' projects. A further 23 per cent has gone into 'future technology' studies while the balance (3 per cent) has been expended on management studies, strategic analysis, tactical warfare studies and the like.[23] As the new fighter-aircraft, the Tornado, is now in the production stage, military R&D is expected to decline somewhat in the early 1980s. The future of military R&D will be determined by the decisions about a future tactical fighter aircraft and a future main battle tank in the early 1980s (see below). If arms production is treated as an independent economic sector, its input of R&D in comparison to total value-added is among the highest within all sectors (13 per cent in 1978). Other sectors with similar R&D intensity include aerospace (21 per cent in 1977/8) and the electrical industry (13 per cent in 1975).[24]

Structural Aspects of Arms Production in the FRG

Arms and Sectors of Industry

Arms production, as has already been mentioned, is not treated as an independent economic sector in West German economic statistics. It is therefore necessary to describe arms production as part of sectors such as machine building or aerospace which *are* treated this way, although the data base one can work with is scanty.

First, there are figures from the federal procurement agency, BWB, which is responsible for government contracts for major weapon systems and some military-related R&D. These have severe limitations. They only cover about two-thirds of all domestic arms production demand, the rest being contracted by local procurement agencies and directly through the Ministry of Defence. Furthermore, these figures only represent payments to prime contractors. Thus, when using these figures, the fact is ignored that subcontractors for a specific weapon system may belong to a different sector than the main contractor. The distribution of arms production derived from these figures can be seen in Table 4.3. Arms production is concentrated in four investment-good sectors: aerospace, machine building, electrical industry and, on a somewhat lower level, road vehicle building. These four sectors, which together acounted for about 31 percent of turnover* in manufacturing, have comprised around 80 per cent of all arms production value-added. In the late 1970s, shipbuilding became somewhat more important because of the frigate production programme.

Somewhat different figures would result if BWB payments could be accurately disaggregated according to industrial sector. A first attempt at this is the sectoral classification of the 30 largest arms producing firms found in Table 4.3. Obviously this carries a bias, since these 30 firms represent only half of the total arms production in the FRG. The sectoral distribution (for 1977) derived from the list of firms places the electrical industry on a higher level (37 per cent of orders) than machine building (21 per cent), aerospace (20 per cent) and road vehicle building (13 per cent). The limited size of 'other industries' (9 per cent) shows that this list of large firms tends to over-emphasize

*The German word is 'Umsatz'. 'Turnover' should be read as 'sales' by the American reader.

Table 4.3: Procurement Payments by Sectors of Economic Activity[a] (percentages)

Sector	1970	1972	1974	1976	1978
Aerospace	32	31	28	24	17
Machine building	17	16	19	23	23
Electrical industry	21	22	23	19	19
Road vehicle building	11	12	9	17	14
Shipbuilding	3	3	2	2	9
Other investment good industries	8	6	6	4	6
Other industries	8	10	13	11	12

Note: a. Payments by the Procurement Agency (BWB) only.
Source: Bundesamt für Wehrtechnik und Beschaffung.

the main sectors of arms production. This second result of sectoral distribution is more congruent with expectations about the sectoral distribution than the first one. Electronics are coming to play an ever more important role in new weapon construction and communication for the armed forces.

As the above-mentioned sectors are of very different size, arms production is of very different importance within these sectors. In aerospace, more than 60 per cent of all economic activity was of a military nature in the 1970s. The ratio lies below 2 per cent for machine and road vehicle building and probably above 2 per cent for the electrical industry. Military-related shipbuilding increased to around 10 per cent in 1980 from about 2 per cent in the early and mid-1970s.[25]

The Regional Concentration of Arms Production

There is one main regional centre of arms production in the FRG: the area around Munich. The reasons for this are not especially clear. While some other areas are attractive because of their abundance of skilled workers (the Neckar Valley and the Ruhr region) or because of lack of alternatives for workers, coupled with state subsidies (Black Forest area, near the GDR border), Munich does not stand out in any respect. This concentration may be due partly to political factors, as one of the first Ministers of Defence, Franz Josef Struass, has a very strong political base in Bavaria, in which Munich is located.

Exact figures on regional concentration are hard to obtain. Again it is necessary to rely on the figures supplied by the BWB. The limitation of the data compiled on the basis of payments to prime contractors is considerable. All payments to subcontractors situated in areas other than that where the main contractor resides will not be correctly classified. The distribution of these payments by states ('Lander') for 1976 can be found in Table 4.4. This is the most recent year for which data are available. These figures are put into perspective in the table by dividing them by the number of people employed in industry in each state. Using this method, Bremen has the highest concentration of arms production of any state, followed by Bavaria, Hamburg and Schleswig-Holstein. Bremen is ranked too high though, since its largest arms producing firm Vereinigte Flugtechnische Werke (VFW) has plants in several other states. Under the BWB accounting methods, however, the production of these subsidiaries is attributed to Bremen, site of the parent company, rather than the states in which the production actually occurred. The characterization of Munich as the 'arms production centre' is obvious when it is considered that 87 per cent of all orders for Bavaria in 1975 went to the Munich area.[26]

At the county level, the picture changes somewhat. As in Bavaria, arms production is concentrated in a few areas in the larger states. Besides Munich, Hamburg and Bremen, Kiel, the area around Kassel, the Rheingau and the Black Forest and Bodensee areas are among the centres of arms production.

The Industry Level of Arms Production

The Largest Firms

The increased political self-esteem of arms producing firms in the second half of the 1970s has had one valuable side-effect for the analysis of arms production: much more information is now available than was previously. It is, for example, possible to list the largest arms producers in the FRG according to their turnover in arms production. Table 4.5 lists all arms producing firms with a turnover of more than 100 million DM (about $43 million) in 1977.

Not surprisingly, two large electronics firms are at the top of the list (Siemens and AEG). They are followed by the aerospace companies Messerschmidt-Bölkow-Blohm (MBB) and VFW-Fokker. In 1979, VFW and the Dutch firm Fokker separated, and VFW merged with MBB in 1981. It is likely that this new combination will lead the arms producers' list for 1981. The only other airframe producer in the FRG, Dornier, can be found in eleventh place while aeroengine builders Motoren- und Turbinenunion (MTU) and Klöckner-Humboldt-Deutz (subcontractors only) occupy sixth and seventeenth places respectively.

The first tank builder is to be found in fifth place (Krauss-Maffei),

Table 4.4: Procurement Payments by States[a]

	1974		1975		1976	
Federal State	million DM	DM per employee in manufacturing industry	million DM	DM per employee in manufacturing industry	million DM	DM per employee in manufacturing industry
Schleswig-Holstein	259.8	1,556	300.1	1,924	275.2	1,609
Hamburg	320.0	1,290	271.6	1,176	313.3	1,852
Niedersachsen	226.8	314	219.2	332	153.9	227
Bremen	509.9	5,367	301.9	3,392	341.2	3,708
NRW	739.3	297	872.3	370	949.5	414
Hessen	389.1	519	369.2	530	555.8	858
Rheinland-Pfalz	140.9	428	147.6	472	164.7	451
Baden-Württemberg	1,091.8	668	1,145.3	750	999.5	712
Bayern	2,146.9	1,532	2,579.4	1,957	2,526.2	2,001
Saarland	94.4	694	84.8	652	85.5	559
Total Bundesrepublik	5,919.0	718	6,291.4	815	6,364.8	856

Note: a. Payments by the Procurement Agency (BWB) only.
Sources for base figures: H. Maneval and G. Neubauer, *Untersuchung über die Wirkung von Verteidigungsausgaben auf die regionale Wirtschaftsstruktur*. Munich: Hochschule der Bundeswehr, 1978, Table 2, and *Statistische Jahrbuch für die Bundesrepublik Deutschland*, Wiesbaden: various years, Table 'Beschäftigte des Produzierenden Gewerbes mit mehr als zehn Beschäftigten'.

Table 4.5: The Largest Arms Producing Firms in the FRG 1977 (Turnover above 100 million DM)

	Firm	Owner	Production Sector	Employment	Turnover million DM	Rank[a]	Arms Production percentage	Arms Production Turnover million DM
1.	Siemens AG, München	Streu	Electronics	319,000	25,198	2	8	2,000
2.	AEG-Telefunken, Frankfurt	Streu	Electronics	158,400	14,286	9	10[b]	1,400[b]
3.	Messerschmidt-Bölkow-Blohm GmbH, München	Siemens, Thyssen, Land Bayern, Land Hamburg	Aerospace	20,700	1,801	87	60	1,080
4.	Vereinigte Flugtechnische Werke – Fokker GmbH Bremen	Krupp Land Bremen	Aerospace	17,500	1,705	91	60[b]	1,020[b]
5.	Krauss-Maffei AG, München	Flick	Tanks	4,600	1,005	155	70	705
6.	Motoren- und Turbinen-Union GmbH, München/Friedrichshafen	Daimler-Benz MAN	Engines	11,000	1,063	150	52	550
7.	Rheinmetall GmbH Düsseldorf	Röchling	Guns, small arms	7,500	779	204	67	520
8.	Maschinenfabrik Augsburg-Nürnberg AG	Haniel	Trucks	62,000	6,329	30	8[c]	500[c]
9.	F. Werner Industrieanlagen	DIAG (Bund)	Small Arms	1,702	441		100[b]	441[b]
10.	Karl Diehl GmbH & Co AG Nürnberg	Diehl	Electronics	13,200	1,220	126	35	430
11.	Dornier GmbH Friedrichshafen	Dornier	Aerospace	6,700	723	212	51	370
12.	Thyssen-Industrie AG Düsseldorf	Thyssen	Tanks, War ships	42,500	4,291	38	8[c]	330[c]
13.	Howaldtswerke – Deutsche Werft AG Hamburg/Kiel	Salzgitter Bund, Land Schl.-Holst.	War ships	13,800	1,215	129	20	250[c]

14.	Blohm & Voss AG, Hamburg	Thyssen, Siemens	War ships	6,500	633	238	40	250
15.	Dynamit Nobel AG Troisdorf/Köln	Flick	Munitions	14,900	1,956	75	12[c]	235[c]
16.	Industriewerke Karlsruhe AG	Quandt	Guns, Small arms	8,100	623	244	33	235
17.	Klöckner-Humboldt-Deutz AG, Köln	Henle	Engines	31,200	4,015	43	5	200
18.	Standard Elektrik Lorenz AG Stuttgart	ITT, USA	Electronics	32,800	2,735	59	6	165
19.	Wegmann & Co, Kassel	Wegmann	Tanks	2,400	250-300		55[b]	165[b]
20.	Zahnradfabrik Friedrichs-hafen	Stadt Friedrichshafen	Transmissions	19,800	1,903	78	7.5[c]	150
21.	Bodensee-Gerätetechnik GmbH & Bodenseewerk Perkin, Elmer & Co., Überlingen	Perkin Elmer, USA	Electronics	1,600	160		95	150
22.	MaK Maschinenbau GmbH Kiel	Krupp	Tanks	3,600	500		30	150
23.	Luther Werke Braunschweig Mainz		Trucks, Tanks	2,600	170		85	145
24.	Industrieanlagenbetriebs-gesellschaft, München	Bund	Studies	1,600	160		80	130
25.	Deutsche Philips, Hamburg	Philips, Niederlande	Electronics	31,000	4,026	12	3[c]	120
26.	Rohde & Schwarz, München	Rohde, Schwarz	Electronics	3,733	360		33	120
27.	Fr. Lürssen-Werft, Bremen	Lürssen	War ships	1,100	100[c]		100	100[c]
28.	Daimler-Benz AG, Stuttgart	Streu	Trucks	169,200	24,723	3	0.4	100
29.	Elektronik-System Gesell-schaft, FEG-Gesellschaft für Logistik,München	AEG, Rohde & Schwarz, Siemens, SEL, Eltro, Honeywell, USA, Litef Teldix	Electronics	900	100		100	100
30.	Heckler & Koch GmbH Oberndorf	Heckler & Koch	Small arms	2,000				

Notes: a. Rank among the 250 largest industrial enterprises in the FRG.
b. Maximum figure. The actual figure is probably somewhat lower.
c. Minimum figure. The actual figure is probably somewhat higher.
Source: Data files of the Arbeitsgruppe Rüstung und Unterentwicklung, University of Hamburg.

followed by Rheinmetall (number seven), Thyssen-Industrie AG (number twelve), Wegmann (number nineteen), Zahnradfabrik Friedrichshafen (number twenty), MaK Maschinenbau (number twenty-two), and Luther-Werke (number twenty-three). (The latter went bankrupt in 1979.) As this subgroup of firms suggests, tank building is a major strength of the West German arms industry. Another major strength is infantry weapon production. Here there are Rheinmetall again (number seven), F. Werner Industrieanlagen (number nine), Dynamit Nobel (number fifteen, especially munitions production), Industriewerke Karlsruhe (number sixteen) and Heckler and Koch (number thirty) among the largest arms producers.

The major arms producers in the vehicle-building industry have varied over time. In 1977, Maschinenfabrik Augsburg-Nürnberg (MAN) was eighth on the list while Daimler-Benz was twenty-eighth. In other years, Magirus Deutz and Volkswagen would have to be included. Shipbuilding firms with large warship-building capacities are Howaldtswerke/Deutsche Werft AG (especially submarines for export), Blohm and Voss (frigates, landing craft) and Lürssen (fast patrol boats for export). The list is completed by a number of smaller electronics firms and one 'think-tank', the Industrieanlagenbetriebsgesellschaft (IABG).

Concentration in the Arms Industry

It is not possible simply to compare the turnover of the arms producing firms and the figures for 'arms production demand'. The latter stands for final demand, and thus all inter-industrial shipments are only counted once, while in the former they are counted as often as transfers between firms occur. The figures for 'value added' must be transformed into figures measuring turnover. One crude, though simple, way to effect this transformation is to use the general ratio between turnover and value-added for firms in the manufacturing sector as a whole. In the late 1970s, this ratio was about 2:1 in the FRG.[27] For 1977, one can thus estimate a figure of 23.4 billion DM (about $10 billion) for arms production turnover. This can be related to the turnover figures in Table 4.5.

The following concentration ratios can be derived from this comparison:

Turnover concentration of the largest firm	9 per cent
– three largest firms	19 per cent
– ten largest firms	37 per cent
– thirty largest firms	52 per cent

More than half the total turnover of arms producing firms, then, is concentrated in those companies which had more than 100 million DM (some $42.7 million) in arms production turnover in 1977. Compared with other countries, the arms industry in the FRG seems less concentrated.[28] Compared with other sectors of industrial activity in the FRG, the 'arms industry' (which is not a sector in terms of statistics) seems to be moderately concentrated. The aerospace, road vehicle building, shipbuilding and electrical industries, for example, are all more concentrated, while the optical, steel and machine-building industries are less so.[29] As mentioned above, the degree of concentration in arms production (as well as in other sectors) is probably much greater than documented here, as there is no, or hardly any, competition for a large number of its products.

Dependence on Arms Production at the Industry Level

Even if, as has been shown to be the case for the FRG, arms production is not a very large factor for the economy as a whole, dependence on arms production by single firms will influence their behaviour towards procurement decisions. If these firms are important regional employers, politicians and bureaucrats will also be heavily influenced in their decision-making.

In order not to fall into this political trap (mixing procurement with employment decisions), the BWB officially follows the policy of limiting individual firms' dependence on arms production.[30] Although the BWB has not been as successful as it might have been, this policy (along with the general trend towards increased concentration in West German industry) may have resulted in a pattern of dependence among arms producing firms in the FRG which is somewhat different from that in other countries, at least in Western Europe. Among the ten largest arms producing firms in the FRG, only one was more than 75 per cent dependent on arms production. In France,the comparable figure is five, and in Italy and the UK, it is four each. Four of the top ten West German firms were less than 50 per cent dependent on arms production.[31] Among the 30 largest firms, seven are more than 75 per cent dependent on arms production and sixteen are less than 50 per cent dependent. If smaller arms producing firms are included, there are a large number of firms totally or almost totally dependent on arms production, especially those in electronics and infantry weapon production and subcontractors for tank building. These smaller firms have between 100 and 1,000 employees. Some of them are located in economically depressed regions.[32]

Another way of measuring the dependence on arms production is to analyze the integration of arms producing firms into industry in general. If some financially strong enterprise is standing behind a smaller arms producer, a drop in the arms business can possibly be dealt with differently than in a firm without such financial resources. Eight of the ten largest arms producers in the FRG are among, or are subsidiaries of, those companies with more than 1 billion DM (about $435 million) turnover in 1977. Three of the companies have more than 5 billion DM ($2.175 billion) turnover. These figures are increased to fifteen and four if the 30 largest firms are considered. In general, it can be said that a large portion of arms production in the FRG is part of the industrial production of the very large corporations. The situation is not quite the same in the other West European countries with large arms industries.[33] At the same time, there is a substantial number of independent, middle-sized firms engaged in arms production in the FRG (Dornier, Lürssen, Rohde and Schwarz among the 30 largest).

Again, these general figures reveal only a small amount of information. Subsidiaries of large corporations working in arms production may be quite independent of the parent firm and heavily dependent on arms production. Sometimes, arms production within one big firm is located at one plant only. Dependence usually accumulates: firms which are very dependent on arms production tend to be in economic branches that are heavily dependent and in regions that are overproportionally dependent on arms production.

Ownership in the Arms Industry

While it is possible to distinguish many kinds of ownership, the discussion here will be limited to four types: stockholder, family, state, and foreign ownership.

The largest arms producing firms are two big public companies (AEG and Siemens) with numerous stockholders, including banks and financial companies. A second group of companies follows, consisting of a large number of smaller arms producers. These companies also contain stockholders, though usually one dominant one with a majority or near majority of the votes. They are usually tied to the names of families that have large industrial interests, not only in arms production (Krupp, Röchling, Haniel, Thyssen, Quandt and Henle).

Private companies in the FRG are not required to publish as much information about their activities as public companies. It seems that among arms producing firms, the tendency to remain a private company is stronger than among firms in general. (So one finds 'GmbH' [one type of private company] instead of 'AG' [a company with stockholders] more often among large arms producers than among large companies in general.) Some of these firms are owned by other large corporations, though, or by the states or the federal government (Bund). The largest real 'family enterprise' among the arms producers seems to be Diehl. Others are Dornier, Wegmann, Luther, Rohde & Schwarz, Lürssen and Hechler & Koch. Obviously, among smaller arms producers, there are a large number of these 'family enterprises'. Those arms producing firms which have gone bankrupt in the last few years (Elac and Luther, for example) have come from this group of companies.

State ownership in arms production in the FRG seems not to be much different from state ownership in industry in general. Among the ten largest arms producers, one is federally owned (DIAG) and states hold shares in two others (MBB and VFW, now merged). Among the 30 largest, public ownership increases to five (HDW, IABG, and Zahnradfabrik Friedrichshafen [where the city of Friedrichshafen has a majority]). Compared with other West European countries, there is very little state ownership in arms production.[34]

Foreign ownership is sizeable if measured in numbers of firms but small if measured by the turnover of these firms. There are three foreign-owned companies among the 30 largest arms producers: SEL (owned by ITT, USA), Bodensee-Werke (owned by Perkin-Elmer, USA) and Philips (owned by Philips, Netherlands). The number of foreign-owned firms is especially large in electronics. Almost all of the large electronics firms of the Western world have subsidiaries in West Germany, particularly US electronics firms. Many of these subsidiaries work mainly in the field of arms production. This situation seems to reflect the technological gap in the advanced field of military electronics. Some other firms seem to use the FRG as a base for exports to third countries (thereby overcoming restrictive arms export legislation of their home countries, for example, the Swiss company Oerlikon-Bührle) but the predominant motive is to participate in the West German arms market.

Activities Outside the FRG

As can be seen from Table 4.1 and Figure 4.1, arms exports from the FRG increased throughout the 1970s and at the end of that decade were an important factor for the arms industry in the FRG. About 20 per cent of total arms production is production for exports. This is so despite frequent claims by the government that West German arms exports are heavily restricted.[35] These claims cannot be taken too seriously in view of the experience since the mid-1970s. At the same time, it has to be acknowledged that in the past there were more restrictions in West Germany than in other West European countries like France, Great Britain, Belgium or Italy.

Arms exports account for approximately 1 per cent of all exports and 3 per cent of exports of higher technology commodities (as listed in category 7 of the Standard International Trade Classification). Some of the large arms producers have been quite heavily dependent on arms exports, for example, F. Werner Industrieanlagen, Howaldtswerke/ Deutsche Werft AG (before the frigate programme started), Lürssen and Heckler & Koch. The arms export market is not dominated by these firms, though, as almost all of the large arms producers are active exporters. This can be seen, for instance, in the arms promotion journals designed for export markets that are put out by German publishing houses.[36]

German firms also participate in arms production outside the FRG. The main reason for this has been to circumvent the restrictions on arms exports in force within the FRG. Cooperation has been especially strong with French firms, where German subsystems are often combined with French subsystems and exported worldwide. Examples of this include Dornier and Dassault on the Alpha-Jet and MBB and Aérospatiale on missiles marketed as products of the French firm Euromissile (a trading company which sells MBB and Aérospatiale products).

There has also been collaboration with Italian and Belgian firms.[37] In addition, companies which are completely or almost completely German-owned produce arms outside the FRG. Such companies are prevalent in the Netherlands, but also exist in Italy and Switzerland. Production by these firms is partly re-imported into Germany, indicating that production outside the FRG may be cheaper or subject to less restrictive laws (for example, environmental regulations). Most of the production in the Netherlands owned by West German firms is munition production.

Profits in the Arms Industry

Profits from arms production are an issue which has been much debated. Some analysts have contended that arms industry profits are probably lower than those for industry in general[38] while others have stated the opposite.[39] Quantitative studies are now available which – despite severe measurement problems – indicate that both arguments are partly correct. As a percentage of turnover, arms production firms tend to have lower profit rates than other companies. The picture changes drastically, though, if profits are related to equity capital. Then profit margins in arms production seem to be substantially above those in civil industry.[40]

The reason for this difference is that the ratio between equity capital and turnover is lower in arms production than is the rule throughout comparable industries. While this ratio is 30 per cent for all public companies, it is around 20 per cent for the largest arms producers that publish relevant information.[41] This situation is made possible through federal financing of production runs which eases requirements for equity capital quite substantially. The low level of equity capital is made possible through pre-financing and very early payment while work is still in progress (for example, in quarterly instalments). The federal state also provides some companies with land and machines. Quality control tests are generally performed by state personnel in the factories.

Arms production could, in theory, occur without any equity capital at all – something which seems impossible outside the arms production sector. As contracts between the state and the arms producer tend to grant a fixed, cost-plus profit margin (which is often below cost-plus margins in other industries), profit on equity capital (the only figure relevant for the capital owner) will depend on the equity/turnover ratio. The risk for the arms producing firms is not one of not making profits while producing (at least not for arms production for domestic demand). It is rather one of not having orders. If there are no orders, alternative civilian production is limited by the lack of equity capital. Therefore, the pressure on the part of the German arms industry to level out the troughs and peaks in arms procurement is very strong.

Arms Production and the State

Some of the issues involved in the relationship between the arms industry and the state have already been sketched out in the preceding discussion: those pertaining to ownership and financing production runs. This state-arms industry relationship is obviously a very special one. The state is by far the largest buyer of weapons. It also controls the export market and, thus, the disposition of these arms that it does not purchase itself. In view of this, some analysts have concluded that the arms industry is or must be less efficient than corporations working in more competitive markets. The arms industry might become more or less part of the state bureaucracy, even if state ownership is limited. Firms may also be able to accumulate more profits because there is no competition, just inefficient financial control by the bureaucracy.[42]

In the case of the FRG, the relation between the state and the arms producers seems to be less special than in other countries, particularly in Western Europe. This results partly from the relatively smaller size of the West German arms industry and partly from its structure, being as a whole less dependent on arms production than many of its Western European counterparts. As the last few years have shown these looser ties have also to a large extent been the result of the economic boom in the FRG in the 1950s and 1960s, especially in arms production. When general and specific employment problems arose in the second half of the 1970s, the arms industry pressed hard for a review of this position, demanding special subsidies and, in particular, a loosening of the restrictions on arms exports. Arms production began to be used as an instrument of regional employment and industrial policy whereas it had previously been one of technology policy only. While the aerospace industry was primarily built up through government arms orders in the 1960s, there had been no special subsidies to other industrial branches. Towards the end of the 1970s, some arms procurement was legitimized because of its employment effects: the Tornado, Leopard II and, especially, the frigate programme. The construction of six frigates was allocated to five different shipbuilding yards although this resulted in extra costs of more than 114 million DM.[43] The reason was one of regional policy: shipbuilding along the German coast was experiencing a crisis and the idea was not to let one yard prosper while others would have to cut employment.

It has been shown, though, that arms procurement expenditure is not a very efficient economic policy instrument. As a fiscal policy instrument, it is much too weak, since arms production represents such a small part of total production and also operates with considerable time lags.[44] As an instrument of regional and structural policy, it has been shown to be less efficient than other instruments. While it does have some short-term effects on employment, in the long run, employment problems will be aggravated if military-related expenditure is not increased steadily.[45] The employment effect of arms expenditures is also less than that of other expenditures.[46]

The fact that arms expenditures have been used as an economic instrument despite these limitations demonstrates the political success of arms lobbyists. The approach of 'simultaneous policies' – that is, to reach several goals with one instrument – has been used with respect to arms expenditure mainly because the arms producers' lobby stressed the possibility of this in specific cases. In the past, this lobby has even been successful in obtaining the collaboration of shop stewards and other representative workers' bodies at the plant level to support demands for the de-control of arms exports.[47]

With weapon systems becoming more complex and more expensive, the government has looked for ways to reduce the financial burden, for example, through joint procurement with other NATO member-countries. The arms industry has been successful in limiting this to areas where production would not have been possible at all in the FRG without foreign collaboration, due to the lack of financing or technology in Germany. Thus, the West German arms industry has willingly participated in co-production schemes for aircraft, where it still lacks some technology, but has been successful in pushing through national programmes for tanks and warships, although collaboration was considered in each case. Pressure on the part of the government to collaborate with other European producers recently decreased when it was discovered that savings from this type of production may be very limited and may be offset by the limited financial control of supranational firms.[48]

The limits of competition in West German arms production are currently enforced by the state. This occurs not only through the practice of not bidding for tenders (see above, p. 117), but also through federal support of jointly owned design bureaus, for example, for electronics and shipbuilding. Here the large firms competing in the market get together with the procurement agency and decide on work shares. One of the 30 largest arms producers, the ESG/FEG, is such a firm. The practice of work sharing for weapon systems has existed from the beginning of arms production in the FRG (for example, in the case of the Lockheed F-104 Starfighter, or the Leopard I tank family). It has, however, become much more important with the economic crisis of the 1970s (with the Leopard II, Tornado, and Frigate 122).

The Future of Arms Production in the FRG

Arms production in the FRG is now in a phase where it has all the requirements – economic, technological (though with some limits), and political – for sustaining a large independent arms industry. Paradoxically, this development has only been possible due to the existence of several factors that may severely limit the future expansion of West German arms production. Among these are the participation in collaborative projects where West Germany has acquired valuable technology, the economic crisis, which strengthened the arms industry's political base, and its connection with a strong civilian industry that made the FRG one of the two largest exporters in the world.

Further expansion is possible only if work sharing with other nations does not limit the possibility of selling them German equipment or components, if the economic crisis in the FRG and on a world scale does not become more severe and thus limit demand drastically, and if civilian industry is not hampered by the diversion of too much of the technologically advanced manpower and too many raw materials into arms production. To further close the technological gap in the defence-related industry with the US or at least keep the gap from growing, more emphasis on R&D as well as gaining production experience in high technology areas is necessary. Despite the newly discovered 'Soviet threat', it is doubtful that the money necessary to finance this can be found at home or in the export market. The next tactical fighter programme is already at stake and the next main battle tank, which was to be produced jointly with France, is being reconsidered. Alternatives are to buy a US fighter aircraft 'off the shelf' and to simply improve the Leopard II.

While this greater reliance on existent and proven weapon systems might be in the interest of the military, which has – though not necessarily in public – complained about some of the German and jointly built weapon systems,[49] it is clearly against the interests of the German arms industry. As the arms producers are part of industry as a whole and especially well-represented among the larger corporations, and might even be supported by some workers' organizations, at least at the factory level, the outcome of these conflicting interests cannot be predicted. The most sensible alternative – to reduce arms production capacity and to finance alternative employment schemes – does not seem to be politically feasible in the short run. It might be the only economically feasible path, however, if the economic crisis continues.

Notes

1. For an overview of the legal aspects of arms production and arms exports before and after World War II, see T. Mammitzsch, *Rechtliche Grenzen von Rüstungsproduktion und Rüstungshandel*, Militärpolitik Dokumentation Heft 18, Frankfurt: Haag und Herchen, 1980.

2. On discussions relating to German rearmament, see Ulrich Albrecht, *Die Wiederaufrüstung der BRD*, Köln: Pahl-Rugenstein, 1974; R. McGeehan, *The German Rearmament Question*, Urbana, Chicago and London: University of

Illinois Press, 1971. Albrecht stresses the German debate while McGeehan concentrates on US interests and moves.

3. For these and other deals, see Ulrich Albrecht, *Politik und Waffengeschäfte*, Munich: Hanser, 1971: C.J.E. Harlow, *The European Armaments Base: A Survey*, London: International Institute for Strategic Studies, 1967.

4. This is documented in G. Brandt, *Rüstung und Wirtschaft in der Bundesrepublik*, Witten and Berlin: Eckardt, 1966. This study is the most extensive overview of arms production in the FRG up to 1965.

5. This is well documented in P. Schlotter, *Rüstungspolitik in der Bundesrepublik, Das Beispiel Starfighter und Phantom*, Frankfurt: Campus, 1975, pp. 15-63.

6. *Jane's Fighting Ships 1977/78*, London: Jane's, 1977.

7. Small arms production facilities were set up by firms from the FRG in Algeria, Burma, Saudi Arabia, Iran and Malaysia among others. The standard rifle of the 'Bundeswehr', the G-3 is used in 36 countries in the third world, cf. *Jane's Infantry Weapons 1979/80*, London: Jane's, 1979.

8. This is documented in: Ulrich Albrecht, Peter Lock, and Herbert Wulf, *Arbeitsplätze durch Rüstung?*, Reinbek: Rowohlt, 1978; and – especially with regard to military related journals and advertisements – R. Saloch, H. Walden, and K. Weihe, *Rüstungswerbung in der BRD*, Militärpolitik, Heft 9/10, Stuttgart: Alektor Velag, 1978.

9. This again is documented in Albrecht, Lock and Wulf, *Arbeitsplätze*. Since 1974 the total employment figure has been declining. Although the active labour force shrank, too, since foreign workers left the country and women and older people retired early, the unemployed figure has stabilized around one million people. This amounts to almost 5 per cent of total employment.

10. On arms exports see Albrecht, Lock and Wulf, *Arbeitsplätze*; for newer developments, see also Michael Brzoska and Herbert Wulf, 'Offensive im Rüstungsexport', in *Aufrüsten um Abzurüsten?* eds. Studiengruppe Militärpolitik Reinbek: Rowohlt, 1980.

11. O. Greve, 'Politik, heute ein bestimmendes Faktor bei der Vergabe von Rüstungsaufträgen', *Wehrtechnik* no. 1 (1979); 16. For more general information, see H.-G. Bode, *Rüstung in der Bundesrepublik Deutschland*, Regensburg: Walhalla und Praetoria- Verlag, 1978, pp. 49-53. This very limited competition has been criticized recently from a conservative standpoint by B. Köppl who argues that through this procedure mismanagement of federal funds by private fims is aggravated. See, for a short account of his arguments, B. Köppl, 'Ist das NATO-Rüstungsmanagement reformbedürftig?' *Internationale Wehrrevue* 13:5 (1980): 658.

12. The two underlying definitions are different in a few minor points. Personal belongings like uniforms etc. are excluded from 'procurement' in the FRG, while they are partially included by ACDA. The same applies to equipment for arms production industries. For the ACDA definition see: US Arms Control and Disarmament Agency, *World Military Expenditures and Arms Transfers 1968-1977*, Publication 100, Washington, DC: October 1979, p. 23.

13. Especially prominent in this is a joint high level study group, the 'Arbeitskreis Rüstung und Wirtschaft'. Details are to be found in the lobby circular 'Wehrdienst', Bonn-Verlag.

14. These figures are also given in the official White Paper for 1979. See Federal Ministry of Defense, *White Paper 1979, The Security of the Federal Republic of Germany and the Development of the Federal Armed Forces*, Bonn: 1979, p. 36.

15. Ibid.

16. See *Metall*, 15 November 1978, p. 13.

17. Deutsches Institut für Wirtschaftsforschung, *Macro-Economic Effects of Disarmament Measures on Sectoral Production and Employment in the Federal Republic of Germany with Special Emphasis on Development Policy Issues*, Berlin: May 1980, mimeo, p. 24.

18. C. Bielfeldt, *Rüstungsausgaben und Staatsinterventionismus*. Frankfurt: Campus, 1977, p. 88.

19. W. Klank, 'Struktur und Entwicklungstendenzen der BRD-Rüstungsindustrie', *IPW Berichte* no. 11 (1979): 23. In another publication by East German authors a figure of 850,000 employees is given, cf. Autorenkollektiv, *Militarismus heute*, Berlin (GDR): Staatsverlag der DDR, 1979, p. 283.

20. The same trend can be found in C. Bielfeldt's figures and also in other estimates with less recent data that are cited by her. Bieldfeldt, *Rüstungsausgaben*.

21. J. Schmidt, *Zur Bedeutung der Staatsausgaben für die Beschäftigung*, Beiträge zur Strukturforschung, Heft 46, Berlin: Deutsches Institut für Wirtschaftsforschung, 1977.

22. Cf. Bundesminister für Forschung und Technologie, *Bundesforschungsbericht VI*, Bonn: 1979, Table 27.

23. Ibid. p. 136.

24. Ibid., Table 15.

25. Comparison of figures in Table 4.3 with *Statistisches Jahrbuch der Bundesrepublik Deutschland*, Tabelle Produktionswert im Warenproduzierenden Gewerbe; see also Federal Ministry of Defence, *White Paper 1979*, p. 36.

26. H. Maneval and G. Neubauer, eds. *Untersuchungen über die Wirkung von Verteidigungsausgaben auf die regionale Wirtschaftsstruktur*, Forschungsbericht 1, Munich: Hochschule der Bundeswehr München, 1978, p. 110.

27. See *Statistisches Jahrbuch der Bundesrepublik Deutschland*, Tabelle Kostenstruktur im warenproduzierenden Gewerbe. This ratio applies to all firms with more than twenty employees.

28. Cf. Michael Brzoska, Peter Lock, and Herbert Wulf, *Rüstungsproduktion in Westeuropa*, IFSH Forschungsberichte, Heft 15, Hamburg: University of Hamburg, 1979, pp. 71-91. For the US, see C. Marfels, 'The Structure of the Military-Industrial Complex in the US and its Impact of Industrial Concentration', *Kyklos* 31: 3 (1978).

29. Cf. Monopolkommission, *2. Hauptgutachten*, Baden-Baden: Nomos, 1978, p. 474.

30. Cf. Bode, *Rüstung in der Bundesrepublik Deutschland*, pp. 48-49.

31. Brzoska, Lock, and Wulf, *Rüstungsproduktion*, p. 46.

32. Tables of smaller arms producers can be found in the appendix of ibid.

33. Ibid., pp. 46-70.

34. Ibid.

35. Federal Ministry of Defence, *White Paper 1979*, p. 36.

36. Prominent among these are: *Military Technology, Technologia Militar*, and *Aerospace International*.

37. Cf. Michael Brzoska, 'Transnationalisation of Arms Production Base in Western Europe and its Consequences for Disarmament Measures', Hamburg: IFSH-Arbeitsgruppe Rüstung und Unterentwicklung, 1979, mimeo.

38. See for example E. Czerwick and E. Müller, *Rüstungskonzerne, ökonomisches Interesse und das Konzept des 'militärisch-industriellen Komplexes'*, Friedensanalysen 6, Frankfurt: Suhrkamp, 1977.

39. Ulrich Albrecht, *Rüstung und Profite*, Friedensanalysen 9, Frankfurt: Suhrkamp, 1979; K. Engelhardt and K.H. Heise, *Militär- Industrie-Komplex im staatsmonopolistischen Herrschaftssystem*, Berlin: Staatsverlag der DDR, 1974.

40. Cf. W. Voss, 'Profite', in: *Rüstungs- oder Sozialstaat?*, ed. J. Huffschmidt, Köln: Pahl-Rugenstein, 1981.

41. Huffschmidt, ed. *Rüstungs-oder Sozialstaat?*

42. In the US, Seymour Melman stresses these points. For the FRG, see B. Köppl, 'Ist das NATO-Rüstungsmanagement reformbedürftig?' and B. Köppl, *Rüstungsmanagement und die Verteidigungsfähigkeit der NATO*, Straubing and Munich: Donau, 1979.

43. 'Fregatte 122: Wichtigstes Programm der schwimmenden Marine', *Wehrtechnik*, November 1980, p. 35.

44. See Bielfeldt, *Rüstungsausgaben*.

45. See Albrecht, Lock and Wulf, *Arbeitsplätze*.

46. An econometric study for the FRG gave the following employment effects (direct and indirect) for each 1,000 million DM in public expenditure: 'defence' – 18,000; 'health' – 20,000; 'construction' – 21,000; 'social work' – 23,000. See Schmidt, *Zur Bedeutung der Staatsausgaben*.

47. See Albrecht, Lock and Wulf, *Arbeitsplätze*.

48. In early 1981 the Minister of Defence, H. Apel, blamed much of the cost-overruns of the MRCA-Tornado programme on its multinational character. See, for example, W. Flume, 'Brief aus Bonn', *Wehrtechnik* no. 1 (1981): 9.

49. In press reports, the MRCA Tornado has been criticized particularly because its terrain-following radar makes control by the pilot very difficult. The automated gun of the Leopard II has not been well received by the troops.

5 SWEDEN

Per Holmström and Ulf Olsson

The Birth of a Modern Swedish Arms Industry

During the Second World War, Sweden had sought to build up an armaments industry which was technically advanced and which could, independent of the outside world, provide the country with well-equipped armed forces.[1] Sweden's isolation from trade with the Allies during the war and the fear of becoming dependent on deliveries of various supplies from Germany were contributing reasons to this policy. The rearmament during the Second World War period did not, however, have a great or lasting influence on the industrial structure of those sectors that were producing traditional armaments.

Great changes did occur in those sectors that manufactured 'new' types of armaments, above all aircraft and tanks, the latter an area in which previous experience was largely lacking. Svenska Aeroplan-aktiebolaget (Saab) was established with the active cooperation of the Swedish government by combining the two largest firms within the aircraft industry and by 1939 it had a virtual monopoly over the production of aircraft. Nydqvist & Holm AB (NOHAB) acquired a similar dominance in the production of aircraft engines. The production of aircraft and tanks during World War II suffered considerable delays because of Sweden's dependence on foreign designs and technology for these products.

It is interesting to note that the rearmament associated with World War II caused no increase in state ownership of arms industries. The possibility of establishing a state monopoly was discussed during the period between World War I and World War II but was rejected for economic and defence reasons. The government wanted Sweden to have an arms industry and this meant that weapons had to be exported in order for the Swedish arms industry to maintain the desired size and level of technical competence. At the same time, Sweden was heavily involved in questions of disarmament and international cooperation, and could not very well export weapons manufactured in state-owned factories.

The Second World War was not followed by disarmament in Europe. Instead, considerable tension between the so-called Eastern and

Western blocs developed towards the end of the 1940s. Sweden ended up standing alone between these two blocs after attempts to organize a Nordic defence alliance failed and Denmark and Norway joined NATO.[2] The Swedish government then pursued a foreign policy based on the doctrine of 'non-alignment in peace aiming at neutrality in war'. The similarities between this stance and the situation during World War II are obvious.

If the policy of non-alignment between the blocs organized by the great powers was to arouse confidence in foreign countries, it was thought necessary that Sweden maintain strong armed forces supplied by Swedish industry. Defence expenditure, which had declined immediately after the war, increased again in the early 1950s, both in relation to total government spending and to Sweden's gross national product (GNP). The Swedish government felt exposed to pressure from the Soviet Union during this period and serious incidents occurred in the Baltic. According to Swedish strategic planning, the armed forces were to be able to meet an enemy beyond or at the country's frontiers and halt an invasion. The possession of an advanced air force therefore became increasingly important for the country's defence, and about half of all the new materiel purchased for the armed forces was equipment for the air force.

Partly as a result of the development of international relations, the structure of the Swedish armaments industry changed strikingly little after World War II. Many of the civilian-oriented firms which had been engaged in the production of armaments reverted to civilian production and began to participate in a world market that was hungry for goods. For these companies, the war served as a springboard for the postwar period. The more specialized arms firms, on the other hand, were, after a short period of diminishing defence orders, again heavily involved in military production. One example is Saab, which tried to free itself in various ways from total dependence on orders from the armed forces when the war ended. Production of a motor car was started and a civil passenger aircraft was developed. However, the rearmament of the early 1950s utilized the entire capacity of the aircraft industry and the production of civil aircraft was stopped.

All the well-known companies from the war years, Flygmotor, Bofors, L.M. Ericsson and so on, took part in this cold-war rearmament. The government's own production of military equipment was maintained within the narrow framework established during World War II. About 10 per cent of the total purchases of military equipment were from state-owned manufacturers. In 1945, 50 surplus fighter planes were bought from the US, but apart from this, very little equipment was imported. The ambitious aims of Sweden's defence led the country to shoulder almost the same burden in terms of military research and technological development as the great powers. Grants for military research increased quickly in the years after the war. Nevertheless, compared with other industrialized countries, a smaller proportion of Swedish private industry's research was financed out of government funds. In the main, the large grants went primarily to the leading defence industries. Basic military research was carried out by industry and those research groups outside industry, above all, at the Swedish National Defence Research Institute (Försvarets Forskningsanstalt, FOA) which was founded in 1945 and works directly under government auspices.

Defence Orders and the Production of Military Equipment 1954-79

The Demand from the Swedish Armed Forces

It was the demand from the Swedish defence that formed the basis for and determined the direction of Sweden's military industry.[3] During the years following the end of the Korean War, no basic changes occurred in Sweden's foreign relations or in Swedish defence policy. Defence costs remained relatively high. From the mid-1950s to the end of the 1960s, defence spending doubled in constant value. As a proportion of total Swedish resources, however, military spending dropped from 25 per cent to about 15 per cent of the national budget and from barely 5 per cent to barely 4 per cent of the gross national product (see Figures 5.1 and 5.2). During the 1970s the share of defence in the total budget declined even further and its share of GNP stabilized around 3.5 per cent by the end of the decade.

During these years, a diminishing portion of total defence resources was used for the procurement of new equipment, while a growing share was spent on personnel and maintenance. This is what can be expected from a non-growing armed forces. In the mid-1950s, about half of the defence grants were used for procurement. At the beginning of the 1970s, about one-third were used for the same purpose (see Figure 5.3). This meant that the purchase of material for the Swedish armed forces grew slowly up to the mid-1960s but then began to diminish.

Simultaneously, the cost of producing weapons rose. The technical complexity of the weapons grew constantly as did, in many cases, the size of individual units. The Swedish armed forces were comparatively

Figure 5.1: Defence Spending in Constant and Current Prices, 1953/54-1979/80 (1959 = 100)

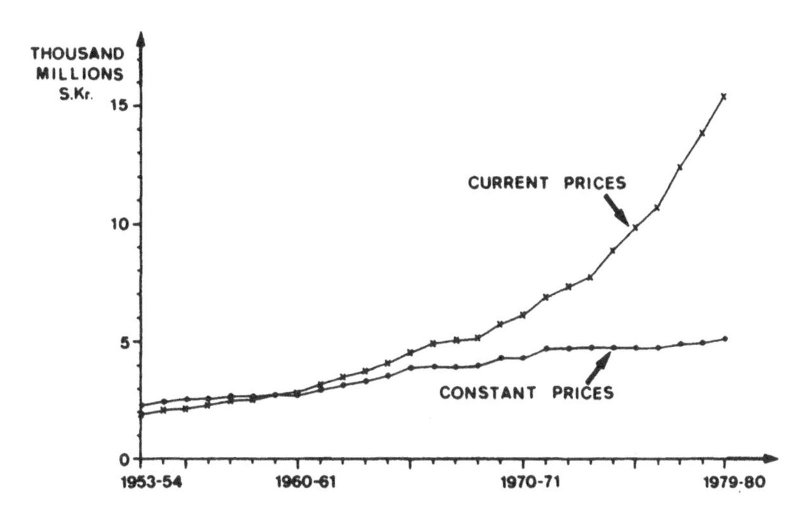

Source: *Riksräkenskapsverkets/Riksrevisionsverkets Budgetredovisning, 1953/54-1979/80*, Stockholm: 1954-80.

Figure 5.2: Swedish Defence Spending as Share of GNP, 1953/54-1979/80

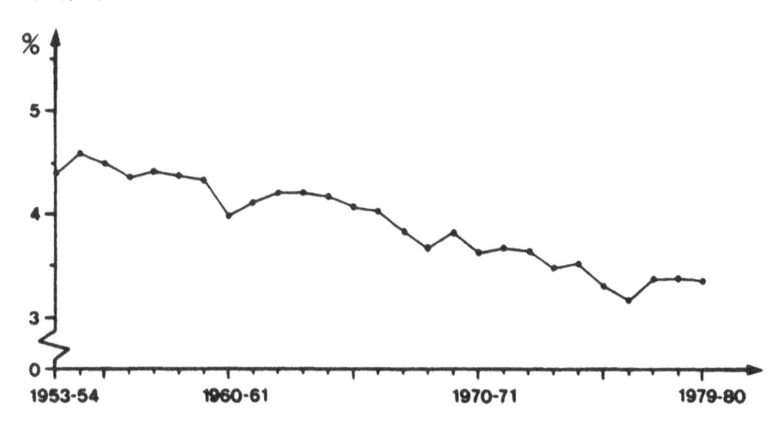

Source: *Riksräkenskapsverkets/Riksrevisionsverkets Budgetredovisning, 1953/54-1979/80*, Stockholm: 1954-80; *Sveriges officiella statistik: SM N*, no. 4.4, Stockholm: 1980; and *Sveriges officiella statistik: Statistisk Årsbok*, Stockholm: 1953-80.

Figure 5.3: Procurement of Equipment as Share of Total Defence Expenditure, 1953/54-1979/80

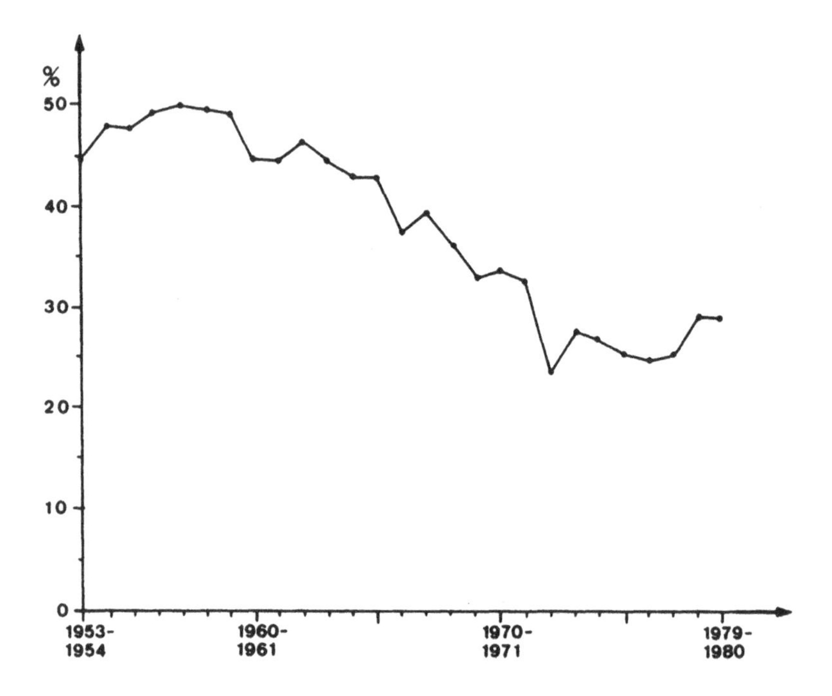

Source: *Riksräkenskapsverkets/Riksrevisionsverkets Budgetredovisning, 1953/54-1979/80*, Stockholm: 1954-80.

large during this period* because there was compulsory military service. The Swedish military services were therefore faced with the problem of equipping the entire armed forces with modern weapons in a period of diminishing economic resources. Considerations of quality were constantly set against considerations of quantity throughout the 1954-1979 period. The basic choice was between buying as much as possible of the most advanced materiel and equipping only a small part of the fully mobilized force, or keeping the maximum of fighting forces and distributing the necessarily less advanced equipment among them all.

*The Swedish armed forces are built around a small peacetime standing force, and a very large rapidly mobilizable army called to service in times of crisis.

For a long time Swedish policy tended towards the first alternative. This meant, for example, that it was no longer thought possible to halt and defeat an invading enemy from more than one direction. In other areas fighting would have to occur within Sweden's borders and without the most advanced weapons. The emphasis on defence at the borders continued to favour the air force, which received almost 60 per cent of equipment grants during the first half of the 1960s. This did not, however, prevent a 50 per cent reduction in the number of squadrons during the 1960s and 1970s. A substantial reduction also took place in the number of fully equipped army units and the number of modern, larger naval vessels. In the latter case, tactical considerations also played a role.

The contradiction between quantity and quality was sometimes weakened or even disappeared, for example, when technical developments led to cheaper alternatives. A change from aircraft to missiles was at times such a step and was made possible by rapid developments within the field of electronics. In the aircraft industry during the 1970s, there was a tendency towards lighter structural materials, such as composites, and lighter engines. In the 1970s, quantity was sometimes favoured over quality. This was evident in the debate over the acquisition of aircraft (which will be touched upon later) but the reversal of the trend was also apparent in the inter-service allocations for material procurement. The air force's share declined to close to 50 per cent at the beginning of the 1970s, partly in favour of the army.

A smaller number of producers received most of the diminishing volume of orders for military equipment. The major producers were Saab (which merged in 1968 with a major truck manufacturer to form Saab-Scania AB), Bofors, Volvo, Ericsson, Hägglunds (which became part of ASEA in 1972), and the state-owned National Defence Industries (Förenade Fabriksverken, or FFV). By the end of the 1960s, the armed forces bought three-quarters of their equipment from these six corporations (see Figure 5.4). By the 1970s, concentration had progressed so far that there tended to be only one or two producers for each category of product. In 1979, there were only 50 separate companies producing military equipment in its strictest sense – arms and ammunition.[4] In many cases, moreover, joint ventures had been formed between different companies for research, development and production of military equipment or components.

This concentration had certain advantages. Continuous contact between the buyer and a seller and more or less guaranteed orders allowed the long-term build-up of competence and the on-going use of

Figure 5.4: Accumulated Shares of the Deliveries to the Materiel Administration of the Armed Forces, 1972/73-1979/80

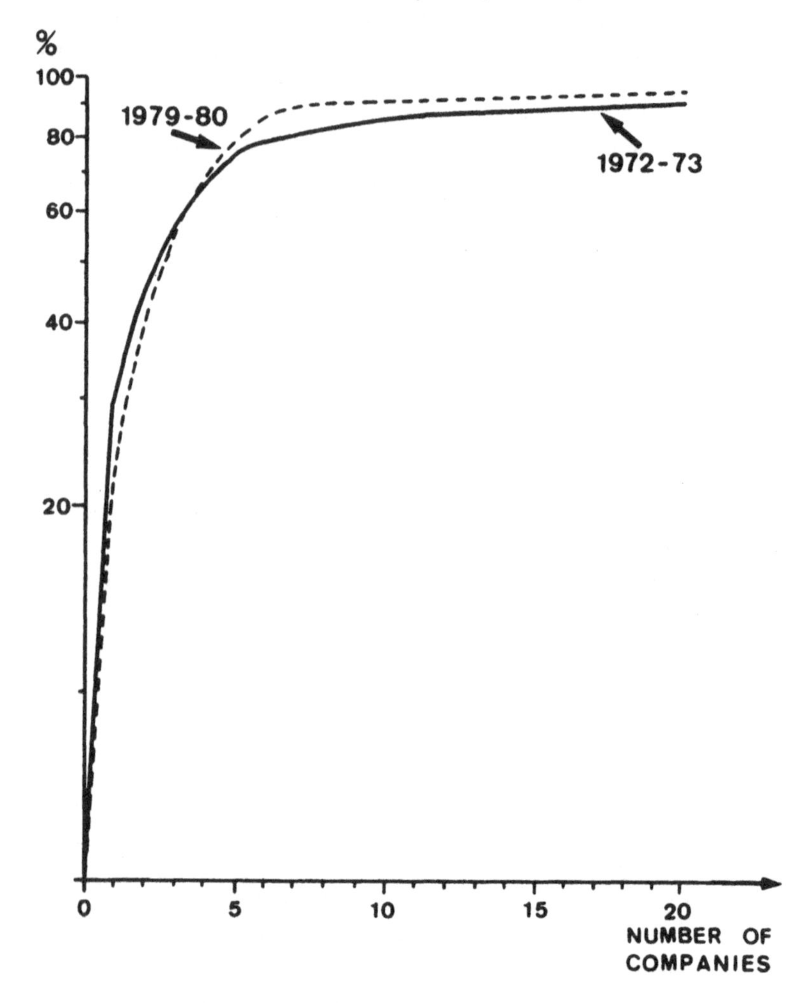

Note: This graph is based on data from Swedish companies and corporations which delivered more than one million Skr. worth of goods: 103 companies out of 479 in 1972/73 and 120 out of 1,431 in 1979/80.
Source: Delivery statistics from the Materiel Administration of the Armed Forces.

research, development and production capacity. Production runs became longer, which theoretically lowered prices. The advantages of concentration, however, had their limits and these were to a great extent reached during the 1970s.

Another problem facing the Swedish government concerning the equipment of the armed forces was the growing contradiction between the need to use funds for the procurement of military equipment as efficiently as possible and the desire to 'buy Swedish'. There were certain arguments in favour of letting Swedish industry produce military equipment.[5] In many cases, it was clearly economically advantageous to produce in Sweden. Often the Swedish armed forces wanted equipment with specific features and it was difficult to find products abroad which could be used without expensive adjustments.

The Swedish armed forces did not take into account the possible effects of spin-off from military R&D when considering where to buy equipment. Their decisions were based on calculations of the defence effects of certain purchases, not on any indirect advantages from these for the Swedish economy or society. The Swedish military was, nonetheless, concerned about the long-term consequences of a diminishing Swedish arms industry on the defence capacity of the country. The armed forces believed that it was important to maintain at the least a sufficient domestic technical capability to be able to test, modify, or manufacture foreign products on licence. It was also suggested that Sweden derived a tactical advantage from the fact that domestically produced equipment differed in some of its technical aspects from the equipment of any of the political blocs and thus was less vulnerable to countermeasures. The main argument for the development and production of war material within the country remained, however, that dependence on defence-related imports could not only threaten Sweden's ability to defend itself but it could also open the door to political pressure and erode confidence in Sweden's policy of non-alignment and neutrality.

Even so, the economic motives for imports grew stronger. The increasing sophistication of defence systems led to higher development costs, especially because the possibility of producing a large number of units and thus lowering the unit costs are strongly limited in Sweden.[6] For political reasons, it was usually not possible to work together with foreign arms manufacturers in order to reduce costs. However, choices were made to import or to produce foreign designed components under licence. Throughout the 1950s and 1960s direct imports of defence equipment were still only 10 per cent of total purchases. By 1968, however, it was estimated that indirect purchases accounted for more than 20 per cent of equipment purchases and the trend was upward. It seems likely that more than one-third of the equipment bought by the military purchasing agency, Försvarets materielverk (FMV), during the 1970s was imported.[7] Missile, helicopter and aircraft production have been particularly dependent on imports.

Aircraft Procurement in the 1970s

The growing problems affecting Swedish procurement of arms were clearly demonstrated in the debate concerning the acquisition of new aircraft during the 1970s. Development costs for each new generation of aircraft (measured in millions of Swedish kronor at constant prices) had risen at a sensational rate since the end of World War II (Table 5.1).

Table 5.1: Development Costs for Aircraft

Decade and aircraft	Development costs total	Numbers of delivered aircraft	Development costs per aircraft
1950s (Tunnan)	324	662	0.5
1960s (Draken)	2,500	550	4.5
1970s (Viggen)	6,800	350	19.6

Source: *Veckans affärer*, no. 42 (1976).

When the deliveries of the Viggen began in the early 1970s, a lively debate started concerning this aircraft and its successor. The debate came to include Swedish defence policy and the arms industry in general. To make room for the expensive Viggen within the defence budget, the number of aircraft produced had been reduced from the projected 800 or so to about 350. To bring down development costs, many imported components were accepted. A large part of the electronic system, for example, was developed in the US, and the jet engine was basically a US design produced under licence in Sweden.

Because of indecision concerning the successor to the Viggen, the Swedish aircraft industry was unable to proceed in the 1970s to develop its suggested aircraft for the 1980s, the advanced B3LA. This threatened both the aircraft industry and its technical personnel. Eventually it was accepted that foreign know-how had to be important in one form or another and that a 30-year tradition of Swedish self-reliance in developing aircraft had to be broken.[8]

Arms Producers in the 1970s

In this section, the focus will be on the single firm, specifically, the major arms producers and exporters. The role of the arms industry in the Swedish economy as a whole will be discussed in a subsequent section.

After a long period where the domestic market was secure and growing and profits were good, the 1970s brought with them uncertainty regarding the direction of defence, and a squeeze on profits for defence firms.[9] The Swedish defence industry was constrained by a number of technical, economic and political considerations from expanding production – which essentially meant exporting more – to bring down unit costs. Foreign military systems were often built upon different technological traditions and infrastructure, and the prices for Swedish-produced arms were often high. Foreign markets were protected, not the least against a country outside the two main military alliances.

Most important, however, were Swedish regulations concerning exports of weapons which have been in force since the 1930s.[10] Swedish industry can export weapons easily only to Nordic and to non-aligned European countries. For each weapon export deal, a special licence is needed, the granting of which depends primarily on the situation of the importing country. In addition, an important distinction is made between exporting typical defensive equipment and arms in general. It is easier to obtain permission for the export of artillery for coastal defence and anti-aircraft artillery than, for example, the export of tanks and aircraft.

Another possibility for the arms manufacturers to counter a crisis without leaving their traditional field of production has been by cooperation or mergers. Such possibilities have been, however, limited by the already very high degree of concentration within the Swedish defence sector. Within many of the defence sub-sectors there was only one domestic producer in the 1970s; an exception to this rule was the electronics industry. Swedish firms have therefore been required to cooperate with foreign producers, a course which has not been entirely without complications.

Most industrial firms try not to be dependent on a single product and to look for new markets when the old ones stagnate. For the arms industry, diversification can take different forms. One is to develop products in which the firm can use its specific technical capability, that is, where positive spin-off effects exist. Another is to combine the production of military equipment with the production of completely different civilian-oriented goods. As will be discussed below, the production of arms is often only a small part of the production of a group of companies formed through mergers. Such conglomerates do not depend on orders from the armed forces even if some of their member companies continue producing military items as long as they are profitable. It is rarely the case that a firm specializing in arms production faces the need of a dramatic 'conversion' into civilian production.[11]

The Aircraft Industry

In the mid-1970s, the Swedish aircraft industry consisted essentially of nine privately owned companies and a few state-owned concerns. It employed around 10,000 people, one-third of whom were involved in development work. The remainder of the work force was involved in the actual production of aircraft, engines, weapons and other kinds of equipment related to air defence. To these 10,000 people should be added 7,000 others involved in aircraft maintenance. Around 90 per cent of the aircraft industry's output was sold to the Swedish armed forces (Table 5.2) while the remaining 10 per cent was exported.[12]

Saab-Scania, Volvo and Ericsson comprise more than 80 per cent of the aircraft industry (50 per cent of deliveries, 20 per cent and 10 per cent respectively) and dominate their particular fields of production. There is only one other company of any quantitative importance, Förenade Fabriksverken (FFV) which is state-owned and accounts for around 8 per cent of total deliveries. Strictly speaking, FFV does not belong to the aircraft sub-sector since it is almost completely engaged in maintenance work for the air force. The rest of this sub-sector consists of smaller companies with, at the most, a few hundred employees each.

Saab-Scania AB. Saab-Scania was formed in 1968, when Saab bought up Scania-Vabis AB, a manufacturer of trucks and buses.[13] During the 1970s this group lessened its dependence on the production of military equipment. Most military production took place in the aerospace division, whose share of the company's total sales decreased from around 15 per cent to around 10 per cent during the course of this decade. Employment in the aerospace division in the 1970s fluctuated between 13 and 18 per cent of total Saab-Scania employment (see Table 5.3). The company also produced cars and electronics but the technical core

Table 5.2: Deliveries to the Materiel Administration of the Swedish Armed Forces, 1972-73 and 1979-80 (millions Skr.)

Aerospace	1972/73	1979/80
Saab-Scania	682	1,066
Volvo	336	670
Ericsson	152	364
Other Defence Industries		
Bofors	250	705
FFV	280	596
ASEA	61	382
Svenska Varv[a]	–	116
Svenska AB Philips[a]	78	52
Statsföretag[a]	54	45
Teleplan[a]	33	37

Note: a. Information is available for these companies but the individual firm is not discussed in the text below due to space considerations.
The following companies are included in each group:

Group:	1972/73	1979/80
Saab-Scania	AB Förenade Flygverkstäder Stansaab El AB, and Svenska Volkswagen AB	Data-Saab AB and Svenska Volkswagen AB
Ericsson	SRA and Sieverts Kabelverk	SRA and Sieverts Kabelverk
Bofors	Lindesbergs Ind AB	Lindesbergs Ind AB, and Norabel AB and Bofors Aerotronks and AB Trelleborgsplast, and 50 per cent of HB Utvecklings-AB
FFV	Telub	Telub and Norma Projektil
ASEA	Hägglunds and Kohlswa Jernverk	Hägglunds, Kohlswa Jernverk and 50 per cent HB Utvecklings-AB
Svenska Varv		Karlskronavarvet, Kockums Ind.
Statsföretag	Karlskronavarvet and Kalmar Verkstad	Kockums Mek, Kockums Landsverk, Eiser and ABAB

Source: Delivery statistics from the Försvarets materielverk, various years.

of the group was the production of aircraft. During the 1970s, Saab-Scania's aerospace division in Linköping, employed over 2,000 engineers, one-quarter of whom had university degrees. This constituted a development capacity outstanding among Swedish companies.

The aerospace division produced primarily military aircraft (during the 1970s, primarily the Viggen). Missiles and military electronics were also produced on lesser scale. In the early 1970s, Saab-Scania was guaranteed a profit of 10 per cent on its deliveries and the work proceeded on current account. This system was subsequently replaced by contract arrangements in which a 'reasonable' profit was calculated.

Table 5.3: Major Swedish Arms Producers, Sales and Employees, 1970-79 (million Skr. and percentages)

	1970	1971	1972	1973	1974	1975	1976	1977	1978	1979
Saab-Scania										
Total sales	3,509	4,111	4,545	5,412	6,552	7,900	9,613	10,796	11,642	13,426
Aerospace div. sales	491	794	630	787	838	858	1,096	1,051	1,114	941
%	14.0	19.3	13.9	14.5	12.8	10.9	11.4	9.7	9.6	7.0
Total Employees	28,588	29,587	30,135	32,436	34,973	37,492	41,386	42,000	39,249	39,000
Aerospace employees	5,226	5,165	5,008	4,920	4,910	5,092	5,483	5,452	6,307	5,930
%	18.3	17.5	16.6	15.2	14.0	13.6	13.3	13.0	16.1	15.2
Volvo Group										
Total sales	5,324	6,104	7,346	8,986	10,537	13,692	15,743	16,168	19,133	23,472
VFA	117	189	255	290	330	372	379	433	504	571
%	2.2	3.1	3.5	3.2	3.1	2.7	2.4	2.7	2.6	2.4
Total Employees	38,866	41,144	44,801	51,408	56,744	63,068	62,441	59,874	61,650	65,054
VFA	2,367	2,398	2,431	2,532	2,720	3,006	2,989	2,850	2,850	2,800
%	6.1	5.8	5.4	4.9	4.8	4.8	4.8	4.8	4.6	4.3
Bofors										
Total sales		1,117	1,050	1,179	1,495	1,709	1,925	2,325	2,788	3,267
Defence-related sales[a]		457	379	412	436	559	588	908	1,286	1,671
%		40.9	36.1	34.9	29.2	32.7	30.6	39.0	46.1	51.1
Total Employees			12,381	12,083	12,645	12,798	13,131	13,557	14,001	13,750
Defence-related employees			3,900	3,900	4,100	4,200	4,300	5,100	5,001	5,700
%			31.5	32.3	32.4	32.8	32.8	37.6	35.7	41.5

ASEA										
Total sales	3,690	4,001	4,952	5,249	6,917	7,863	8,400	9,718	9,814	11,830
Hägglunds		176	186	209	271	346	514	516	426	465
%		4.4	3.8	4.0	3.9	4.4	6.1	5.3	4.3	3.9
Total Employees	36,591	37,911	38,651	39,154	41,217	43,604	44,246	43,233	43,071	43,404
Hägglunds			1,833	1,994	2,189	2,357	2,372	2,328	2,253	2,258
%			4.7	5.1	5.3	5.4	5.4	5.4	5.2	5.2
Ericsson[b]										
Total Employees	61,900	66,900	70,600	75,600	80,600	84,100	71,100	66,400	65,100	59,500
Mil-rel. Employees	2,300	2,300	2,400	2,500	2,600	2,700	2,800	2,800	2,900	3,000
%	3.7	3.4	3.4	3.3	3.2	3.2	3.9	4.2	4.5	5.0

Notes: a. Includes internal deliveries.
b. Figures for military related-sales unavailable. Figure for military-related employment are estimates.
Sources: Saab-Scania, *Annual Reports*, Linköping: 1970-79; Volvo, *Annual Reports*, Gothenburg: 1970-79; Bofors, *Annual Reports*, Karlskoga: 1970-79; ASEA, *Annual Reports*, Västerås: 1970-79; Director Christer Ericsson, M1 Division, Mölndal.

The long-term and secure orders from the armed forces were particularly advantageous since they were to a great extent paid in advance. The profit rates of the aerospace division were very high during the 1970s, although they did decrease towards the end of the period (see Table 5.4). The portion of total Saab-Scania profits derived from the aerospace division correspondingly decreased from more than 50 per cent to around 33 per cent.

If, however, only the figures showing turnover and profits are considered, the importance of the aerospace division for Saab-Scania will be underestimated.[14] It is only by producing military aircraft that the corporation as a whole has been able to maintain its technological leadership. Saab-Scania was, therefore, very alarmed when the future of the Swedish aircraft industry was debated in the 1970s. Different measures were undertaken in an attempt to safeguard the economic position of the group. One involved the creation of joint ventures with Bofors and Ericsson. Another was to attempt to bring down the growth in development costs for aircraft by using existing foreign-designed components.

Table 5.4: Saab-Scania, Profitability 1975-79 (percentages)

	1975	1976	1977	1978	1979
Corporation	8	6	9	10	13
Flygdivision	21	16	15	10	7

Source: Saab-Scania, *Annual Reports*, Linköping: 1975-79.

Great efforts were also made to export aircraft and a degree of success was recorded in the 1960s and early 1970s. The Draken, the predecessor of the Viggen, was sold to Denmark and Finland. The Saab 105 trainer/attack aircraft was exported to Austria and the Saab Safir/Supporter to Pakistan and other countries. At the beginning of the 1970s, the Saab-Scania aerospace division exported almost 20 per cent of its production. During the 1970s, however, the export share declined and Saab-Scania never succeeded in exporting the Viggen despite several major attempts. Although the Swedish aircraft industry – in the form of a consortium of Saab-Scania, Volvo and Ericsson – attempted to win a contract in the early 1970s from Holland, Belgium, Denmark and Norway for a fighter aircraft to replace their F-104 Starfighter system, it was unable to compete with the US aerospace industry. The latter was able to offer more aircraft at a lower unit price.

A similar offer to sell 50 training aircraft to Finland was rejected for the same reasons. In both cases, the offers included generous compensation in the form of investment related to the manufacutring of the aircraft in the purchasing country plus investment in the civilian sector. A subsequent attempt to sell the Viggen to India also failed. These major attempts to export illustrate important aspects of the international aircraft market in the 1970s. First, it was a buyer's market; sellers were prepared to compensate buyers generously in order to obtain large production runs for their aircraft. Second, not only aircraft industries were sellers. States were as well. The Swedish Social Democratic government took an active part in marketing Swedish military aircraft. The degree to which international political considerations could affect these deals was thus increased.[15]

The Swedish aircraft industry is comparatively very small; most of the aircraft industries in Western Europe are five times its size. This reduces the possibilities open to the Swedes of offering tempting compensation.[16] Furthermore, Saab-Scania is legally prohibited from doing business with the potentially most attractive clients, Middle Eastern governments. For different reasons, therefore, Saab-Scania was unable to expand its foreign sales as it wanted to. By the late 1970s, exports by the aerospace division accounted for only 10 per cent of its total sales.

Faced with the threat of reduction, the Saab-Scania aerospace division sought to diversify its production. Its innovative capacity in the field of civilian products was impressive. Some of these products were marketable, but none of them reached a satisfactory volume of sales or profitability. Work within the field of civil aircraft was more successful and took the form of acquiring orders as a subcontractor for components such as wing flaps for foreign aircraft, as well as new passenger aircraft. Reductions in the Swedish military aircraft industry in the 1970s were felt primarily by the development personnel while the production of the Viggen kept the manufacturing personnel occupied. All in all, attempts at diversification in Saab-Scania's aerospace division affected the composition of sales very little. In 1979, 83 per cent of the sales still were of military products.

The Volvo Group. The manufacturing of aircraft engines, by Volvo Flygmotor (VFA) in Trollhättan in southwest Sweden, declined in importance during the 1970s and by 1979, it accounted for only about 2.5 per cent of the group's total sales (see Table 5.3). Of VFA's production, 70 per cent was for military purposes. In addition to aircraft engines, Volvo delivers cars, trucks and tractors, rocket engines and tanks to the Swedish armed forces. During the first half of the 1970s, when sales doubled and profit margins rose, VFA was in a strong position. Later, the volume of sales stagnated but profits were maintained. For the Volvo group as a whole, profitability was around 6 per cent between 1975 and 1979. For VFA, it was around 10-11 per cent (see Table 5.5). In the early 1970s, VFA began to investigate the possibilities of diversifying into civilian products. Probably of greatest long-term economic importance were contracts signed towards the end of the decade for the development of civil aircraft engines together with the American companies General Electric and Garret Corporation.

Ericsson. Like the Volvo group, Ericsson, is, as a whole, not heavily dependent on arms orders. Its M1 division manufactures technically advanced military equipment. Ericsson also owns a 79 per cent share in Svenska Radio AB (SRA), a company which delivers a considerable amount of its products to the Swedish armed forces. Deliveries to the Swedish military at the beginning of the 1970s accounted for only about 2 per cent of the Ericsson group's total sales (including those of SRA). At the same time, the development and production of defence materiel has played an important role in building up the technological base of the corporation.

By the end of the 1970s, the production of miliary materiel had become increasingly important and, in 1979, came to 6 per cent of total corporate sales (see Table 5.3). A considerable part of this production was exported, especially radar equipment. The M1 division, which was completely dependent on orders from the Swedish armed forces, began to investigate different civilian markets during the 1970s. Its major difficulties lay in the fact that the company's organization was designed to manage big, complicated and advanced systems while most civilian products are of a completely different character.[17] The most promising field has been radar for air-traffic control. SRA has sought civilian applications for its technological know-how within the fields of road and rail traffic control.

Looking at the aircraft industry's efforts to move into civilian markets, two points must be taken into consideration. First, the period under study is short while it takes many years to develop and market new products. Second, the 1970s were characterized by international economic stagnation and the fortune of Swedish industry tends to be closely tied to that of the international market.

It is very hard to judge how serious the attempts to diversify have in fact been. Up to the end of the 1970s, it seemed as if orders from the

Table 5.5: Volvo: Profitability and Profit Margins, 1970-79 (percentages)

	1970	1971	1972	1973	1974	1975	1976	1977	1978	1979
Profitability[a]										
Corporation	9	8	10	10	9	6	6	5	5	7
Flygmotor				11	11	11	10	13	10	9
Profit Margins[b]										
Corporation		8	10	10	6	4	4	3	3	5
Flygmotor		10	11	15	15	15	19	21	18	12

Notes: a. Profitability: profits before deductions of interest payments and taxes and allocations to internal funds in relation to total assets shown on the firm's balance sheets.
b. Profit margins: profits as a percentage of the value of sales.
Source: Volvo, *Annual Reports*, Gothenburg: 1970-79.

armed forces would be maintained at a normal level. There has been a reluctance to run down the Swedish aircraft industry not only on the part of the military and the companies themselves but also on the part of the unions and the government. This is one area which still is a show-case for Swedish technological capability. The hope is that this will lead indirectly to orders for other industrial goods. Technical know-how is thought to spill over into other parts of the economy. While some have argued that, in the long-run, a small nation can hold a leading position in a specific field of technology only if state support is forthcoming,[18] the Swedish government has chosen to provide financial support for the diversification of production. The aim has been to find projects where the companies could use their technical skill, but where old traditions could be replaced by a new consciousness of the market. A special committee in the Department of Industry investigated these questions during 1979 and this resulted in support for some projects at Saab-Scania and Volvo.[19]

State support for different industries became rather common in Sweden during the latter half of the 1970s. Normally, as with the textile, steel or shipbuilding industry, the aim was to help industries in trouble to survive, or at least to continue to function while employees sought new jobs. The aircraft industry is a unique case in that it was subsidized while it was still financially solvent.

Other Industries

The aircraft industry dominates the Swedish defence sector. The largest non-aircraft arms producer in Sweden, AB Bofors, delivered only one-third to one-half as much materiel (in terms of cost), to the armed forces during the 1970s as did Saab-Scania (see Table 5.2). Some of the manufacturing of traditional arms still takes place in state-owned factories. *Förenade Fabriksverken* (FFV) was mentioned previously as an important aircraft servicing organization.[20] During the 1970s it also produced about an equal amount (in terms of value) of small arms, anti-tank weapons, bombs, ammunition and other mechanical military equipment. FFV employed around 7,000 workers and is about 80 per cent dependent on military orders. Approximately 10 per cent of FFV's total sales were exported. FFV experienced typical problems when trying to find civilian products suited to its engineering capabilities. The first projects were too technically unsophisticated and the company's technical capability was not fully used. Eventually, suitable projects were found, notably the Stirling engine which was developed together with, among others, American firms.

The Swedish state has long been engaged in the construction of warships. As a result of a general crisis in the shipbuilding industry in 1976-7, the government subsidized the industry and formed *AB Svenska Varv*. Svenska Varv is comprised of all the bigger Swedish ship yards, including AB Kockums, the biggest private shipyard and a traditional producer of warships. By the end of the 1970s, therefore, all warship production took place in state-owned yards. The Swedish navy ordered only a few ships dring the 1970s, mainly patrol boats and submarines. The level of exports has been low as well, although sales have been made, mainly to Malaysia, Chile (before 1973), and Bangladesh.

AB Bofors. The production of traditional weapons in Sweden has been dominated by AB Bofors for many years. During the 1950s and 1960s, however, Bofors developed a very wide range of civilian products, including toothpaste, water turbines, anaesthetics, and thermal cameras. To some extent, it was able to draw upon its expertise as an arms producer to develop the new products. The attempts to diversify were motivated by the belief that it was no longer possible for Bofors to grow within the armaments sector since orders from the Swedish military were stagnating and exports were curtailed by Sweden's export restrictions.[21] It turned out, however, to be easier for Bofors to innovate than to sell these new products. The product mix was soon found to be too diverse and the items produced too expensive. For example, using the technology employed in tank production, a very advanced farm tractor was developed. Unfortunately, the tractor was extremely difficult to sell on the competitive market. Manufacturers with a broader product mix, lower prices and better marketing techniques easily outsold Bofors.[22] It was only thanks to arms production that Bofors avoided losses (see Table 5.3).

In 1972, Bofors was reorganized. The number of products was reduced and grouped into eight profit centres. The most important industrial divisions were: (1) Defence, where guns, tanks, missiles and ammunition were dominant; (2) Steel, in size as important as the defence division; (3) Chemicals, with plastics as the most dynamic product; (4) Engineering (NOHAB), which produces locomotives, diesel engines, turbines and printing presses.[23] There are obvious connections between production for the Swedish armed forces and for the civilian market. Steel, chemicals, and engineering all have their counterparts in arms production and, within the corporation, research and development serves as a natural bridge between the production of defence items and

and the production of civilian goods.

Arms production has been a vital profit generator for Bofors (see Table 5.6). It is thus not surprising that Bofors expanded its military production and purchased several smaller producers of ammunition, military instruments and chemicals. Bofors also formed subsidiaries in the 1970s with Saab-Scania for missiles and Hägglunds for combat vehicles in order to strengthen its market position. There have been several reasons for the success of Bofors' arms production. Domestic demand has been greater for the traditional military products than for aircraft. Nor have traditional arms, or tanks and missiles, been exposed to such dramatic increases in development costs as have aircraft. Bofors has been able to have fairly large and regular production runs and it has developed important new weapons such as a new field howitzer and the Robot 70 missile.

Bofors has also been able to expand its export markets. There probably were purely commercial – technical and economic – reasons for this, but Bofors also benefited from the fact that its weapons are normally classified as 'defensive weapons' by the Swedish authorities which has meant that export licences are more easily acquired. In the late 1970s, Bofors sold half of its output abroad. In contrast to the majority of Swedish arms producers, Bofors could compensate the loss of domestic orders through exports.[24] Many countries buy Bofors' products – mainly anti-aircraft artillery. Bofors weapons are also produced under licence in five countries, India, Netherlands, the UK, Italy and Spain. Several of these countries in turn export the licence-produced materials and they do not in any way need to comply with Swedish export restrictions.

Hägglunds: Begun as a family business Hägglund & Söner was acquired in 1972 by Allmänna Svenska Elektriska AB (ASEA), one of Sweden's biggest enterprises. Hägglund & Söner had concentrated on heavy engineering with military materiel as its core. About 30 per cent of its sales in the early 1970s were defence orders. After its merger with ASEA, Hägglunds concentrated on military products. It accounted for some 4-5 per cent of ASEA's production and, in general, somewhat more of the corporation's profit. (For production and employment data, see Table 5.3.) Thus ASEA, the parent corporation, like Volvo and Ericsson, purchased a military division which was valuable for the group of companies but was not big enough to create a serious dependence on defence orders for the corporation as a whole.

Table 5.6: Bofors: Profits 1971-79 (million Skr.)

	1971	1972	1973	1974	1975	1976	1977	1978	1979
Corporation	38	28	42	110	110	66	37	92	167
Defence materiel			32	38	38	62	84	111	172
Net profit corporation	12	12	15	26	20	20	23	24	30

Source: Bofors, *Annual Reports*, Karlskoga: 1971-79.

This survey of Swedish arms producers has shown that individual companies have had different opportunities for coping with decreasing orders from the Swedish armed forces and for expanding beyond the home market. In many cases, the threat to the domestic market never materialized. In conclusion, two points should be emphasized. The first concerns the concept of an arms industry. No Swedish company of any importance has completely specialized in the production of war materiel. Rather, there has been a tendency, particularly in recent years, for arms production to become a small part of the total activity of a larger corporation. It is important to remember this when discussing the Swedish arms industry or 'military-industrial complex'.

There are, however, problems of definition involved. Here, the term 'military materiel' is used to indicate such equipment which performs to the specifications of the military purchaser. That is, products bought by the FMV have been regarded as military materiel. The distinction, however, between, say, radar equipment for a military airfield and radar for a civilian one is not clear. This problem of definition will be explored further in the discussion of the export of military materiel.

The second point that should be emphasized concerns ownership.[25] In 1973-4, state-owned companies delivered only about 13 per cent of the FMV's total purchases. The government clearly still finds it desirable that private companies manufacture the equipment for the Swedish armed forces. Since arms production is closely linked to the Swedish engineering industry as a whole, the private corporations which dominate the Swedish economy as a whole also are the leading arms producers. These corporations are tied to each other by family relations or through commercial banks. Thus, the Wallenberg group is involved in half of the deliveries to the FMV through its dominant position in Saab-Scania, Ericsson and ASEA. Bofors, in conjunction with certain subsidiaries, accounted for around 10 per cent of deliveries and the Volvo group for around 13 per cent. These three groups of owners are the only ones of any significance.

Foreign influence exhibits itself in different ways. Foreign companies have subsidiaries in Sweden, as in the case of Svenska Philips (completely Dutch-owned) and in addition, Sweden buys some equipment directly from abroad. In 1973-4, purchases by the FMV via these two mechanisms, amounted to 7 per cent of total deliveries. In several cases, foreign companies also hold minority shares in Swedish companies and their deliveries are not included in the 7 per cent figure. Furthermore, as was pointed out earlier, the 'volume of import' in the procurement of the armed forces was considerably larger due to the use of foreign subcontractors. Finally, it should be emphasized that maintenance work performed by the different industries under discussion has not been dealt with here. During the 1970s, the value of maintenance work doubled from around 500 million Skr. to about 1,000 million Skr.

Industry's Influence over Military Decision-Making

The question of the extent to which the arms industry influences the armed forces and military decisions has been much discussed in Sweden. During World War II, private industry increased its influence over military decision-making with the creation of a centralized procurement agency, the forerunner of the FMV,[26] and government-industry relations were generally cooperative in spirit. When the war ended, much of the cooperation concerning general industrial policy disappeared, although there was less contention between industry and government over matters relating to national defence and weapons procurement than over other issues. This is related to Sweden's position as a non-allied country near the frontline of the cold war. During the 1950s and 1960s, there was little discussion of defence questions and the need for a strong defence was never questioned. The important decisions were taken by Parliament only after the groundwork for them had been undertaken by Parliamentary committees. The Communist Party was not allowed to participate in this committee work, and compromises were worked out between the Social Democrats and the parties to the right of them.

In the course of this preparatory work, representatives of the armed forces as well as the defence industry played important roles. Many factors contributed to keeping the tradition of cooperation alive. The political system was very stable, the relations between government and private industry were very friendly, and personal contacts between the individuals involved were very close. The more complex the weapon systems and defence planning became, the longer the development of a new weapon required. As a consequence, collaboration had to start at an early stage, many years before the product could be used. With rising prices it became prohibitively expensive to ask more than one manufacturer to develop a certain item. The process of concentration in the Swedish arms industry made the armed forces dependent on the remaining companies. In general, importing weapons was not considered a viable alternative. Under such conditions, it was easy for industry to take the initiative regarding when and in what direction Swedish weapons should be developed and produced. Thus, general considerations – the concern to maintain a particular domestic production capability – rather than a specific identified military requirement for a particular weapon system was able to determine how the armed forces would be equipped.[27]

In the late 1960s, Swedish defence policy and the defence industry were criticized more strongly than they had been before. This was, of course, connected to the decline in political tension in Europe. Criticism came from many quarters but was fuelled largely by the radical debate in the US concerning the military-industrial complex.[28] The basic argument here is that power is held by a combination of the military, the arms producers and the state bureaucracy who create power, profits and national prestige for themselves by building unnecessarily big and expensive armed forces.

There is no simple way of determining if a military-industrial complex has existed in Sweden during the postwar decades. The situation is complex and the concept both abstract and ambiguous in nature.[29] Empirical investigations of the question have tended to deal with only part of the armed forces, primarily the development of aircraft. A book on the Viggen project, published in 1973, showed that the means by which the defence industry in Sweden exerted influence on government decision-making has taken the same course as in other industrially and scientifically developed countries, particularly the US and West Germany.[30] Critics of Swedish defence procurement have argued that there has been no real democratic influence on the decision-making process in Sweden and they have demonstrated that individuals have moved back and forth between important positions in industry and in the Ministry of Defence and the armed forces. They have also argued that defence has become too technologically advanced and that this actually degrades defence capability.

Supporters of Swedish defence efforts in the postwar period have denied that the technological level of the armament industry negatively effects Swedish defence and point to other, positive effects of the arms industry such as technological spin-off, employment generation and so on. These arguments obviously were not good enough. The diminishing economic resources allotted to defence by Parliament required the air force and the aircraft industry to reduce their expectations. It can be argued that the Viggen project was an example of the political power of a military-industrial complex in Sweden in the 1950s and 1960s. At the same time, developments in the 1970s showed that the objective conditions for its influence had changed.

The Role of the Arms Industry in the Economy: Some Basic Facts

A useful point of departure for discussing the role of the development and manufacture of arms in the Swedish economy during the 1970s is the level of total defence expenditure for the financial year 1977-8. This was 13.3 thousand million (t.m.) Swedish kronor. If one subtracts the costs of protecting civilians and of maintaining production and providing for people during war and emergencies, 12.3 t.m. kronor remain for purely military defence: command and units, 7.5 t.m. kronor; acquisition of equipment, 3.2 t.m. kronor; acquisition of buildings, 0.5 t.m. kronor; and research and development, 1.0 t.m. kronor.[31] These 12.3 t.m. kronor were equivalent to about 3.5 per cent of the country's GNP and about 9 per cent of the state budget. These percentages have been declining since the mid-1950s (see Figures 5.2 and 5.5).

It can be argued that the cost of defence might be somewhat underestimated if measured as a percentage of the budget rather than as real costs for the total economy since conscripted personnel would have been better paid in the civilian labour market.[32] This is, however, a marginal question which will not be developed here. Instead, the focus will be on the quantitative aspects of arms production, including production for export.

Table 5.7 shows total production of arms and aircraft in Sweden during the 1970s. Weapons production accounted for between 1 and 2 per cent of the total value of Swedish industrial production (see Figure 5.6). Some 80-90 per cent of the total value of arms was produced by the engineering subsector.[33] This accounted for about 5 per cent of the output of the engineering industry. The military component of production in the chemical industry – mainly explosives – was also approximately 5 per cent, although the size of the Swedish chemical industry, and hence the value of its output, is smaller than for the engineering industry.

Altogether, about 40,000 persons worked with the production of arms in the 1970s; 36,000 of these were directly involved in production. To this figure can be added about 18,000 people working for defence subcontractors. The companies in which these workers are employed were discussed in the preceding section. Geographically, arms production has been concentrated in Sweden's traditional engineering-

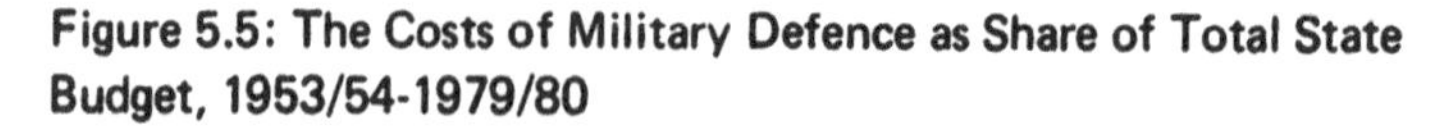

Figure 5.5: The Costs of Military Defence as Share of Total State Budget, 1953/54-1979/80

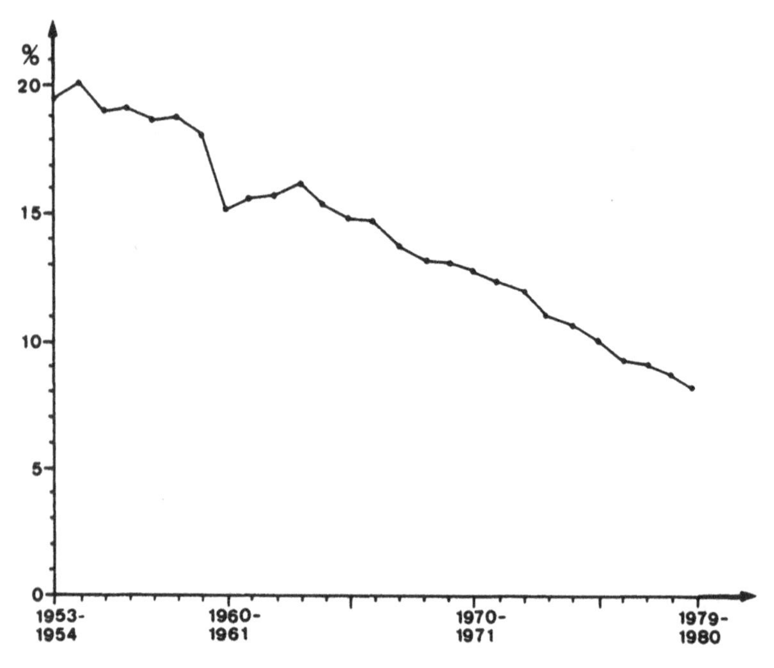

Source: *Riksräkenskapsverket/Riksrevisionsverkets budgetredovisning, 1953/54-1979/80*, Stockholm: 1954-80.

industry belt, which stretches from the Lake Mälaren region towards the iron-making districts in western Sweden and to Gothenburg in the southwest (see Map 5.1). Hägglunds in Örnsköldsvik is a northern outpost while Saab-Scania's aircraft production has been situated in eastern Sweden. Normally the arms industry has played no important role in the employment situation in the different regions, but there have been some exceptions. Saab-Scania in the Linköping region is the most important of these. The area around the city of Linköping is not an industrial region and the Saab-Scania group employed half of the industrial workers in this area during the 1970s. The production of military materiel, mainly aircraft, during the last decade provided work for one-quarter of all engineering employees in the region.[34]

The interest on the part of trade unions and political parties in discussions about the aircraft industry and the possibility of finding alternative civilian production can be explained by the effects on

Table 5.7: Production of Arms and Aircraft, 1970-79[a] (million Skr and percentages)

	1970	1971	1972	1973	1974	1975	1976	1977	1978	1979
Aircraft	676	830	850	936	1,012	1,057	1,175	1,404	1,362	1,429
%	49	56	56	57	54	51	44	43	38	34
Warships	–	34	3	35	55	58	122	44	84	299
%	–	2	0	2	3	3	5	1	2	7
Tanks	232	134	78	74	33	71	148	231	176	102
%	17	9	5	5	2	3	5	7	5	2
Weapons	128	147	190	186	257	262	383	446	558	633
%	9	10	12	11	14	13	14	14	16	15
Ammunition	340	340	411	412	500	627	846	1,169	1,412	1,753
%	25	23	27	25	27	30	32	35	39	42
Total	1,376	1,485	1,532	1,643	1,857	2,075	2,674	3,294	3,592	4,216
%	100	100	100	100	100	100	100	100	100	100

Note: a. Distributed according to customs tariffs with a statistical commodity list, based upon the Brussels nomenclature. Aircraft includes parts and engines but not radar and computer systems. Ammunition includes bombs, grenades, torpedoes, mines and missiles and war ammunition of similar kinds.
Source: *Sveriges officiella statistik. Industri*, Stockholm: 1970-79.

Figure 5.6: Aircraft and Arms Production as Share of Total Swedish Industrial Production, 1970-79

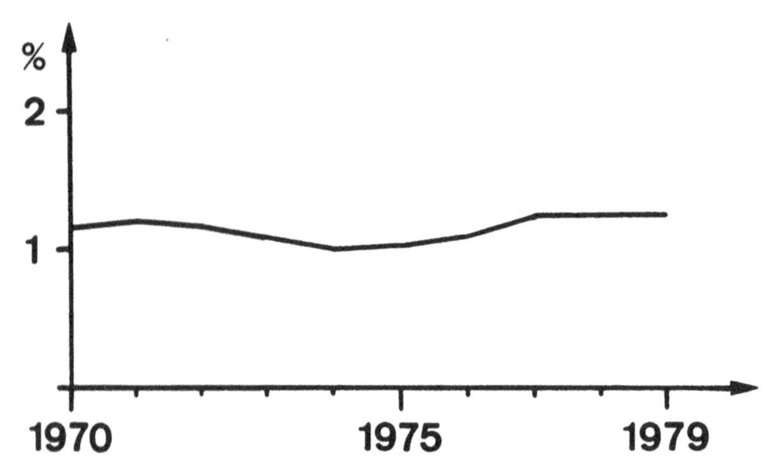

Source: *Sveriges Officiella Statistik: Industri*, Stockholm: 1970-79; *Sveriges Officiella Statistik: Statistisk Årsbok*, Stockholm: 1970-79; and *Sveriges Officiella Statistik: SM I*, no. 7, Stockholm: 1981.

Linköping of a reduction in military orders during the 1970s. The problem was aggravated by the fact that so many of the employed were highly specialized and qualified technicians who had little chance of finding new jobs. This was especially true for those engineers in the 'middle' of the production process, who were concerned with project development or questions of material and the construction of tools. The normally more advanced technology used in the early stages of the production process had more general applicability, as did the skills used in the later stages of control, marketing and servicing.[35]

Volvo Flygmotor did not dominate the town of Trollhättan or its district so heavily. In these areas, industry is well-developed and the manufacturing of aircraft engines only employed 20 per cent of the area's industrial workforce in the 1970s. Nevertheless, for some groups of highly specialized employees, the situation was sufficiently gloomy. For Bofors, the situation was somewhat different. The company heavily dominates the little town of Karlskoga and employs one-quarter of the engineering workers in the region. On the other hand, Bofors expanded during the 1970s and questions such as diversifying production or government involvement on a larger scale to secure employment in the region never arose.

Map 5.1: Main Swedish Arms Producers

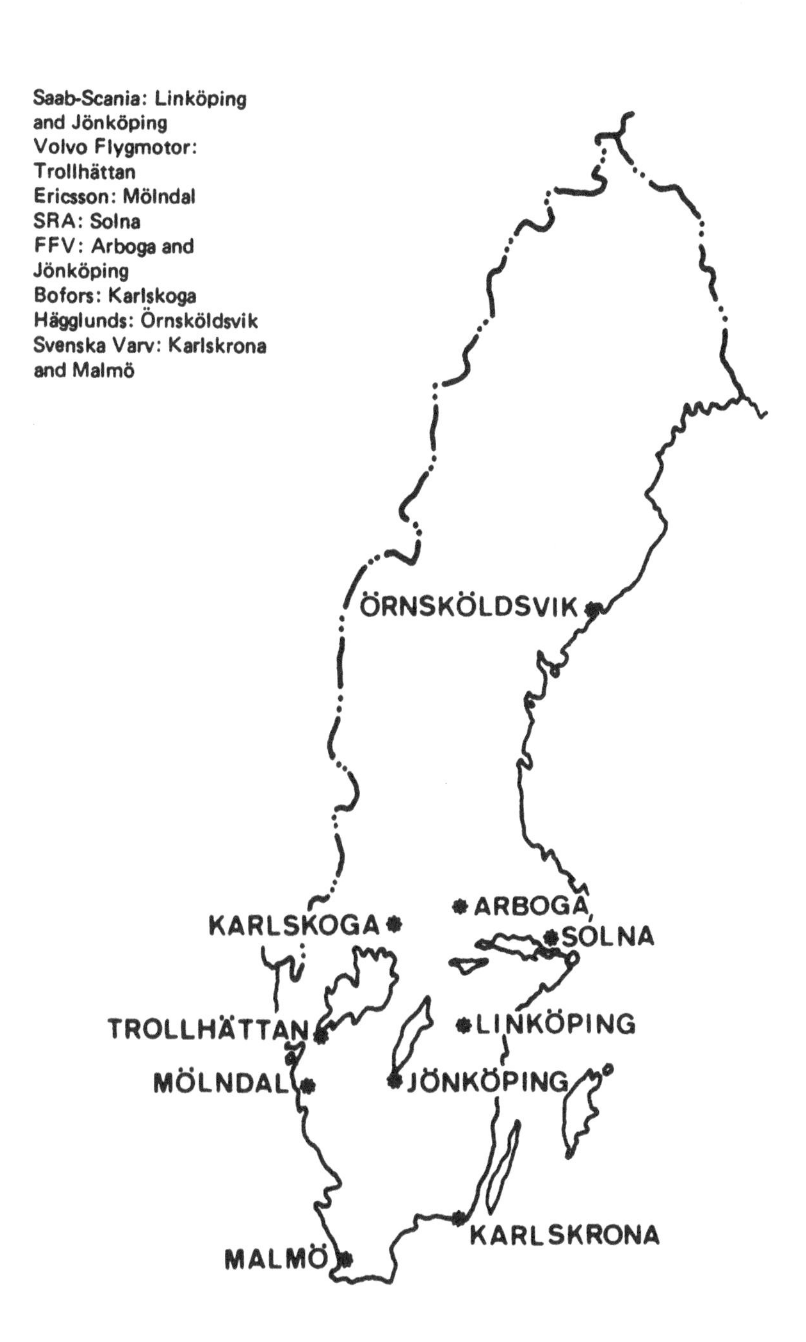

The export of arms during the 1970s amounted to about 1 per cent of Sweden's total exports and nearly 3 per cent of exports of the engineering industry (machinery, apparatus, and transport equipment) [see Figure 5.7]. Exports of military materiel increased during the 1970s and by the end of the decade they accounted for some 21 per cent of total output from the defence industry (Figure 5.8). Some questions regarding the definition of 'arms' have been raised earlier. They are defined here as military equipment which is either constructed, equipped or modified so that it clearly diverges from corresponding

Figure 5.7: Export of Arms as Share of Total Swedish Exports, 1970-79

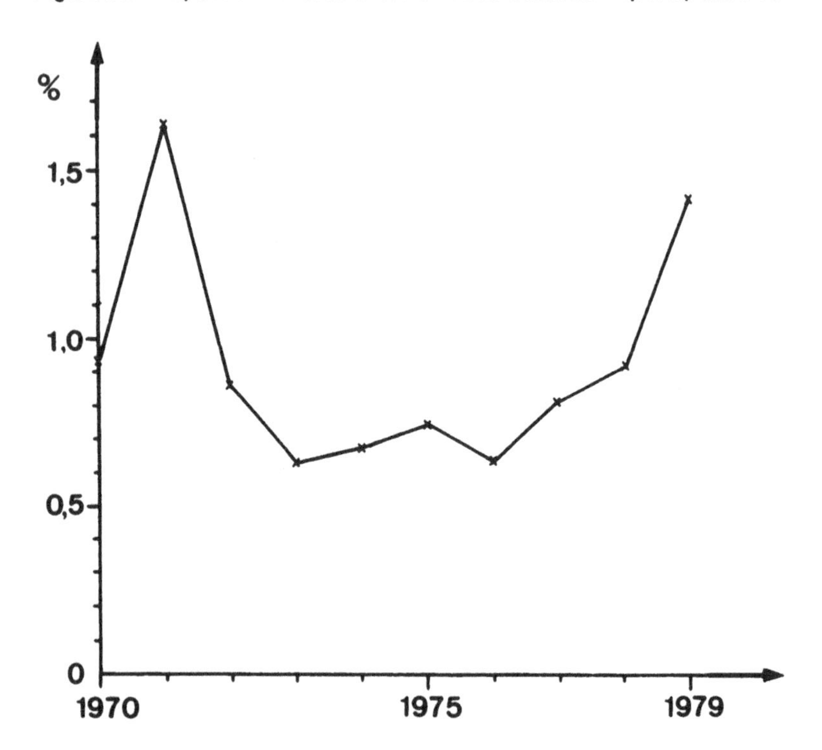

Source: Ministry of Commerce and Industry, *Krigsmaterielinspektörens berättelse*, Stockholm: 1970-79, and *Sveriges Officiella Statistik: Utrikeshandel*, Stockholm: 1970-79.

civilian equipment. Thus, radar used to supervise Sweden's air-space and territorial waters is not included. Neither is equipment for radio communications or terrain vehicles without weapons.[36] During the 1970s, between 30 and 40 companies were exporting arms. In 1978, Bofors alone accounted for more than 51 per cent of such exports, while Bofors, Karlskronavarvet (ships) and FFV together accounted for 87 per cent of such exports.[37] It is normal for the export of arms to vary considerably over time due to the dominance of single orders in certain years. In the beginning of the 1970s, the export of aircraft was important while weapons and ammunition played a greater role towards the end of the decade (see Table 5.8).

Figure 5.8: Composition of Swedish Defence Industry Output, 1979-80

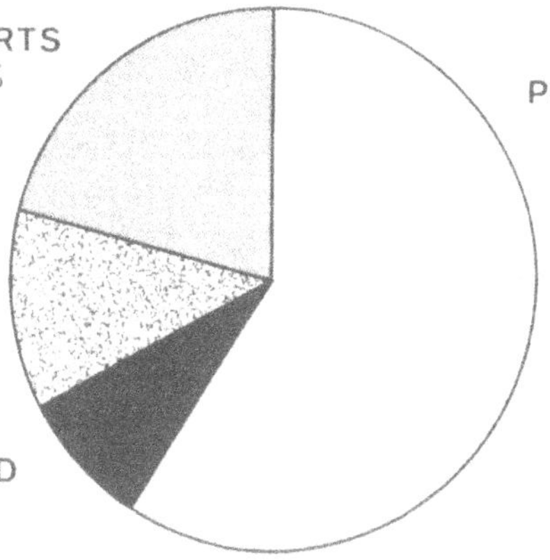

Source: Försvarets forskningsanstalt, Huvudavdelning 1, Sektionen för samhällsekonomi, *Svensk Försvarsindustri: Struktur, kompetens, utvecklings-betingelser*, FOA report C-10200-M5, Stockholm: February 1982, p. 201.

Great changes also take place in the destination of Swedish arms exports from one year to another. None the less, about two-thirds of arms exports normally have gone to the Nordic countries and the non-allied countries in Europe. It has been calculated that Sweden's share of the arms trade is 0.3 per cent of the total and that in the late 1970s Sweden was the twelfth largest exporter of arms.[38]

Military research and development (R&D) accounts for a higher percentage of total Swedish R&D than military production does for total industrial production or the export of arms does for total exports. The military field was the first where the state engaged in large-scale R&D efforts after the Second World War. In the 1970s, as much as one-quarter of total state R&D expenditure still concerned military projects. Most of this money went to the Ministry of Defence and only a small

Table 5.8: Export of Military Equipment from Sweden, 1970-79 (millions Skr and percentages)

Item	1970	1971	1972	1973	1974	1975	1976	1977	1978	1979
Weapons and ammunition										
kronor	127	207	175	189	219	262	225	341	639	1114
percentages	39.6	33.0	48.6	56.4	45.9	48.9	44.8	49.0	70.5	66.7
Explosives and certain detonating equipment										
kronor	14	15	17	14	22	32	36	32	45	40
percentages	4.4	2.4	4.7	4.2	4.6	5.9	7.2	4.6	5.0	2.4
Tanks and other armed vehicles with equipment										
kronor	0	–	1	–	1	21	24	22	4	1
percentages	0	–	0.3	–	0.2	3.9	4.8	3.2	0.4	0
Naval ships with equipment										
kronor	–	30	–	–	–	–	–	–	–	326
percentages	–	4.8	–	–	–	–	–	–	–	19.5
Military aircraft, parts and parachutes										
kronor	132	339	92	96	125	47	110	83	28	50
percentages	41.1	54.2	25.6	28.7	26.2	8.7	21.9	11.9	3.1	3.0
Other military equipment										
kronor	48	35	75	36	110	175	107	218	190	140
percentages	14.9	5.6	20.8	10.7	23.1	32.6	21.3	31.3	21.0	8.4
Total (kronor)	321	626	360	335	477	536	502	696	906	1671

Note: Weapons and explosives category includes such products destined for non-military uses also.
Source: Ministry of Commerce and Industry, *Krigsmaterielinspektörens berättelse*, Stockholm: 1970-79.

part to other ministries (see Table 5.9).[39] From the Ministry of Defence more than three-quarters of the R&D grants have been transferred to industry via the FMV, while relatively small sums have been used by the National Defence Research Institute (FOA), by central military headquarters or for medical research.

Table 5.9: Grants from the Ministry of Defence for R&D 1972/73-1979/80 (millions Skr.)

Year	Amount
1972/3	536
1973/4	587
1974/5	644
1975/6	626
1976/7	663
1977/8	830
1978/9	572
1979/80	665

Source: *Riksräkenskapsverkets/Riksrevisionsverkets budgetredovisning, 1953/54-1979/80*, Stockholm: 1954-80.

Most R&D grants have gone through the FMV to state-owned and private industries or consulting companies. Some three-quarters of all R&D for military purposes has been performed by Swedish industry.[40] This, in turn, has been concentrated in the engineering subsector, especially in aircraft and automobile companies. More than two-thirds of total military R&D has been performed by the aircraft industry (see Figure 5.9).

It is not easy to give a precise figure for the military-financed share of industrial R&D. The main problem is the identification of R&D costs in the production process. Official statistics such as those given in Table 5.10 indicate that the figure was 13 per cent in 1973. According to industry itself, the proportion of state-financed R&D in the same period was 16 per cent which fits well with Table 5.10 since almost all state financing for industrial R&D goes for defence purposes. Other calculations based on FMV's costs show that defence R&D in industry amounted to 20 per cent of total R&D in the middle of the 1970s.[41] Such a high figure can be explained by the fact that FMV uses R&D money to pay for work which industry itself regards as pure production costs, not as R&D. Whichever way it is calculated, the share of defence R&D in industry's total decreased somewhat during the 1970s. Industry's own R&D expenditures for military purposes were

Figure 5.9: Grants for Military R&D, Service Shares, 1972/73-1979/80

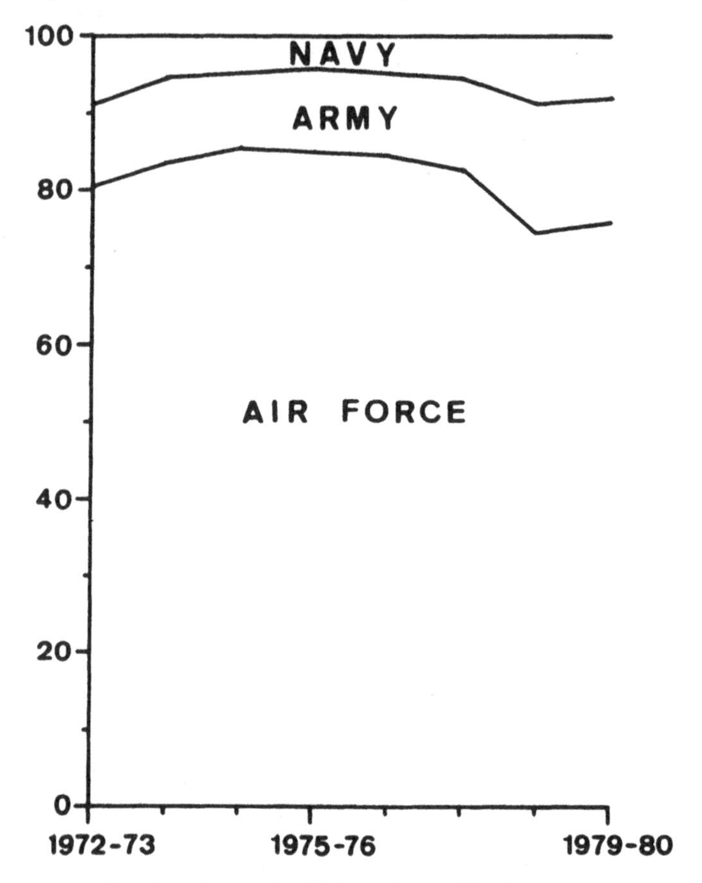

Source: *Riksräkenskapsverkets/Riksrevisionsverkets budgetredovisning, 1953/54-1979/80*, Stockholm: 1954-80.

small compared to the government's during the 1970s. However, a small part of the R&D funds allocated by private industry concerned military projects which were outside the interest of the Swedish defence authorities. Both Saab-Scania and Bofors developed some products directly for the export market. It is very hard to calculate these R&D costs. Despite the problems involved in calculating the military share in Swedish R&D, a reasonable estimate would be that between 15 and 25 per cent of all Swedish R&D – almost 5 thousand million kronor in the late 1970s – went for military projects. Most of this R&D was financed by the government through the FMV and performed by private Swedish companies, mainly in the aircraft industry.

Table 5.10: The Financing by Military Authorities of R&D of Swedish Industry 1973, 1975, 1977 and 1979 (millions Skr and percentage of total industrial R&D)

	1973		1975		1977		1979	
	mSkr	%	mSkr	%	mSkr	%	mSkr	%
Military authorities	280	13	358	12	499	12	467	9
Total	2,100	100	3,060	100	4,153	100	5,339	100

Source: *Sveriges officiella statistik:* SM U, 1975:19, 1977:23, 1979:25, 1981:18, Stockholm, 1975, 1977, 1979, 1981.

A purely quantitative presentation of the military sector is, of course, only part of the truth of its economic role. It is therefore necessary to touch upon the indirect effects of military production and R&D. Some popular hypothetical questions will not be dealt with here. One such question is whether an investment of the same magnitude in some other sector would have had a more favourable effect on employment and welfare in Sweden. This might be so, but it is not possible to discuss the political prerequisites for and consequences of such alternatives here.

In the debate over arming the Swedish defence forces, it has been suggested by some analysts that a different level of technology or another balance between the services might be more favourable in a number of ways. A less sophisticated technology, it has been argued, would have provided a better defence, had a more positive effect on the Swedish economy in general and led to more technological spin-offs for industry.[42] Those who have made this argument believe that military technology has developed more quickly than civilian technology, thereby reducing spin-off. The costs of developing civilian applications of military technology has, according to these critics, escalated rapidly and the products developed in this way have been very difficult to market. As discussed above, Swedish defence industries have had great difficulties in diversifying into civilian markets but the obstacles to diversification have lain in the structure and organization of the companies rather than in any technological gap.

It has also been denied, for instance by representatives of the aircraft industry, that civilian and military technology are drifting apart. In their opinion, there still is close contact between the two and influence operates in both directions. The French-British Concorde project is used as an example of how the civil sector has taken the lead in the field of heavy aircraft. According to representatives of the aircraft industry, wing profiles, engines, electronics and material for military aircraft have often been adopted from the civil sector.[43] In this debate, there probably has been a tendency to underestimate the very important common base of purely scientific research which is mainly financed by civilian funds and performed outside the industry.

The basic question in this connection is to what extent defence research and production have – through spin-offs – in fact stimulated Swedish science, technology and industrial capacity. Unfortunately, this question is very hard to answer. Great problems exist concerning both the data and the methods which should be employed in an investigation of this kind. Nonetheless, it is possible to present some of the results from two studies of the Swedish aircraft industry.

One possible spin-off effect is that new products and new production processes are created as a result of state orders for military goals. This effect can be measured by comparing the 'spin-off increase' in turnover with the volume of military orders. One of these studies calculated that the aircraft industry had increased its turnover by 60-85 per cent of the value of the FMV's orders.[44] In these calculations pure savings of costs are reckoned as a tenfold increase in turnover. Most valuable was access to a qualified R&D unit which was financed by state orders and was able to work for civilian portions of the corporate group at very low and otherwise unprofitable prices. The effects for the aircraft industry's subcontractors were calculated in the same way and resulted in somewhat lower percentages.[45]

In a related study, an attempt was made to measure the spread of technological know-how from the state-financed R&D unit within the aircraft industry. According to this study – which is complicated and difficult to appraise – 300 million kronor spent on aircraft research generated R&D in the civilian sector, that is, universities and technical schools, worth 200 million kronor. One-third of this R&D had taken the form of people shifting from military to civilian employment and one-third was the type of subsidized R&D service mentioned above. Both investigations were based on data from the aircraft industry itself. Keeping in mind the general circumstances during the 1970s when the studies were made, one must be somewhat sceptical of the very high estimate of spin-off from aircraft production.[46] These results no doubt correspond to widely-held views on the indirect economic benefits of investment in arms production, views which derive to a large extent from the importance that the aircraft producer places on having access to state resources for research, development and manufacturing of military materiel.

Conclusion

The rearmament of the Second World War was the point of departure for this survey of the Swedish arms industry. Both the industrial structure and the general direction of Swedish defence policy have changed very little during the postwar years. The working conditions of the Swedish arms manufacturers have been set by the state to a considerable degree. Defence policy has largely determined the amount of weapons produced and their character. This is not to deny that there has been room for influence by the military and the arms industry when important decisions have been taken. Political bodies have concerned themselves not only with the acquisition of materiel for the Swedish military but also with the terms under which Swedish military equipment can be exported. The laws regulating export have constantly been discussed and have been changed on several occasions. The problem has been to balance ideological demands against economic and political requirements. Politicians have tended to accept some morally questionable export of arms in order to secure cheaper and better equipment for their country's own armed forces.

The Swedish arms industry has been exposed to two main external forces during the later part of the postwar period. The first is the trend towards détente in Europe. This has recently weakened the consensus on Swedish defence policy and has led to a relative decrease in grants allotted to the military. The share for materiel procurement has been most affected. The second main external influence stemmed from the trend towards the production of more expensive military equipment with rapidly escalating development costs. These trends have pushed Sweden in two different directions which have been hard to reconcile with Sweden's traditional defence policy. It has become necessary to import an increasing amount of defence materiel and to provide the armed forces with technically less advanced equipment. Such tendencies have most clearly been seen in the debate over the procurement of aircraft during the 1970s.

Swedish arms producers have been very concerned to obtain new orders from the Swedish armed forces. Such orders are obviously very profitable both directly and indirectly through the access they provide to R&D capacity. Simultaneously, the companies have had to secure themselves against a real or potential reduction in military orders. This could be accomplished either by enlarging the military market, which meant exporting, or by going into new markets, which meant diversification. Different companies have experienced different degrees of success with these two strategies. On the whole, Swedish military industry slowly expanded its volume of production during the 1970s. It is interesting to note that a growing proportion of Swedish arms manufacturing has been occurring within engineering conglomerates, which can profit from the R&D capacity built up in their defence branches but which do not become dependent for their existence on state orders.

Defence orders have been of greatest value for the Swedish economy in terms of developing an R&D capacity. In quantitative terms, such as the number of workers employed or its export value, this sector of industry has not been very important. Attention is none the less focused on it because of its technical sophistication and its specific implications for domestic and foreign policies.

Notes

1. Research on rearmament and the engineering industry in Sweden has previously been published in Ulf Olsson, *The Creation of a Modern Arms Industry: Sweden 1939-1974*, Gothenburg: Institute of Economic History, University of Gothenburg, 1977, and in Ulf Olsson, 'The State and Industry in Swedish Rearmament during the Second World War', *Scandinavian Journal of History* no. 4 (1979). These works are the sources for statements made in this essay where no other reference is given. A very new publication which unfortunately has been impossible to take full account of here is, Försvarets forskningsanstalt, Huvudavdelning 1, Sektionen för Samhällsekonomi, *Svensk Försvarsindustri: Struktur, kompetens, utvecklingsbetingelser*, FOA report C-10200-M5, Stockholm: February 1982.

2. G. Lundestad, *America, Scandinavia and the Cold War, 1945-1949*, New York: Columbia University Press, 1980, *passim*.

3. The general sources for this chapter are different official statistics and publications from the parliamentary defence committees, especially *Säkerhets- och försvarspolitiken*, SOU 1972:4, Stockholm: 1972; *Säkerhetspolitiken och totalförsvar*, SOU 1976:5, Stockholm: 1976; *Vår säkerhetspolitik*, SOU 1979:2, Stockholm: 1979 [hereafter SOU 1979:1] as well as publications from the Swedish Ministry of Defence. Among the latter, of particular importance are, S. Hellman, *Säkerhetspolitik och försvarsindustri*, Stockholm: Försvarsdepartementet, 1976, mimeo, and S. Hellman, *Försvarspolitikens utveckling och nuvarande problem*, Stockholm: Försvarsdepartementet, 1978.

4. *Svensk Krigsmaterielexport. Betänkande av 1979 års krigsmateriel-kommitté*, SOU 1981:39, Stockholm: 1981, p. 55. (Hereafter, SOU 1981:39.)

5. For example, Hellman, *Säkerhetspolitik och försvarsindustri*.

6. *Försvarsindustriella problem. Rapport till försvarsutredningen*. Stockholm:

1975, mimeo, pp. 75-90.

7. Ibid., pp. 56-8; *Flygindustrikommitténs betänkande*, Part I, Ds Fö 1978:8, Stockholm: 1978, pp. 66-9 (hereafter Ds Fö 1978:8) and *Flygindustrikommitténs betänkande*, Part II, Ds Fö 1979:1, Stockholm: 1979, p. 54 (hereafter Ds Fö 1979:1).

8. G. Jonazon, 'Nu avgörs Saab-Scanias öde: Ny ÖB-order enda räddningen', *Veckans affärer* no. 42 (1976): 22-3; *Försvarsindustriella problem, passim; B3LA-beredningens rapport*, Ds Fö 1977:7, Stockholm: 1977, pp. 24-6, 60-1; *Samhällsekonomisk analys av framtida flygplansanskaffning. Rapport på uppdrag av 1974 års försvarsutredning*, Stockholm 1976, mimeo; *Totalförsvaret 1977-1982*, SOU 1977:1, Stockholm: 1977; and Ds Fö 1978:8, pp. 227-9.

9. *Försvarsindustriella problem*, pp. 109-25.

10. Ibid., pp. 59-61; *Kungl.Maj:ts proposition nummer 146/1971*, Stockholm: Riksdagen, 1971.

11. An interesting Swedish study in this field is O. Frensborg and Peter Wallensteen, *New Wine and Old Bottles. Product versus Organization: Swedish Experiences in Changing from Military to Civilian Production*. Report no. 21, Uppsala: Department of Peace and Conflict Research, Uppsala University, 1980.

12. *Samhällsekonomisk analys av framtida flygplansanskaffning*, pp. 63-6.

13. The presentation of the companies is primarily based on the annual reports of their boards of directors and articles in economic journals, for example, *Affärsvärlden* and *Veckans affärer*. See also, *Flygindustridelegationens betänkande*, Ds I 1980:2, Stockholm: 1980 (hereafter Ds I 1980:2).

14. Jonazon, 'Nu avgörs Saab-Scanias öde'.

15. Concerning 'the deal of the century', see Anthony Sampson, *The Arms Bazaar. The Companies, the Dealers, the Bribes: From Vickers to Lockheed*, London: Hodder, 1977, pp. 260-70; *Samhällsekonomisk analys av framtida flygplansanskaffning*, pp. 48-62; *Veckans affärer*, 1974-75, *passim*.

16. *Försvarsindustriella problem*, p. 61.

17. Ds I 1980:2, p. 60.

18. Försvarets Forskningsanstalt, *Svensk flygindustri i ett samhällsekonomiskt perspektiv – resurser och alternativutnyttjande*, Stockholm: FOA, 1978.

19. 'Bofors och flygmotor: Stöd i nya former', *Affärsvärlden*, nos. 29-30 (1979):17.

20. Förenade Fabriksverken, *Betänkande av 1972 års FFV-utredning*, SOU 1974:38, Stockholm: FFV, 1974; *Försvarsindustriella problem*, p. 34; Statliga Företag, *En redogörelse för affärsdrivande verk och statligt ägda aktiebolag*, Stockholm: 1970-1979; G. Jonazon, 'Försvarets fabriksverk blir civiliserat och lönsamt', *Veckans affärer*, no. 43 (1977): pp. 62-3.

21. M. Hallvarsson, 'Efter storstädning i Bofors: Plats (och pengar) till nya företag', *Veckans affärer*, no. 3 (1975): 18-19.

22. 'Bofors rensar vidare', *Affärsvärlden*, no. 48 (1979): 15-21.

23. Hallvarsson, 'Efter storstädning i Bofors'.

24. G. Jonazon, 'Därför måste Bofors fördubbla sin vapenexport', *Veckans affärer*, no. 8 (1979); 35-7.

25. *Försvarsindustriella problem*, Annex 3.

26. Olsson, *The Creation of a Modern Arms Industry: Sweden 1939-1974*, *passim*.

27. Ds Fö 1979:1, pp, 69-78, shows how this schema operated in the case of missile production.

28. H. Schiller and J. Phillips, eds. *Superstate: Readings in the Military-Industrial Complex*, Chicago: University of Illinois Press, 1970, *passim*; and K-E Schulz, ed., *Militär und Ökonomie, Beiträge zu einem Symposium*, Göttingen: Van den Hoeck und Ruprecht, 1977.

29. Ulrich Albrecht, 'Theoreme vom "militärisch-industriellen Komplex" – Eine kritische Bestandsaufnahme', in ibid.

30. Ingemar Dörfer, *System 37 Viggen. Arms, Technology and the Domestication of Glory*. Norwegian Foreign Policy Studies, no. 6, Oslo: Universitetsforlag, 1973.

31. SOU 1979:2, p. 94.

32. Ibid.; Hellman, *Försvarspolitikens utveckling och nuvarande problem*, p. 31.

33. *Försvarsindustriella problem*, p. 35.

34. Ibid., pp. 36-9; and *Samhällsekonomisk analys av framtida flygplansanskaffning*, pp. 112-31.

35. Försvarets Forskningsanstalt, pp. 32-5.

36. SOU 1981: 39, pp. 64-8. On the other hand, weapons, ammunition, explosives and certain detonating equipment for non-military uses are included.

37. Ibid., p. 57.

38. Stockholm International Peace Research Institute, *World Armaments and Disarmament, SIPRI Yearbook 1980*, London: Taylor & Francis Ltd., 1980, pp. 82-5.

39. *Forskning för försvarssektorn*, SOU 1970:54, Stockholm: 1970, pp. 15-45.

40. Försvarets Forskningsanstalt, p. 49.

41. Ibid., p. 50.

42. Bo Persson, *Samhällsnyttan av militär högteknologi*, Skillinge: Forskningsstiftelsen eco, 1977, mimeo.

43. I. Olsson, '"Gapet" mellan civil och militär teknologi en perssonsk myt', *Affärsvärlden*, no. 47 (1977): 35-9.

44. C-G Regårdh, *Teknologispridning från flygindustriell till annan verksamhet. Studie för flygindustrikommittén*, Fö 1978:01, Stockholm: November 1978, p. 45.

45. H. Lindgren, 'Flygindustrins betydelse för dess underleverantörers teknologiska standard', in ibid., Annex.

46. O. Ljungström, *Flygindustrins FoU-spridning till den civila forskningsvärlden och näringslivet. Studie för flygindustrikommittén*, Fö 1978:01 Stockholm: November 1981.

6 CZECHOSLOVAKIA

Stephen Tiedtke

The Czechoslovakian armament industry is the second most important such industry in Eastern Europe. Only the Soviet defence industry is larger and more technologically advanced. At the same time, the CSSR, which is situated at the border line between East and West within the so-called 'strategic triangle' of the Warsaw Treaty Organization (WTO) (together with Poland and the German Democratic Republic), is politically and strategically one of the most important allies of the Soviet Union. No analysis of the arms industry in Czechoslovakia can be undertaken in isolation from certain external political considerations.[1] In view of the confrontation between East and West, two questions are of particular importance.

First, Czechoslovakia, which once was foremost in Eastern Europe in terms of technological capability and productivity, suffered an economic crisis during the 1950s, the severity of which was unparallelled in Eastern Europe by 1962-64.[2] One can ask if the development and extension of an arms industry of high international standard contributed to this crisis (which in turn led to the economic reforms instituted during the 'Prague Spring'). Or, to phrase the question differently, one can ask whether the economic decline of Czechoslovakia resulted from the structural rigidity inherent in the planning process and the political systems, or whether it was also a result of the increased military confrontation between East and West during the cold war. This question, transposed to the era of détente, leads one to ask if the effects of arms production on the Czech economy were so severe – compared to those of other East European WTO members – that they may also have given rise to a specific Czech interest in military détente in Europe and that these experiences may today continue to condition Czech attitudes.

The second point to be raised here is related to the fact that the conflict between East and West has created far-reaching though non-comparable dependencies among the allies on each side. For instance, in contrast to NATO countries, the Eastern European members of the WTO cannot decide by themselves how much and what items their arms industries will produce. (As a result of the Potsdam agreement, the FRG and the GDR constitute a special case; Romania could partly make itself independent.) It is therefore important to ask how the framework of the WTO influences national arms production in Eastern Europe. As can be expected, the focus here is on the possibility of differences between the interests of the Soviet Union and of the other WTO countries, which need not to be identical.

The supremacy of the Soviet arms industry considerably limits the opportunities for arms production in the other WTO countries. The West tends to see this as another way the USSR has of controlling the military policy of its allies.[3] Soviet armament and military policies regarding the WTO are no doubt affected by the USSR's interest in controlling its allies but this is only one aspect of the division of labour between the WTO countries in the arms sector. Military and economic factors must also be considered. When West European governments came to realize that the development and production of modern weapons was so expensive that only supranational arms production was justifiable, the creation of a West European arms industry was suggested. Rather than asking if a West European arms industry might not be suboptimal, discussion has focused on how US standards can be reached or at least how the distance can be diminished.[4] In Eastern Europe, the situation is different. The economic potential of the East European states is so small – compared to that of Western Europe – that it is economically unrealistic for them to seek the sort of independent arms industry that West Europeans want. Western specialists can see economic and military, if not political, advantages for the WTO because of its division of labour in the arms sector which includes all East European countries. Economically, cooperation offers the possibility of larger production runs which diminish the unit cost of the product. Most important militarily is the fact that the large degree of standardization of equipment among East European armies is considered an advantage for the WTO, partly because it diminishes the problem of logistics. The political disadvantage of standardisation, however, is that the Soviet Union, as the chief producer of heavy weapon systems, operates in accordance with its *own* political and economic interests and does not have to take its allies' interests into consideration. One example of this is that the Soviet Union gives preference to third-world countries instead of its European allies regarding the delivery of advance weapon systems.

The national and international situation of the Czech arms industry will be analyzed in three stages. First, the general development and structure of the Czech arms industry from the end of the Second World War until the so-called 'Prague Spring' will be discussed. Second, the output of the Czech arms industry will be examined (particularly for the period after 1967 when a re-equipment programme was begun for the Czech army). Finally, the Czech arms industry will be considered within the framework of the East European alliance.

Development of the Czechoslovakian Armament Industry

Of the Soviet WTO allies, only the CSSR can look back on a considerable tradition of armament production. The internationally most well known armament enterprise in Czechoslovakia, the Skoda Works in Pilsen, began producing weapons in the 1880s.[5] It was not, however, until Czechoslovakia became independent in 1918 that the armament sector became important, and it was the increasing strength of German fascism after 1933 that led to a considerable growth of the armament industry.[6] In anticipation of serious international conflict, by 1936 the Czechoslovakian army was reorganized and modernized. The experience of the Zbrojovka Brno (Brno Arms Factory) can serve as an example of the general trend. In 1918, at the end of the war, it employed about 600 workers; by 1938 that number had risen to more than 25,000 blue-collar and 18,000 white-collar employees.[7] During the Second World War, Czechoslovak arms producing capacity was used by the German Wehrmacht which also converted a large number of civilian Czechoslovak enterprises to arms production.

The Czech arms industry that existed at the end of World War II was none the less not used, at least not to re-equip the Czech army with new weapons. Czech sources have pointed out that, 'At the beginning of the 1950s the Czechoslovakian People's Army was equipped predominantly with obsolete and worn-out technology dating from the end of the Second World War.'[8] Whether this was due – as has been claimed – to the large-scale conversion of the arms industry to civilian production[9] or to the possibility that the capacity created between 1939 and 1945 was moved to the Soviet Union or was producing for foreign needs remains an open question. It is known that the Army High Command considered it necessary in 1950 to ban the export of war materials to capitalist countries because, it was argued, 'tendencies' had arisen in the armament industry to favour 'market interests . . . sometimes even at the expense of political and military interests.'[10] What the High Command was obviously referring to were Czech arms shipments to Israel during the war in Palestine in 1948-9.[11]

The revitalization and expansion of the Czech arms industry started in 1950. From the time that the communist government had come to power, part of the capacity of heavy industy (about 25-30 per cent) had not been used but had been reserved for future arms production. The establishment of NATO, the war in Korea, and, especially, the expectation on the part of the Soviet leadership of a military conflict in Europe in 1953/4 marked the post-1950 expansive stage of the arms industry.* The short period of time which was available both for preparing for the expected European war and for adjusting to the needs of the Korean War led to a situation where economic planning in the CSSR got out of control and the economy as a whole developed considerable problems.

From a Czech point of view, the development of the arms industry during these years can be described as follows. At the beginning of 1950, 'military agencies in the national planning office' were formed and they took over 'all economic obligations for the defence of the country.'[12] In February 1950, the Presidium of the Central Committee of the Czech Communist Party demanded that within the framework of the

> accelerated re-equipment of the army . . . the Czechoslovakian People's Army should be equipped with new military technology predominantly produced under licence from the Soviet Union. Until the end of 1953, pre-requisites for a rapid reorganization of the production for war purposes in case of war should be created, so that [Czechoslovakia] could help the people's democracies to equip their armies.[13]

It was possible to attain these objectives quickly only by giving the army 'preference regarding deliveries from industry';[14] in other words, rearmament had been given top priority.

The important effects of this programme on the Czech economy are shown in Tables 6.1 and 6.2. In the space of three years, arms production as a share of total machinery production increased more than six-fold. This rate of increase is even more serious if one takes into account the fact that of all industrial sectors only machinery production was able to raise its production share substantially, from 13 per cent in 1948 to 22.7 per cent in 1958.[15] The average annual growth rate of industrial production in general between 1948 and 1953 was 14.1 per cent.[16]

*Washington also anticipated a military conflict in 1953/4.

Table 6.1: Arms Production as a Share of Total Machinery Production

Year		Percentage
1950		4
1951	about	10
1952		20
1953	about	27

Source: Autorenkollektiv, *Die Tschechoslowakische Volksarmee*, Berlin: Militärverlag der DDR, 1979, p. 127 ff.

Table 6.2 shows that in the same three years investments for defence purposes tripled.[17] These figures include the construction of such important armaments enterprises as Dubnica, Adamov, and Martin. Martin, previously the J. Stalin machine works, had about 4,000 inhabitants in 1949. By the middle of the 1950s, it had some 30,000 residents.

Table 6.2: Volume of Investments for Defence Purposes (1950 = 100)

Year	Investment Level
1951	137
1952	237
1953	221

Source: Autorenkollektiv, *Die Tschechoslowakische Volksarmee*, Berlin: Militärverlag der DDR, 1979, p. 128.

The export of military technology to the 'other people's democracies' was also of considerable importance. It 'temporarily' accounted for the largest portion of Czech exports to these countries. In 1952, it amounted to 29 per cent of total exports to the East European nations under Soviet influence.[18] This figure does not, however, accurately represent the share of military-related products in total exports to the WTO as deliveries (of, for instance, uranium) to the Soviet Union are excluded. In 1953, the share of the people's democracies in Czech foreign trade turnover was 43 per cent; the share of the Soviet Union was 35.5 per cent.[19] While estimates such as those by J. Sláma claiming that arms production in Czechoslovakia in 1950-1 rose 800 per cent are – to say the least – exaggerated,[20] there were large increases in Czech arms production and their effect on the economy should not be underrated. The situation has been described in very general terms by Czech sources:

> The success [in fulfilling the programme of 1950] was achieved at the cost of great sacrifices on the part of society, and in the following years this led to not inconsiderable problems.[21]

The effect of significantly expanded arms production can be seen more clearly in the development of the first five-year plan (1949-53), the chief goal of which was the expansion of heavy industry. During the five-year plan, 'the plan's index figures were raised three times; in other words, the growth rate of the total economy was increased considerably, investment quotas were pushed up and the growth rate of heavy industry was raised to the extreme.'[22] In view of the timing of the first plan revision and the decision by the Presidium of the Central Committee to accelerate the re-equipment of the Czech army (both in February 1950), it seems likely that the first five-year plan was adjusted at that time to conform to the requirements of the arms industry. There was a similar and more obvious pattern in the development of industry in Poland. Between 1950 and 1953, the main branches of the Polish arms industry were established, in direct contradiction to 'the original orientation of the plan', which strongly emphasized the development of light industry, the food industry, agriculture and the construction industry.[23] Up to 1955, 75 per cent of the output of the Polish machine industry and 50 per cent of the output of the chemical industry consisted of military-related items.[24]

At the end of 1954, that is, a year later than anticipated, '. . . the goals and tasks that had been laid down were essentially accomplished.'[25] This meant that the People's Army of Czechoslovakia was re-equipped and that the armament industry was ready for war. The year's delay and the difficulties which developed because of the failure to coordinate the processes of industrialization and rearmament in a way that sensibly corresponded to the possibilities of the CSSR clearly contributed to the postponement of the second five-year plan. In 1954, a one-year plan was issued. The last-minute decision on the part of Czech military leaders, in December 1954, to further enlarge the armed forces similarly contributed to the fact that there was another one-year plan in 1955. It was only on 22 December 1954 that a report was decided upon which regulated the further re-equipment of the Czech army for the period 1955-60. One reason for the lateness of this report was that it was preceded by discussions at the 'First Conference of European Countries for the guarantee of Peace and Security in Europe' in Moscow at the beginning of December 1954. In Moscow, it had been declared that the East European countries would respond to the ratification of the treaty of Paris (which enabled West Germany to join NATO) by establishing a military alliance of their own. The report summarized the results of the negotiations that the Czech government delegation had carried on in Moscow 'on questions regarding the coordination of production in the defense industry.'[26] Among the questions decided upon at the conference were the import of military technology and the provision of raw materials for the arms industry from the Soviet Union plus the arms trade with the allied people's democracies. Also decided upon were the scope of the Czech armament programme for the following years and the costs of the re-equipment of the Czech forces.

It is doubtful, however, that the decisions reached in Moscow, which were confirmed once more in Moscow in October 1955, brought any relief to the Czech economy during the following years. The cooperation between arms industries in Eastern Europe did, however, result in the CSSR once again designing some weapons of its own and not limiting itself to licence production. In 1959, Czechoslovakia started the development of a jet aircraft for training purposes, and an armoured personnel carrier (APC) was developed in cooperation with Poland. (See page 193.) None the less, the technological level of Czech industry declined continuously. Officials announced in 1965 that only 39 per cent of various machine products were internationally competitive.[27] Other studies dating from the same period show that important branches of industry (including arms production) lagged ten to fifteen years behind the West. The growth of the arms industry and its concentration on licence production of Soviet weapon systems are undoubtedly responsible for the declining level of technical proficiency of Czech industry at least until the 1960s. Armament production had no spinoff effects which could have at least partly countered the decline in technological capability. In effect, the negative consequences of the integration of Czechoslovakia into the CMEA are important to note. The then existing clear technological superiority of Czechoslovakia in many branches of industry compared to other member-states and the latter's immense need of industrial products from the CSSR hardly required any technological innovations on the part of Czech industry.[28] Therefore, there was no requirement for such innovations in order for Czech industrial products to be saleable on the international market.

In contrast to the early 1950s, there are no official figures or estimates of the economic burden of the Czech arms industry after the establishment of the WTO. The behaviour of the Czech bureaucracy during 1968, however, makes it clear that Czech military expenditure at the end of the 1960s was still considered almost unbearable both in the national economic context and compared to the outlays of other allies. While East German defence expenditure was planned to increase by 62 per cent in 1968 (over 1967 levels) – itself an unparalleled increase – the Czech government planned to increase the defence budget by only 6.8 per cent. This plan was revised in July 1968. The Czech Minister of Defence declared that the CSSR did not intend to increase defence expenditure until 1970,[29] which in fact meant a reduction of the defence budget. Because, however, defence outlays are not restricted to the budget of the Czech Ministry of Defence but can also be found in the budgets of other ministries, these figures do not reflect the real intentions of the government. The July declaration should rather be seen as a symbolic action directed above all against the WTO.

Table 6.3: Total State Investment and Defence Investment, Selected Years, 1950-68 (billions Kronen)

Year	Total state investment (1)	Defence investment (2)	(2) as a percentage of (1)
1950	14.3	4.9	34.3
1952	20.5	10.4	50.7
1953	21.3	18.0	84.5
1954	21.0	10.4	49.5
1955	22.6	17.6	77.9
1956	25.6	16.8	65.6
1957	28.1	21.0	74.7
1958	31.9	28.5	89.3
1959	38.2	32.2	84.3
1960	43.0	33.4	77.7
1961	46.1	37.3	80.9
1965	62.6	23.7	37.9
1966	68.6	30.0	43.7
1967	70.4	64.0	90.9
1968	67.2	32.5	48.4

Source: J. Hodic.

The importance given to decreasing military expenditure during the 'Prague Spring' is underlined by an unpublished estimate from 1967/8 produced by the State Planning Commission which compared total state investment with defence investment over an eighteen-year period. (See Table 6.3.) These estimates are so high that it seems very unlikely that they could be explained even by the extremely liberal Eastern European interpretation of the concept of 'securing the economic basis of national defence'.[30] It is not unreasonable to assume that these data were meant to support demands within the bureaucracy for reductions of arms expenditure, rather than to give the share of defence investment in total investment as accurately as possible. However, the fact that the authors considered their figures at least politically justifiable shows that many people in the bureaucracy were ready to accept the notion that defence-related investment was such a huge burden.

On the Structure of the Czech Arms Industry

As was mentioned above, the basis of the contemporary Czech arms industry was laid in the days of the Austrian monarchy and the Czechoslovakian republic. Apart from the Soviet economic model with its central administrative planning, introduced to Czechoslovakia after 1948, the structural elements which were new to the Czech armament industry following World War II derived primarily from the division of labour among the arms industries of Eastern Europe and the WTO's military planning based on the likelihood of an imminent war. Compared to the Soviet Union, where the vastness of the nation's territory allowed a regional distribution of the arms industry directed by military considerations, the arms industry of Czechoslovakia retains more of its traditional economic structure. The Czech arms industry is distributed throughout the country and, unlike the Soviet Union, is not largely isolated from other branches of industry. Many enterprises take part in the production of a weapon system, for instance a tank, and the political leadership appoints one of them as chief entrepreneur. This enterprise then becomes responsible for the production and quality of the product. In the case of T-model tanks, the chief enterprise at the beginning of the 1950s was CKD in Prague. Since about the middle of the 1950s, it has been the machine works in Martin. Under the leadership of this enterprise, production of the most essential tank parts was distributed among the following companies: CKD – transmission; Skoda – engine; Martin – body; Vikovica – armour; Dubnica – turret and gun; Pal Magneton – electronic components. To these six companies, which also produce for the civilian market, should be added more than 100 minor enterprises which are subcontractors, producing both civilian and military products. For this reason, all enterprises in the CSSR are not and have never been completely dependent on arms production. There is always a civilian product to which production can be transferred. In addition, enterprises have little interest in arms production because the state, which carries the burden of investment and guarantees demand, does not give them any influence on the formation of prices and allows a 'profit rate' of (at most) 5 per cent.

Regional Concentration

Military aspects were of considerable importance during discussions on the location of arms enterprises created after World War II. Expectations concerning the course of a military conflict in Central Europe and the functions of the Czechoslovakian arms industry in the event of such a war were the decisive considerations. Soviet military strategy, which is also the strategy of the alliance, aims, as it did in earlier years, at achieving a decisive victory as soon as possible in case of a European war. This means that 'efforts to guarantee the economic capacity of industry to supply the national defence with arms are concentrated on the period preceding the outbreak of war.'[31] At the same time, however, and independent of military-strategic notions of a rapid military victory, economic conditions for a long war are also developed. For the Czech arms industry, this means that it must be able to function after the outbreak of war. The Slovakian armament enterprises are assigned the task of guaranteeing that war materiels damaged on the southwestern front will be repaired.

This wartime mission considerably burdens the Czech economy in peacetime as it requires the build-up and maintenance of the necessary capacity in Slovakia. (Parenthetically, in 1939, Soviet-Czech projects already foresaw 'an evacuation of most of the arms industry in Western Bohemia . . . to Eastern Slovakia, indeed, to Soviet territory.'[32]) Slovakia's ability to produce arms during a war is to be secured by the dispersal of armament enterprises, which precludes the destruction of more than one arms enterprise for every nuclear explosion – a goal which is facilitated by the mountainous terrain of Czechoslovakia.

The following statistics outlining Slovakia's share in the industrial development of Czechoslovakia suggest that the arms industry has played an important role in Slovakian industrialization. Slovakia's share of total national industrial *production* amounted to 7 per cent in 1936, grew to 18.4 per cent by 1963 and to 21 per cent by 1968. Slovakia's share of total national *investment* rose from 29.1 per cent in 1946 to 36 per cent in 1968.[33] Even if it cannot be demonstrated statistically, it seems likely that armaments are a comparatively heavier burden to the Slovakian portion of industry than to the Czech portion and that the economic benefit of total national investments in Slovakia thus cannot be compared to those made in the rest of the CSSR. This may also be a reason for the economic tension existing between the Slovakian and the Czech parts of the country.[34]

The Production Capacity of the Czech Arms Industry

As an extensive and detailed analysis of the Czech arms industry is not possible due to lack of sources – especially for the period after 1960 – an attempt will be made to analyze the output of weapons, even though this exercise may prove to be of limited significance. In doing this, the following qualifications must be taken into consideration. First, the output of the Czech arms industry can only be ascertained by an analysis of the worldwide stock of Czechoslovakian armament products, since so much of its output was destined for foreign users. The available figures are, however, very uncertain; even information on the equipment of the Czech army itself is contradictory in the West. Second, rough estimates of production figures must be translated into some number which can supply an idea of the economic importance of the arms industry. One way of doing this is to compare Czech data with production figures from other countries where the weight of arms production in the economy as a whole can be estimated more accurately. The problem is that such estimates can only be made with the necessary degree of accuracy for capitalist states. The differences between capitalist and socialist countries in terms of economic system, industrial productivity and weapon systems design are considerable, so again it is only possible to convey a general impression.

Czech weapons production will be compared with that of the FRG.* An examination will be made of the output of three sectors of Czechoslovakia's arms industry: tank production, truck production, and the aircraft industry. The production of small arms (pistols, machine-guns, etc.) and ordnance is no doubt important (in terms of earning hard currency) to the Czech economy but it is impossible to estimate its magnitude and it will not be discussed here.

Production of Tanks and Armoured Personnel Carriers

The tank has been and remains the chief weapon in WTO armies. The

*The total population of the FRG is nearly four times that of the CSSR and, according to the World Bank, its labour force is 4.4 times larger. The percentage of the total labour force employed in industry in the two countries is nearly the same, 48 per cent for the FRG and 49 per cent for the CSSR, which means that the FRG has four times as many industrial workers as Czechoslovakia.

number produced in Eastern Europe and the Soviet Union since the end of World War II greatly exceeds the number produced by NATO countries. The latter refrained from competing with the Soviet Union and its allies in tank production during the 1950s, partly for financial reasons. In 1952, Czechoslovakia started licence production of the Soviet T-34 which had performed well during World War II.[35] The negotiations between the CSSR and the Soviet Union relating to this licence production had been concluded by the end of 1949.[36] Subsequent models, the T-54 (which appeared for the first time in the armies of Soviet allies in 1958/9) and the T-55 were also produced in the CSSR under Soviet licence. It is not entirely clear whether the T-62 was also produced in the CSSR.[37] The small number of this model in the Czechoslovak People's Army – and in other Eastern European WTO armies – argues against Czech licence production. According to the *Military Balance*, the Czechoslovak People's Army possessed only slightly more than 100 T-62 tanks in 1980 and there appears to be no indication in the relevant reference works of Czech export of these tanks to the third world.[38]

The peak years for Czech tank production were between 1952 and 1968. During the 'Prague Spring' an end to tank production was reportedly discussed and, in fact, tank production was continued only on a limited scale. These discussions grew out of a general reorientation of the Czech army which started in 1967. The T-55s, which were introduced into WTO divisions beginning in 1968 as a result of structural changes in the forces of member-nations, may not have been manufactured in Czechoslovakia but are likely to have been turned over to the Czech forces by Soviet troops stationed in the CSSR. Soviet troops were simultaneously being equipped with the T-62, and passing the older material (T-55) on to the militaries of the minor WTO members is a normal way of operating in Eastern Europe. This practice has been given further impetus by the difficulties experienced in returning heavy equipment to the Soviet Union.[39] The railway line leading into the USSR is extremely overtaxed. Through the 18 per cent of the Czech rail system which is directly connected to the Soviet rail system must pass about 60 per cent of Czech rail traffic.

In 1952 licence production of the Soviet 'Jagdpanzer' (assault gun) SU-100 began along with production of tracked APCs.[40] The OT-810, which was first produced in the 1950s, is said still to be used by the Czech army today.[41] The OT-62 was obviously produced in the CSSR in large numbers. This APC – produced in the CSSR since 1964 – is another version of the Soviet BTR-50 PK, but is said to be technically superior to the Soviet model.[42] It was delivered by Czechoslovakia to its WTO allies Bulgaria, Hungary, Poland and Romania. Among developing countries, Egypt, Libya, India and Morocco have received the OT-62.[43] Since the mid-1970s, the OT-62 has been replaced by the Soviet BMP which most probably is not produced in the CSSR.[44] This would mean that by the mid-1970s at the latest, most of the production of tracked APCs and main battle tanks had ceased in the CSSR.

At the same time as they started the production of the OT-62, the Czechs also began producing a wheeled APC, the OT-64 (SKOT). The first production-ready designs of this vehicle were available in the CSSR by 1959 but mass production started only in 1964, after an agreement with the Polish government that the OT-64 (SKOT) would be developed and produced jointly.[45] The widespread adoption of the OT-64 is evidence that Poland and the CSSR succeeded in constructing an excellent weapon (with the main responsibility apparently resting with the Czechs). Apart from the Polish and Czech armies, this vehicle is also in service in the Hungarian and the Soviet armies.[46] According to *Jane's Weapons Systems*, it has been acquired by the armies of Egypt, India, Libya, Morocco, Sudan, Syria and Uganda.[47] The OT-65 is another wheeled APC produced in Czechoslovakia. It corresponds to the Hungarian FUG-65A. As regards this four-wheeled APC, it is not clear whether it is uniquely of Hungarian construction or a joint Hungarian-Czech project. It is also unclear to what extent production is divided between the two countries.[48]

Since the early 1970s, Czech production of tracked combat vehicles (tanks and APCs) has without doubt decreased considerably. Since the production of wheeled APCs has also diminished, it is legitimate to enquire how the enterprises that were previously engaged in this production, responsible for a not insignificant portion of Czech industrial output, were able to adjust to the new conditions. The likelihood of output being concentrated more on civilian products (agricultural or construction machinery, for example) is something that can only be assumed.

In view of the general running-down of Czech tank production, the licence production of the most modern Soviet tank, the T-72, something which has been mentioned in Western journals,[49] would signal a very costly resumption of Czech tank production. There are, however, two reasons why this resumption of production is possible, if the T-72 were to become the new standard tank of the WTO, which seems likely. First, a large demand for this weapon would justify – from an economic standpoint – the start-up of a new production line in Czechoslovakia. Second, the CSSR would have to construct factories for the maintenance and repair of these new tanks in Slovakia anyway.

How many APCs and main battle tanks (T-54 and T-55) were produced in the CSSR? These two models were in production for roughly twelve years, between 1958 and 1968/9. In this period, about 3,000 tanks were produced for Czechoslovakia's own forces.[50] To this must be added those tanks that were exported, the exact quantity of which, however, is unknown. In all the material published by SIPRI, only 210 tanks are recorded as having been exported to developing nations by Czechoslovakia. In a 1974 study, the US Arms Control and Disarmament Agency estimated that 760 tanks and self-propelled guns had been exported by the CSSR to developing nations between 1967 and 1971.[51] How many tanks were exported to the WTO allies is an open question. Czech exports to the WTO began in the early 1950s and were important. That tanks were delivered to WTO allies even after 1955 was confirmed by the East German Minister for Defence, Hoffmann, in May 1972. 'The tank crews in the National People's Army, to mention only one example, appreciate the tanks produced for them by the Czechoslovakian workers.'[52]

A conservative estimate of the number of T-54s and T-55s produced in the CSSR would be about 4,000. This figure is certainly too small, rather than too large. It corresponds roughly to production figures for the West German Leopard I tank, of which 3,541 were produced between 1965 and 1977.[53] This means that the annual production of tanks in West Germany and Czechoslovakia is about the same, both in terms of absolute output and on average.[54] The average annual production figures (around 340 tanks) are no doubt considerably lower than Czechoslovakia's production figures for the T-34. During the 1950s, some 700 T-34s were produced annually. Production reached a peak in 1954, when 844 tanks were produced. A total of about 4,000 T-34s had been produced by 1958. This suggests that the estimate of 4,000 tanks for total output of the T-54 and T-55 models is too low. It has been estimated that Soviet output of T-72s is 'at least' 2,000 units per year.[55]

It is difficult to estimate how many jobs are dependent on tank production in Czechoslovakia. When the procurement programme for 1,800 Leopard II tanks was approved by the German parliament (Bundestag), it was justified by the fact that 20,000 jobs in the armament industry would be secured. As Czech industry is not as productive as West German industry, the number of jobs dependent on the production of tanks should be larger than in West Germany.

It is almost impossible to make a correct estimate of the number of APCs produced in the CSSR. If one is to believe the *Military Balance*, about 3,000 OT-62s and OT-64s have been produced in Czechoslovaka for Czech forces since the mid-1960s. To this figure must be added 680 OT-65s.[56] This total production figure corresponds approximately to the number of APCs in the West German Army (Bundeswehr) in 1974.[57] How many APCs have been exported from Czechoslovakia cannot, however, be established, even approximately.

Trucks

All WTO countries produce trucks which can be used both for civilian and for military purposes.[58] Of the East European members of the WTO, the CSSR is the major truck producer. This is partly due to the fact that the Czech arms industry has succeeded in combining truck production with that of wheeled combat vehicles. Apart from imports, demand on the part of the Czech army for trucks and wheeled combat vehicles is met primarily by the Praga and Tatra Works in Koprivnice Probably the most well known product of the Praga Works is Praga V-3-S, which has been produced since 1960.[59] This medium-sized truck is still used, for example, by the East German and the Bulgarian militaries.[60] The Tatra factories, however, are more important in terms of the standardization of truck construction within WTO countries, a goal which they have been striving to attain since the mid-1960s. A loan of 77.5 million (transferable) rubles from the international Bank of the Comecon will enable the Tatra Works to meet the total demand from member countries for trucks with a loading capacity of 12 to 16 tons, starting in 1980. At least 15,000 units of the model Tatra 815 are to be produced annually.[61] This means that the output of this kind of vehicle by Tatra will approximately double. The old target was 8,500 trucks a year. (Two-thirds of these were T-148s, which have been used in the East German National People's Army [NPA] in a variety of versions. One-third have been T-813s.[62]) This Comecon-sponsored project will play an important role in the armament of WTO countries.[63] It evidently represents a decisive step towards the long-neglected standardization of the truck-force of the WTO armies.

How important it is for the Czech arms industry to put an already proven and approved model to the broadest possible military use is evident from the case of the Tatra-813. Besides its ordinary transport function, it serves as the basic vehicle for the APC OT-64 (SKOT). In addition to this, it can also be converted for use as a multiple rocket launcher. This particular model which was officially shown for the first time in 1972 – with launchers produced in the Soviet Union – is considered superior to the Russian ones by East European specialists.[64] According to the *Military Balance*, the Czechoslovakian army had 300 Tatra-813s equipped in this way in 1980, and this model is used by the East German NPA (180 vehicles) and by the Soviet Red Army as well.[65] In comparison, the West German Bundeswehr had 209 multiple rocket launchers at its disposal in 1973.[66] Furthermore, the T-813 serves as a carrier for pontoon elements.[67]

The development of wheeled combat vehicles is quite in line with the technological trends in the international arms industry, but the armoured self-propelled gun built on the chassis of the Tatra-813 which was unveiled in 1980 is definitely an innovation. This self-propelled gun is the heaviest indigenously developed piece of army weaponry ever produced in Czechoslovakia.[68]

Aircraft Industry

The production of airplanes in Czechoslovakia started immediately after the birth of the republic in 1918. By 1939, this industry had grown considerably and it made an important contribution to the German war effort following the occupation of Czechoslovakia.[69] When World War II ended, the production of airplanes ceased, but only for a short period. In 1947, the first ZLIN TRENER appeared. It was considered an 'excellent plane for training and stunt flying'.[70] The Czechoslovak aviation industry can produce light aircraft, transport and passenger planes, fighters, and jet trainers.

In a military context, the light airplanes are used solely for liaison purposes. The ZLIN 43, which was produced up to 1977, still serves in this capacity in the East German armed forces.[71] Up to the beginning of the 1970s, the Czechoslovakians had developed some twenty different varieties of light aircraft and had produced about 2,500 machines of this kind.[72]

The first medium-sized transport plane built by the Czech aviation industry was a copy of the Soviet IL-14 which is still used by transport regiments within the WTO.[73] The light transport and passenger plane L-410, which is used by the Czech airforce as well, is a genuine domestic product.[74] An agreement among Comecon members has given the L-410 a crucial role in Czechoslovakia's aviation industry. According to sources in Prague, thousands of L-410s will be produced during the period up to 1990. Up to 1979, the Soviet Union alone had placed orders for 2,000 L-410s.[75]

As early as 1951, the MIG-15 fighter – at that time one of the most modern airplanes of its type – was manufactured in Czechoslovakia, and the licence under which it was produced stipulated that these fighters were to be produced for Hungary, Bulgaria, Romania and Poland as well. It has been reported that the MIG-17, MIG-19, and MIG-21 have been, and still are, produced in the CSSR as well. As far as the MIG-19 is concerned, however, this seems unlikely. According to the *Military Balance*, the CSSR and the other East European allies of the Soviet Union have only a few MIG-19s and licence production would therefore seem economically untenable. The MIG-21, on the other hand, is widely in use in the Czech airforce (319 planes in 1981),[76] which makes licence production more probable. Important sources, however, mention the import of the MIG-21 from the Soviet Union,[77] and this alternative is plausible in view of the burden that production of MIG-21s would have placed on the capacity of the Czech aviation industry. Production of the MIG-21 would have coincided with the mass production of the Czech jet trainer L-29. It seems probable that the fabrication of Soviet fighters in Czechoslovakia ceased sometime between (very roughly) 1965 and 1975. Polish production of modern military aircraft ended in 1969.[78]

The most important single product of the Czechoslovakian aviation industry is definitely the jet-powered training plane L-29 (Delphin) and its follow-on, the L-39 (Albatross). The L-29 won a competition against Soviet and Polish models and became the official training airplane for the WTO, with the exception of Poland.* This meant that the Soviet mode of procuring material was transposed to the organization as a whole. According to this model, different design bureaus develop the prototype for a weapon system and the most successful of these is singled out for multiple production.[79] Between 1963 and 1974, more than 3,000 L-29 aircraft were manufactured and an average annual output of about 300 planes can therefore be posited.[80] Unlike its follow-on, the L-39, the L-29 is equipped with a Czech engine. As this engine proved to be defective in several respects, a new one was developed for the L-39.[81] When, however, the L-39 was finally ready for multiple production (in 1972), it was equipped with a Soviet engine. For training purposes, this airplane can carry suspension attachments for bombs, rockets and other weapons and it is reported to be suitable for ground attack as well.[82] The coordinated programme for airplane production in Eastern Europe predicts a total of 3,000 units up to 1990[83] and this figure does not take into account any export to

*With regard to Poland, see pp. 200-1 below.

the third world. (The L-29 was quite successful in this market.) The economic importance of the Czechoslovakian production programme can be illustrated by a comparison with the demand for military training airplanes which is expected by the Western aviation industry during the next ten to fifteen years, that is, up to 1990-5. (This comparison concerns markets open to Western producers in which the CSSR also competes.) D.A. Brown has estimated this demand to be 6,000 units of which only 4,000 would be light jet aircraft.[84] The Czechoslovakian output of light jet trainers will, in other words, be only slightly lower than the total production of the entire Western airplane industry (combining US, UK, French, West German and Italian production of such types).

The electronic enterprise Tesla holds a special position within the Czech aviation industry, as it is Eastern Europe's leading producer of radar (ground and on-board equipment). The first development work in the field of radar technology occurred in 1946-8.[85] Many airfields in Eastern Europe have been equipped with radar systems from Tesla.

For aviation and for other armaments sectors (for example, communications) the low-tension electric power industry plays an important role. In this branch, which has expanded quickly from its beginning in 1953, civilian and military interests are heavily intertwined. This is evident from the fact that all forms of communication that serve civilian needs in Czechoslovakia in time of peace can be administered and built only in close cooperation with the military authorities.

According to Czech sources, there are 30,000 wage-earners employed in the aviation industry (exclusive of Tesla). Of these, some 10 per cent are engaged in research and development work.[86] (The air- and spacecraft industry in the FRG employed 47,495 people in 1966 and 55,910 people in 1976.[87]) At present, the Czechoslovakian aviation industry is working at full capacity and the expansion that has been discussed within Comecon (specialization in light passenger, transport and training planes) could occur without raising the number of employees.[88] In the mid-1960s, Tesla employed 50,000 people. Of these, almost 20 per cent were engaged in research and development.[89] In connection with these figures, it must be noted that the aviation industry does not produce exclusively military items and that it also produces for markets other than civilian and military aviation.

Czechoslovakia's Role in Armament Cooperation in Eastern Europe

In Czechoslovakia, as in the other smaller WTO member-states, the development of the arms industry as it has been formed by intra-WTO cooperation can be divided into three phases.

1950-5

The aggravation of the East-West conflict, which produced fears of war in both camps around 1953-4, led to a dramatic reorientation of economic planning towards arms production in Czechoslovakia and Poland. The resultant militarization of economic and social life seriously impeded the economic development of these countries.

From the Founding of the WTO to the End of the 1960s

The massive, but uncoordinated, first phase of the rearmament process made Poland and Czechoslovkia almost self-sufficient in terms of military materiel. In 1962, the Polish general Bordzilowski announced that '. . . the majority of the modern weapons used by our armed forces are produced domestically.'[90] One disadvantage with this development was that the Polish and Czech arms industries were not complementary. In fact, they were often in competition with each other.[91] After 1955 efforts were made to eliminate this shortcoming. The smaller WTO states began to sign bilateral treaties of cooperation. The most striking example of this new trend was the joint Polish-Czech development and production of the APC OT-64 (SKOT).The competition between the arms industries of Poland and Czechsolovakia was brought to an end through a process of divergent specialization and elimination of overlapping production capacities. This tendency within the WTO towards a stricter industrial division of labour in the field of armaments was accompanied by a growing military importance of the smaller armies within the WTO. It was these trends that also made it possible for Romania, for example, to win a certain degree of independence in the fields of armament and military policy.

The Period after 1969

At the end of the 1960s and at the beginning of the 1970s, the production of heavy weapon systems in Poland and Czechsolovakia came to an end, at least temporarily, when the existing licences for fighter aircraft and tanks expired. One reason for this was that the Polish and Czech armies had filled their immediate needs for these weapons and no new follow-on systems were under development. At the same time, the

WTO armies were reorganized and a series of institutional reforms were implemented as a result of decisions made at the meeting of the WTO's advisory political committee in Budapest in 1969. Due to these changes, the economic difficulties experienced by WTO members due to the rapid development of the arms industry in the early 1950s were, to a large extent, surmounted and a more comprehensive and efficient division of labour was established for the WTO arms industry. The aviation industry is a striking example of this renewal. Poland has specialized in helicopters – the production of the first entirely Polish helicopter was expected to start by the middle of 1982. Similarly, the production of a domestically designed Polish medium-sized transport plane is planned.[92] In addition to very light aircraft, the Czech aircraft industry will concentrate on jet trainers and light transport aircraft.

As discussed earlier, Western analysts often interpret this cooperation in the field of arms production as an instrument of political control in the hands of the Soviet Union. The French military theorist F.O. Mischke has stated that the members of the Warsaw Treaty Organization 'are chained to Moscow in matters of strategy and military economy' through the long-term agreements on licence production, weapons export and production for their own forces. According to Checinski, this dependence is increasing rather than decreasing.[93] At first sight, this interpretation seems logical. The more a country can meet its military needs with indigenous products, the more indepependent it is, at least in terms of its arms industry. The more it must rely on imports – especially if these derive almost solely from one country – the deeper it falls into political bondage. The trend during the 1970s towards waning independence for the arms industries of the smaller WTO countries therefore means that the Soviet Union has strengthened its influence over these countries. (The exception here is Romania which together with Yugoslavia developed and built its own fighter aircraft.[94]) this interpretation is true only if patterns of arms transfers within NATO are used as a frame of reference. Interstate relationships between East European countries follow a different logic.

The decision to develop the arms industry in Poland and Czechoslovakia at the beginning of the 1950s was not made by these countries. It was made by the Soviet Union alone. That was the character of the political power structure in Eastern Europe, and it did not change much in the following years. Poland did seek to give its arms production a new and partly independent profile without consulting Moscow. Up to the beginning of the 1960s, for example, Poland is said to have boycotted all agreements dealing with economic cooperation for armament purposes,[95] and it refused to use the Czech training plane that had proved superior to the Polish TS-11 Iskra. Nonetheless, this was neither an attempt to expand Poland's political independence nor a reflection of a growing autonomy based on the expanding arms industry. The causes for Polish recalcitrance were reportedly purely economic. At that time, Poland followed the practice of making the importation of a weapon system dependent on a comparison of the import cost and the amount of investment that would be necessary if the weapon system were to be produced domestically. Checinski argues that Soviet weapons are expensive because they are sold at world market prices (at least to WTO member states), and it was economically desirable for Poland to produce as much as possible on its own. Checinski reports that the Polish military industry accounts for 16-17 per cent of Polish industrial production in terms of value of annual output. The crucial fact is that this deviation was tolerated by the Soviet Union which it could afford to do since equipment for its own army was not affected.[96] The Polish forces had little military significance in terms of the East-West confrontation in those years. Furthermore, Poland could easily be brought to heel whenever Moscow chose to do so. The frequently presented view that an independent arms industry in a WTO member-country could represent a threat to the Soviet Union in a conflict within the organization is highly questionable considering the prevailing power structure in Eastern Europe.[97]

One interesting question is how far the Soviet Union is prepared to let its allies go astray in matters of military-industrial policy. Romania's defection from armament cooperation is the most serious deviation of this kind that the Soviet Union has been confronted with up to now. It is quite in line with Romanian foreign and military policies and is meant to strengthen them. One must ask if Soviet dominance in the WTO arms industry as a whole failed as an instrument of political control or if there are other factors in Soviet military policy that can account for the lenient position in the conflict with Romania. As in the Polish case it can be assumed that Romania's military industry was of negligible importance to the Soviet army. Furthermore, Romania's geographic position is peripheral to the central area of conflict as defined in the military strategies of the WTO. Therefore, its dissidence could be tolerated. A similar situation in the 'strategic triangle' of the WTO would elicit a quite different response. If Czechoslovakia decided to follow the Romanian path towards independence in matters of arms production the Soviet Union, and other WTO members to the north, would hardly tolerate such a decision, considering the function that

Slovakia has as a key military-industrial area in case of war. The Czechoslovakian defence-industrial infrastructure is most probably more important to the Soviet military planning in the event of a European conflict than the Czech armed forces as such.

Since the mid-1960s, political control has played a modest role, if any at all, in arms production cooperation within the WTO. Of course, this does not exclude the possibility that the Soviet Union might influence the development of the arms industry of its allies or that the form taken by arms cooperation increases the allies' dependence on Moscow. This dependence works both ways, however. The growing complexity of weapon systems, which makes a distinction between military and civilian industrial production more and more obsolete, brings about a greater dependence of the Soviet armament industry within Comecon on the industrial capacity and know-how of its WTO allies. Soviet military industry today is clearly more dependent on supplies from the other WTO countries than it was in the 1950s and the 1960s, even if this is difficult to prove empirically.

The GDR provides a striking example of the increasing interdependence of Eastern European military industries. The only indications available of the development or magnitude of a GDR contribution to WTO defence production derive from official GDR statements of the most general sort, lacking any quantification or specifics. This contribution apparently consists of the delivery of components to the arms industries of other WTO member-nations. The value or size of these deliveries is not known. In contrast to its neighbours Poland and Czechoslovakia, both of which have a flourishing arms industry, the GDR produces only a few items that can be classified as military hardware in a strict sense.[98] Up to the beginning of the 1960s, of the GDR's role within the overall armament schema was peripheral even according to its own statements. In the mid-1960s, however, the GDR claimed that it was making a 'not insignificant direct contribution to the total production of military technology and equipment',[99] and three years later (in 1969-70) that 'The GDR has a special responsibility to safeguard the economy of the national defence within the framework of the socialist defence coalition.'[100] In the middle of the 1970s, East German producers of components are described as making '. . . an increasing contribution to the strengthening of the defence power of socialism'.[101] The same can without a doubt be said of the CSSR, with the difference that Czechoslovakia, as a producer of complete weapon systems and military equipment, is a far more important factor for WTO arms production than is the GDR as a subcontractor for the production of components.

As arms-production cooperation intensified, both the material structure and the institutional framework of this cooperation changed. With the creation of the 'Committee for Coordination of Military Technology' in 1969, the minor WTO members-states were given the opportunity to exert some influence over the decision-making process. According to Western sources, this committee deals with the rationalization of research and development and the improvement of military equipment. It also decides '. . . what kinds of weapons and components should be produced in each country'.[102] The committee is apparently composed of '. . . the directors of the military departments of the planning committees . . .' of member-countries.[103] It is impossible to estimate the extent to which the influence of the minor countries really has increased due to the structural changes. In the older system, responsibilities overlapped and the institutions were inefficient. It would nonetheless seem likely that the leading power of the WTO – the USSR – would find it harder to encroach on the autonomy of the individual members in a multilateral institution than in bilateral ones.

That production for armament purposes put an enormous strain on the Czech economy, especially in the 1950s and the 1960s, has been hinted at in a variety of official and unofficial statements and data. Comparisons with other countries in the WTO and studies of the development of living standards and industrial production provide a more tangible picture of the effects of arms production on the Czech economy. (See Table 6.4.) That there is a tension between the armament industry and heavy industry in general as well as consumer industry in Eastern Europe is evident from the frequent, but futile, attempts in the Soviet Union to switch priority to the production of consumer goods. The effects of the large armament-producing sector on Czech living standards has been clarified by the exiled Polish military specialist Michael Checinski by comparing Czechoslovakia with the GDR, a country without a traditional military industry:

> In those Sociaist countries where the armament industry and production for the civilian market do not compete for scarce resources, the quality of the latter is definitely higher. In East Germany, for example, there was virtually no production of final (ready-for-use) armaments for many years. That country was able to distribute its resources and capabilities in a relatively more even way between the various branches of economic activity. That is one reason why civilian products from East Germany have been the best

Table 6.4: Estimates of Per Capita Gross National Product, Czechoslovakia, GDR, Poland and the USSR, 1960-77

Country	1960		1965		1970		1975		1976		1977	
	World Bank	CIA	World Bank	CIA	World Bank	CIA	World Bank	CIA	World Bank	CIA	World Bank	CIA
Czechsolovakia	2,600	2,900	2,800	3,000	3,300	3,600	3,800	4,000	3,800	4,100	3,900	4,200
GDR	2,700	2,500	3,100	2,900	3,700	3,400	4,400	4,000	4,500	4,200	4,700	4,400
Poland	1,500	1,500	1,800	1,800	2,100	2,100	2,800	2,700	2,900	2,800	3,000	2,900
USSR	1,600	2,200	1,900	2,700	2,300	3,300	2,700	3,800	2,800	3,900	2,900	4,000

Source: P.S. Shoup, *The East European and Soviet Data Handbook*, New York: Columbia University Press, 1981, p. 386.

in the whole Soviet Bloc.

The experience of Czechoslovakia was different. Following World War II, Czechoslovakia's technological potential was not inferior to that of East Germany. Since 1950, however, a considerable armament industry, based on Soviet licences, has been developed there and the quality of production for the civilian market has suffered in consequence. Developing armament production required allocating a significant proportion of the available highly skilled workers, modern tools and the most valuable raw materials to that sector of the economy. As a result the relative quality of Czechoslovak products for the civilian market declined. At the same time Czech armament production continues to reflect a high level of modern technology and workmanship.[104]

That the same view is taken in the GDR, though without any direct reference to other WTO countries and without any direct comment on the supply of consumer goods for the home market, is evident from many East German commentaries. The fact that the weapon systems necessary to the NPA can be imported instead of being produced within the country have made a thorough-going transformation of the industrial capacity of the GDR unnecessary and have made it possible '. . . to save the means for necessary long-term development work and to avoid tying up considerable productive capacities'.[105] The Potsdam Treaty, which explicitly dictated that all weapon factories in the former Third Reich be closed down, enabled the GDR to develop its economy unaffected by the economic straight-jacket of arms production that fettered its allies in the 1950s. Later, when the GDR came to contribute its share to WTO arms production, it maintained its favourable position, since the GDR produced military hardware only in accordance with the structure and profile of its national economy, that is, military products produced by civilian industry.[106]

It is commonly assumed that arms industries have a positive effect on the technological level of a country's industry. In the case of the Polish aircraft industry, the licence production of Soviet fighters and helicopters has without a doubt contributed to the development of the modern civil aviation industry.[107] The same is true for the Polish automobile industry which did not exist until after World War II. The Strachowiche plant, reported to be the foremost supplier of military vehicles in Poland, was built in the 1950s.[108] As long as the Federal Republic of Germany did not have a substantial armaments industry and the aim of Soviet policy toward Germany was to prevent the rearmament of West Germany, the USSR did not permit the build-up of an armaments industry of greater proportions in the GDR. The continuation of this major choice by the USSR beyond the mid-1950s may require additional explanation.

As for Czechoslovakia, however, the post-World War II production of weapons has contributed very little to its technological development since the country already had a long tradition of advanced arms production. The licence production of Soviet weapon systems during the post-war period did not provide much 'know-how'. On the contrary, it prevented technological innovation.[109] These conflicting tendencies, which have served to diminish the technological gap between Poland and Czechoslovakia, negatively affected Czechoslovakia prior to the introduction of stricter specialization within the WTO. The abolition of licence production and the orientation towards indigenous development within a specific sector of the arms industry brought progress to Czechoslovakia, at least in the technological sphere.

Summary

Of the countries that fell into the Soviet sphere of influence after World War II, Czechoslovakia was the only one with an arms industry worth mentioning. The cold war and the prospect of a hot war in Europe at the beginning of the 1950s induced the Soviet Union to require Czechoslovakia to reactivate its dormant military-industrial capacity and to expand those arms concerns which were in use. There were two reasons for this. The first was to equip the militaries of the East European countries without straining Soviet industry. The Soviet licences to Czechoslovakia in this period, for example, were accompanied by an obligation to supply the other East European states with military equipment.[110] The second aim was to create a military infrastructure in Czechoslovakia (airfields, repair facilities, depots, lines of communication and so forth) to meet the expected war, and this required substantial investment.

These aims were to be attained within a short period of time. The resultant strain on the Czech economy led to a breakdown of economic planning and caused a series of long-term problems for the overall economic development of the country. By the beginning of the 1960s, Czechoslovakia, a country which once had an international reputation for its sound economic system, became the first nation in Eastern Europe to experience an economic decline (see Table 6.5). This was not, however, caused by arms production and the military apparatus alone. A highly industrialized Czechoslovakia had been integrated into an economic alliance which was at a much lower level of development. Czech industry seldom met any challenges that called for technological innovation and it was gradually brought to a technological standstill. The arms industry could not change this situation even if its products were of a high international standard from the very beginning. The licence production of Soviet weapon systems left no room for the self-development that was the only way to technological renewal.

Table 6.5: Growth Rate of the Czech Gross National Product[a] 1960-66 (percentages)

1960	1961	1962	1963	1964	1965	1966
8.1	6.8	1.4	-2.2	6.6	3.4	10.8

Note: a. Gross national product measured in constant values.
Source: F.-L. Altmann *et al., Die Wirtschaft der Tschechoslowakei und Polens*, Munich and Vienna: Olzog Verlag, 1968, p. 30.

Around the mid-1960s, the situation gradually began to change. There was an increasing awareness in some sectors of the bureaucracy of the excessive strains put on the Czechoslovak economy and attempts were made (during the 'Prague Spring') to reduce military expenditure, or at least to curb its growth, and to reduce arms production. These aims, and particularly the second one, could only be attained within the multilateral framework of the military alliance in Eastern Europe. The following tendencies can be discerned in Czech arms production since the mid-1960s.

First, the licence production of heavy Soviet weapon systems (tanks, fighter aircraft) was reduced and had practically ceased in the 1970s. Second, the CSSR concentrated its production of heavy, indigenously developed, weapons systems in sectors with the greatest possible connection to civilian products (lorries, for instance). Third, there was a similar development in the case of the jet training aircraft which are the Czech contribution to WTO armament production. Fourth, in accordance with the technological trends in the international armament industry, the distinction between military and civilian production has become increasingly blurred in Eastern Europe.[111] One effect of this development is that civilian industrial sectors in the non-USSR member-nations of the WTO – for example, the low-tension electrical power industry in the CSSR – has become more and more important to Soviet armament production.

These four trends point to a reorientation of Czech arms production. Czechoslovakia no longer has the same importance as a producer of final products for military use. The new role of the Czechoslovak military industry within the framework of WTO armament cooperation is to maintain a military infrastructure that is necessary in case of war.

Notes

1. It is not necessary to go into the difficulties associated with the analysis of the military and armament policies in Eastern Europe once again. The lack of a systematic Western study of the Czech arms industry says enough. The author would like to thank Dr J. Hodic, formerly of the Austrian Institute of International Affairs, for his expert help and advice.

2. J. Kosta, 'Veränderungen des tschechoslowakischen Wirtschaftssystems nach dem Zweiten Weltkrieg (1945-1966),' p. 147, in *Die Tschechoslowakei 1945-1970*, eds. N. Lobkowicz and F. Prinz, Munich and Vienna: Oldenbourg Verlag, 1978.

3. Compare this with Michael Checinski, 'The Costs of Armament Production and the Profitability of Armament Export in Comecon Countries', *Osteuropa-Wirtschaft* no. 2 (1975): 130.

4. Michael Brzoska *et al.*, *Rüstungskooperation in Westeuropa*, IFSH-Forschungsberichte, Heft 15, Hamburg: University of Hamburg, 1979, p. 2.

5. The first products of the arms industry were naval guns. See also, W. Himmelberger, 'Die Rüstungsindustrie der Tschechoslowakei 1933 bis 1939', p. 313, in *Wirtschaft und Rüstung am Vorabend des Zweiten Weltkrieges*, eds. F. Forstmeier and H-E. Volkmann, Düsseldorf: Droste Verlag, 1975.

6. Ibid., pp. 308 ff.

7. Ibid., p. 318.

8. Autorenkollektiv, *Die Tschechoslowakische Volksarmee*, Berlin: Militärverlag der DDR, 1979, p. 125.

9. Ibid., p. 127. At a minimum, those companies which had been converted to military production during World War II did return to civilian production once the war ended.

10. Ibid. In 1953, Generals Drgac and Drenec were sentenced for having 'sold weapons to the imperialists, thereby diminishing the strength of the Czechoslovakian army.'

11. See Stockholm International Peace Research Institute (SIPRI), *The Arms Trade with the Third World*, Stockholm: Almqvist & Wiksell, 1971, p. 529. Available evidence indicates, however, that the Czech arms shipments to Israel had the full approval of the USSR when they were arranged and very probably even involved direct negotiations with the USSR. See Gavriel D. Ra'anan, *The Evolution of the Soviet Use of Surrogates in Military Relations with the Third World, With Particular Emphasis on Cuban Participation in Africa*, P-6420, Santa Monica, California: Rand, December 1979, pp. 2-4. A more plausible interpretation of the High Command comment is that the politics of the USSR on the question of Israel may have been changing by 1950. The communist side in the Greek civil war is also said to have received weapons from Czechoslovakia, as did Syria at the same time as shipments went to Israel.

12. Autorenkollektiv, *Die Tschechoslowakische Volksarmee*, p. 126.
13. Ibid., p. 127.
14. Ibid.
15. Apart from the machine industry, only the chemical industry (which also produced for the arms sector) was able to raise its share, although only by about 1 per cent. See Kosta, 'Veränderungen des tschechoslowakischen Wirtschafts-systems', p. 156, Übersicht 8.
16. Ibid., p. 165. diagram 2.
17.The total volume of state investment rose on average by 18.8 per cent each year between 1948 and 1953. Ibid., p. 168, diagram 9.
18. Autorenkollektiv, *Die Tschechoslowakische Volksarmee*, p. 130.
19. Kosta, 'Veränderungen des tschechoslowakischen Wirtschaftssystems', p. 168, diagram 9.
20. J. Sláma, 'Die Tschechoslowakei zwischen dem besiegten Deutschland und der Sowjetunion: Eine sozio-ökonomische Analyse', p. 117, in Lobkowicz and Prinz, *Die Tschechoslowakei*. Serious armament and the decisive development of the arms industry only began in April 1951 as a result of the armament plans for individual countries decided upon at the conference of defence ministers in January 1951.
21. Autorenkollektiv, *Die Tschechoslowakische Volksarmee*, p. 129.
22. Kosta, 'Veränderungen des tschechoslowakischen Wirtschaftssystems', p. 141.
23. T. Grabowski, 'Militärische Aspekte der wirtschaftlichen Entwicklung Volkspolens', *Militärwesen* no. 10 (1969): 1443, 1446.
24. S. Ciaston, *Ekonomiczne aspekty obronnosci*, Warsaw: 1968, cited in Checinski, 'The Costs of Armament Production', p. 137.
25. Autorenkollektiv, *Die Tschechoslowakische Volksarmee*, p. 129.
26. Ibid., p. 274. There are other aspects of the negotiations in Moscow that the authors do not find worth mentioning, for example the admonition against ratification of the Paris Treaty.
27. Compare this with O. Šik, *Fakten der tschechoslowakischen Wirtschaft*, Vienna, Munich and Zürich: Molden Verlag, 1969, p. 56.
28. Compare this with Sláma, 'Die Tschechoslowakei zwischen dem besiegten Deutschland', p. 114.
29. *Süddeutsche Zeitung*, 8 September 1968.
30. More about the concept 'securing the economic basis of national defense' -(das tendenziell die gesamte Volkswirtschaft umfasst) can be found in T.M. Forster, *Die NVA*, Cologne: Markus Verlag, 1979, pp. 200 ff.
31. S. Schönherr, 'Uber den Zusammenhang zwischen Wirtschaft und Landes-verteidigung beim Aufbau des entwickelten gesellschaftlichen System des Sozialismus in der DDR'. *Wirtschaftswissenschaft* no. 8 (1969): 1164.
32. Compare this with Himmelberger, 'Die Rüstungsindustrie der Tschecho-slowakei', p. 327.
33. J.K. Hoensch, 'Entwicklungstrends und Entwicklungsbrüche in der Tschechoslowakischen Republik seit 1945', p. 32, in Lobkowicz and Prinz, *Die Tschechoslowakei*.
34. Ibid., for a comparison.
35. A more detailed description of the production of armour in Czecho-slovakia is found in O. Holub, *Československe tanky a tankiste* [Czechoslovakian Armour and Armoured Soldiers], Prague: Naše voysko, 1980.
36. Information from J. Hodic.
37. Up to 1977-8, SIPRI and the *Österreichische Militärische Zeitschrift* (ÖMZ), which is usually well-informed on military policy in Eastern Europe, left open the question of licence production of the T-62. (See the *Österreichische Militärische Zeitschrift* no. 1 (1977): 21, and Stockholm International Peace

Research Institute, *World Armament and Disarmament, SIPRI Yearbook 1978*, London: Taylor & Francis, 1978, p. 201.) In 1979, SIPRI reported that the licences had been granted to the CSSR in 1970. (Stockholm International Peace Research Institute, *World Armament and Disarmament, SIPRI Yearbook 1979*, London: Taylor & Francis, 1979, p. 144.) This does not, of course, say anything about when production began. In 1980, the *ÖMZ* (no. 5 [1980] : 427) stated that the T-62 is produced in Czechoslovakia. Also in 1980, the West German periodical *Soldat und Technik* mentions only licence production of the T-54/55. (See *Soldat und Technik* no. 9 [1980] : 501.)

38. For data on the number of T-62s in the Czech army, see International Institute of Strategic Studies, *The Military Balance 1980-1981*, London: 1980, p. 15. According to the 1977 SIPRI Yearbook, the CSSR exported a few T-62s to Bulgaria but it simultaneously imported T-62s from the Soviet Union. (Stockholm International Peace Research Institute, *World Armament and Disarmament. SIPRI Yearbook 1977*, Stockholm: Almqvist & Wiksell, 1977, pp. 276-8.) The Institute of Strategic Studies in London seems to have had trouble with the T-62 also. Up to 1976, it reported that 'a few' T-62s were found in the Czech army. Later, up to 1980, it found *no* T-62s at all. The reports in *Military Balance* on the T-62 in the East German military are also contradictory.

39. Compare this with F. Wiener, ed., *Die Armeen der Warschauer-Pakt-Staaten*, Munich: Lehmanns Verlag, 1974, p. 42.

40. Compare this with Holub, *Československe tanky*, p. 339. This tank is based on the T-34.

41. This APC is developed from the German APC Sd-Kfz-251 which was also produced in Czechoslovakia during World War II. Wiener, *Die Armeen der Warschauer-Pakt-Staaten*, p. 264. On its use today, see International Institute for Strategic Studies, *The Military Balance 1980-1981*, p. 15.

42. R.T. Pretty, ed., *Jane's Weapon Systems 1979-1980*, London: Jane's, 1979, p. 310.

43. Ibid.

44. It is only *Soldat und Technik* which believes in the possibility of Czech licence production.

45. Compare this with *Soldat und Technik* no. 4 (1970): 188.

46. Stockholm International Peace Research Institute, *SIPRI Yearbook 1977*, p. 274. Since 1979, *The Military Balance* has registered this APC as a product of the Soviet Union.

47. Since 1970, India has produced the OT-64 (SKOT) on licence. Stockholm International Peace Reseach Institute, *SIPRI Yearbook 1977*, p. 298. See also, *Jane's Weapon Systems 1979-1980*, p. 309.

48. Compare this with *Soldat und Technik* no. 11 (1970): 620 and no. 1 (1978).

49. *Österreichische Militärische Zeitschrift*, no. 5 (1980): 427, and *Soldat und Technik* no. 8 (1980): 501.

50. *The Military Balance* lists 2,700 tanks in 1969-70. Most of these are T-55s.

51. In 1962-3, 130 T-54s were exported to Egypt and in 1967-8, 80 of them were exported to Morocco. Stockholm International Peace Research Institute, *The Arms Trade with the Third World*, pp. 840, 854; ACDA data from United States Arms Control and Disarmament Agency, *The International Transfer of Conventional Arms. A Report to the Congress*, Washington, DC: April 1974, pp. A-15, A-19 ff. Compare this with the data in Stephan Tiedtke, *Die Warschauer Vertragsorganisation, Zum Verhältnis von Militär- und Entspannungspolitik in Osteuropa*, Munich and Vienna: Oldenbourg Verlag, 1978, p. 151.

52. H. Hoffman, 'Sozialistische Landesverteidigung', p. 333, in *Reden und Aufsätzen 1970 bis Februar 1974, Part III*, Berlin: Militärverlag der DDR, 1974.

53. W.J. Spielberger, *Von der Zugmaschine zum Leopard, Geschichte der Wehrtechnik bei Krauss-Maffei*, Munich: Bernard F. Graefe, 1979, p. 171.
54. Here the bridge-laying tanks and the tank recovery vehicles are not accounted for. Krauss Maffei manufactures them on the Leopard I chassis while the Czechs manufacture them on T-54 and T-55 chassis. See *Militärtechnik* no. 3 (1976): 133, and *Österreichische Militärische Zeitschrift* no. 1 (1977): 71.
55. *Jane's Weapon Systems 1979-1980*, p. 335.
56. International Institute of Strategic Studies, *The Military Balance 1980-1981*, p. 15.
57. In Federal Republic of Germany, Ministry of Defence, *Zur Sicherheit der Bundesrepublik Deutschland und zur Entwicklung der Bundeswehr*, Bonn: 1974, a whitebook on West German security and the development of the armed forces, the number of APCs is reported as 3,941. The outmoded APC HS-30 is not included in this number.
58. A more detailed description of the production of trucks in the WTO can be found in L.L. Whetten and J.L. Waddell, 'Motor Vehicle Standardisation in the Warsaw Pact: Problems and Limitations, *RUSI Journal* 124:1 (1979): 55-60.
59. *Militärtechnik* no. 3 (1980): 131.
60. *Militärtechnik*, no. 1 (1977): 36, and Whetten and Waddell, 'Motor Vehicle Standardization'. p. 56.
61. *Militärtechnik*, no. 9 (1976): 419 ff.
62. Ibid.
63. Ibid.
64. *Militärtechnik*, no. 10 (1976): 477.
65. The International Institute of Strategic Studies, *The Military Balance 1980-1981*, p. 15, and *Jane's Weapon Systems 1979-1980*, p. 410.
66. Federal Republic of Germany, Ministry of Defence, *Zur Sicherheit der Bundesrepublik*, p. 184.
67. *Militärtechnik*, no. 12 (1975): 570, and *Militärtechnik*, no. 1 (1979): 40 ff.
68. *Soldat und Technik*, no. 9 (1980): 500.
69. Compare this with E. Peter, 'Die Luftstreitkräfte der Tschechoslowakei: Ein Beitrag zu ihrer Geschichte und Entwicklung', *Österreichische Militärische Zeitschrift* no. 2 (1978): 129. This source also provides a detailed description of the Czech aviation industry prior to World War II.
70. Ibid.
71. *Militärtechnik* (1975), folder.
72. *Österreichische Militärische Zeitschrift* no. 1 (1977): 71.
73 In the 1950s, East Germany tried to build up an aircraft industry of its own through licence production of this model, but in 1958 this project had to be abandoned. See also Ulrich Albrecht and Stephan Tiedtke, 'Rüstungswirtschaft in der DDR', p. 86, in *Die Nationale Volksarmee*, Studiengruppe Militärpolitik, Reinbek: Rowohlt Verlag, 1976.
74. From the beginning of the 1970s to the end of 1977, 90 planes of this type were produced. J.W.R. Taylor, *Jane's All the World's Aircraft 1978-1979*, London: Jane's, 1978, p. 32.
75. 'Czechs Gear for East European Sales', *Aviation Week & Space Technology*, 11 June 1979, p. 282.
76. International Institute of Strategic Studies, *The Military Balance 1980-1981*, p. 15.
77. Autorenkollektiv, *Die Tschechoslowakische Volksarmee*, p. 164.
78. A.R. Johnson *et al.*, *East European Military Establishments: The Warsaw Pact Northern Tier*, Santa Monica, Calif.: Rand Corporation, December 1980, p. 12, and *Jane's All the World's Aircraft 1978-1979*, p. 140. Both these sources

indicate that licence production of MIG-fighters had ended by 1979.

79. J.B. Köppl, 'Probleme des multinationalen Rüstungsmanagements und deren Auswirkungen auf die Verteidigungsfähigkeit der NATO-Staaten under dem Aspekt der wachsenden sowjetischen Bedrohung', Munich: Inaugural Dissertation, 1979, pp. 141 ff.

80. See also *Jane's All the World's Aircraft 1978-1979*, p. 30.

81. *Militärtechnik*, no. 8 (1976): 375, and E. Bau, 'Neue Aspekte für die Luftfahrtsindustrie der CSSR', *Interavia* no. 5 (1967): 715.

82. *Militärtechnik* no. 8 (1976): 375; *Jane's All the World's Aircraft 1978-1979*, p. 31.

83. 'Czechs Gear for East European Sales', *Aviation Week & Space Technology*, p. 282.

84. D.A. Brown, 'Trainer Vies for International Market', *Aviation Week & Space Technology*, 9 July 1979, p. 43.

85. Baur, 'Neue Aspekte für die Luftfahrtsindustrie', p. 717.

86. 'Czechs Gear for East European Sales', *Aviation Week & Space Technology*, p. 282.

87. See G. Krude, 'Military Production and Unemployment in the Federal Republic of Germany', unpublished manuscript, 1979, pp. 18 ff.

88. 'Czechs Gear for East European Sales', *Aviation Week & Space Technology*, p. 282.

89. Baur, 'Neue Aspekte für die Luftfahrtsindustrie', p. 717.

90. Quoted from *Wehrkunde* no. 12 (1962): 683.

91. Compare this with Checinski, 'The Costs of Armament Production', p. 137.

92. Compare this with *Aviation Week & Space Technology*, 12 May 1980, pp. 57 ff.

93. F.O. Miksche, *Rüstungswettlauf*, Stuttgart: Seewald Verlag, 1972, p. 207, and Checinski, 'The Costs of Armament Production', p. 138.

94. N. Cherikow, 'Orao- "Jukom-Fighter", Ein jugoslawisch-rumänisches Mehrzweckflugzeug', *Internazionale Wehrrevue* no. 4 (1975): 488-90.

95. Checinski, 'The Costs of Armament Production', p. 129.

96. Ibid.; personal communication from Checinski, September 1982. Checinski formerly served in the Polish Institute for War Economy.

97. Compare this with Curt Gasteyger, 'Probleme und Reformen des Warschauer Paktes', *Europa-Archiv* no. 1 (1967): 4.

98. Compare this with Tiedtke, *Die Nationale Volksarmee*, pp. 131 ff.

99. G. Peter and H. Einhorn, '500 Betriebe bauen an einem Schiff. Anforderungen der Landesverteidigung an die Wirtschaft', *Neues Deutschland*, 22 May 1966.

100. Schönherr, 'Uber den Zusammenhang zwischen Wirtschaft', p. 1167.

101. Ibid.

102. Compare this with Tiedtke, *Die Nationale Volksarmee*, p. 127.

103. Checinski, 'The Costs of Armament Production', p. 129. It is possible that this committee is the same as the 'Technischen Komitee der Vereinten Streitkräfte' [The Technical Committee of the United Armed Forces] which has about the same functions according to East German sources. Compare this with *Militärwesen* no. 2 (1980): 93.

104. Checinski, 'The Costs of Armament Production', p. 119.

105. The first quotation is from K. Greese and H. Seeling, 'Zur Entwicklung der Gefechtsbereitschaft der Luftstreitkräfte/Luftverteidigung der NVA (1957/58)', *Militärgeschichte* no. 4 (1972): 414. The second quotation is from K. Greese *et al.*, 'Zur Aufstellung moderner Landstreitkräfte der NVA im Jahre 1956', *Zeitschrift für Militärgeschichte* no. 4 (1967): 394 ff.

106. Compare this with G. Glasser *et al.*, 'Zur Geschichte der Nationalen Volksarmee der DDR-Thesen', *Militärgeschichte* no. 4 (1973): 17 ff. (Appendix).

107. Compare this with *Militärtechnik* no. 7 (1974): 325.

108. Whetten and Waddell, 'Motor Vehicle Standardization', p. 57.

109. Wiener, *Die Armeen der Warschauer-Pakt-Staaten*, p. 44.

110. The same was also true for Czech arms deliveries to the third world which began in 1955 with shipments to Egypt. For tactical reasons, this agreement was made between the CSSR and Egypt, rather than the Soviet Union and Egypt. See, W.Z. Laqueur, *The Soviet Union and the Middle East*, London: Routledge and Kegan Paul, 1959, p. 227, note 1; D. Dallin, *Sowjetische Aussenpolitik nach Stalins Tod*, Köln-Berlin: Kiepenheuer und Witsch, 1961, pp. 464 ff.; Uri Ra'anan, *The USSR Arms the Third World: Case Studies in Soviet Foreign Policy*, Cambridge, Mass., and London: MIT Press, 1969, Part 1, pp. 13 ff.; and Oles M. Smolansky, *The Soviet Union and the Arab East Under Khrushchev*, Lewisburg, Pa.: Bucknell University Press, 1974, pp. 23 ff. S.P. Gibert and W. Joshua, *Arms for the Third World, Soviet Military Aid Diplomacy*, Baltimore, Md. and London: Johns Hopkins University Press, 1969, p. 98, estimate that about 10 per cent of the combined military assistance of East European countries between 1955 and 1968 came from Czechoslovakia.

111. Brezhnev's statement at the 24th Party Congress that 42 per cent of the total output of the Soviet defence sector was destined for civilian uses can be taken as confirming this trend in the USSR despite the fact that the statement is somewhat ambiguous and is open to multiple interpretations. See *Materialy XXIV s-ezda KPSS*, Moscow: 1973, p. 46.

7 ITALY

Sergio A. Rossi

Introduction

Anyone who tries to undertake a scientific survey of the Italian defence industry or of anything else related directly or indirectly to contemporary Italian military problems will have to contend with several major difficulties. The first is the general lack of comprehensive, published studies on this subject. Partial attempts have periodically been made in the annual survey of IAI, the Institute of International Affairs in Rome.[1] Only in 1980 was the first book on the Italian defence industry published by a young sociologist at the University of Rome. This book was written from a marxist perspective.[2] However, no publication contains much more than aggregate data on total estimated military production. At most, data are provided for the main subsectors (electronics, naval, engineering, etc.). In particular, there are no detailed data (such as total sales or numbers of employees) for the major Italian companies engaged in defence production.

This unsatisfactory situation is due primarily to the traditional separation of military studies from civil studies in Italy. This reflects, on the one hand, the separation of the armed forces from society as a whole and, on the other hand, the excessive secrecy adopted by Italian military authorities concerning the release of data which is routinely made public in other major industrialized countries of the West. A classical example of this is the publication of the first White Book on Italian defence, which occurred only in 1977. Although the White Book's publication represented a remarkable break with tradition, it did not contain, for instance, any quantitative data on the composition of the armed forces or their major weapon systems. The latter were indicated only in general terms, by class or type. An additional reason for the climate of secrecy surrounding military affairs has been the domestic political situation. A genuine, in-depth debate on national defence and its links with industry would only further exacerbate the existing cleavages between the political parties, both in government and in opposition (mainly the Communist Party). Hence, the choice has been to attempt to sweep the entire issue under the carpet.

Another factor inhibiting the study of the Italian armed forces and

Italian military industry is that, apart from a traditional school of military historians, there is no sizeable and independent group of scholars in Italy researching military affairs and contemporary strategy. It was only in the 1970s that certain institutes, like IAI in Rome, started the first surveys in this field.[3] In the last two years, other centres and institutes have been established to carry out defence-related research.[4] On the whole, however, it can be said that, in Italy, the 'state of the art' is only in its infancy.[5]

Business circles and companies engaged in defence production are exceptionally reluctant to release quantitative data on their sales, profits, and so on, even when they widely advertise their major products in specialized defence journals and magazines. Although the situation has improved somewhat with regard to certain major companies, there is still a deeply rooted diffidence towards outside survey and inquiries. This is fed, it is true, by the frequent manipulation or distortion of the little information that is available by politicians and by the media's biased presentation of the role and function of the Italian 'military-industrial complex'. In the last five years, however, even leftist parties and trade unions have started to acknowledge the importance of the defence industry for industrial production and employment in Italy.

There is thus a scarcity of clear and usable statistical data on defence industries, even in the corporate Statements of Accounts and Balance Sheets that must be issued every year in compliance with Italian law. The reluctance of certain companies to disclose these documents (which should be made public) and the ineffectiveness of the state bureaucracy are well known. Endless, messy piles of official yearly statements of acounts from defence producers lie in cellars or in semi-forgotten archives of Italian courthouses, where by law at least one copy is to be available for public consultation. These files are poorly managed by overworked and underpaid clerks. In addition, the EEC IV Directive concerning international standards of accountancy, especially 'the true and fair view' (the correspondence between official company data and the true financial and economic situation of the company), is being adopted very slowly in Italy.

As a result of all this, many figures provided in this study are, inevitably, estimates. In most cases, however, the margin of error has probably been reduced to less than 10 per cent. In part, this has been possible because certain firms provided information after having received adequate guarantees of objectivity and proper use of relevant data.

The Postwar Years

The development of the Italian defence industry in the postwar period, following the destruction suffered during World War II, went almost hand-in-hand with the entrance of Italy into the Atlantic Alliance and its signature, in 1949, of the North Atlantic Treaty. Beyond its political significance in terms of the East-West dispute during the Cold War years, this event implied the subsequent cooperation with and gradual integration of Italian national production into, first, the Western economic system (naurally led by the United States after the inauguration of the Marshall Plan) and, second, the economic system created by recovering West European countries. This latter development was symbolized by the participation of Italy in the foundation of the European Economic Community in the 1950s.

In so far as defence industry is concerned, Italian membership in NATO and Italy's preferential bilateral ties with the United States were most important in Italy's first stage of postwar development. This stage began in the 1950s and lasted at least until the mid-1960s. In the aircraft industry, after the early years of licence production, mainly assembly of parts (for instance of the British Vampire fighter-bomber by FIAT and Macchi under licence from de Havilland), a period of increasingly sophisticated production under American licence started around 1956 (220 F 86-K Sabre fighter-interceptors were built by FIAT). In those years, the electronics and missile industries were still in their infancy, while the naval industry, mainly under the aegis of the state, built a host of light and medium-sized warships, both for the Italian and NATO navies and for export to developing countries.

By the end of the 1950s,with the design and production of the light tactical fighter G-91 (which used a Bristol Siddeley engine), the Italian defence industry was ready to undertake its first largely domestically conceived ventures. The G-91 was subsequently adopted as a standard aircraft by NATO. Also, Macchi and Piaggio started to export their own light trainer and tactical support aircraft, such as the MB 326, the P-146, and the P-149. The electronics industry (led by Finmeccanica-Selenia) started serious development in 1959 with its participation in the European consortium for the Hawk anti-aircraft missile. Then, in the early 1960s, the whole Italian aircraft industry participated in the NATO consortium for the production of the F-104 fighter-bomber. (This was, however, still produced under licence from the US firm, Lockheed.) This event was of particular relevance for the formation of an autonomous European consortium in the aircraft industry which is currently producing the advanced Tornado fighter-bomber after many years of design and testing in the 1970s.

The creation in 1968 by Italy, the United Kingdom, the Federal Republic of Germany and the Netherlands, of the MRCA consortium for the construction of a multi-role combat aircraft of entirely European design marked a new stage in the development of the Italian defence industry. This stage, which still continues, has been characterized by the maintenance of the somewhat preferential link with the American defence industry as codified in September 1978 by the 'Memorandum between the US and Italy on cooperation in the field of armaments and armaments-related industry'. This cooperation, however different the industrial size and the technological capabilities of the two countries, is slowly coming to favour Italy. One example of this trend is the installation, on about 80 American ships (and 50 West German ones), of the 76/62 Integrated Naval Gun System built by Oto-Melara.

More recently, the Italian Beretta pistol was ranked first in performance tests by the US Air Force in view of its possible adoption as a standard weapon. However, later tests conducted by the US Army, with more stringent requirements, were pronounced inconclusive. Considering the heavy pressure exerted by US manufacturers, the prospects for the Beretta have declined somewhat. At the same time, in a noteworthy speech at NATO SHAPE in the spring of 1981, the US Undersecretary for Policy in the Department of Defense during the Carter Administration, Robert Komer, both acknowledged and criticized the US Armed Forces' reluctance to adopt foreign-designed weapons. This attitude persists inspite of the official NATO 'two-way street' policy relating to arms procurement and production.

Nonetheless, cooperation with the US has allowed the Italian defence industry to build up a technological base on which to rely for developing new designs and projects of entirely indigenous weapon systems. Such indigenous development is a reality as the case of Agusta demonstrates. After a period of cooperation and licence production with Bell and Sikorsky, Agusta is now able to build its own helicopters for the Italian army, for example, the new Mangusta anti-tank model. Indeed, in 1978, thanks to the experience and know-how gained through this cooperation, Augusta was able to sign an important agreement with three other European firms (Aérospatiale of France, Messerschmidt of West Germany, and Westland of Great Britain) for the joint construction of a new generation of military helicopters.

This means that there is a willingness and capability on the part of the Italian defence industry to become involved in – when it is judged to be in its own interests – a growing cooperation with other European defence firms. Apart from the example of the MRCA Tornado aircraft, Oto-Melara cooperated with Kraus Maffei of Germany in the production of the Leopard tank for the Italian armed forces. Here again, Oto-Melara has recently developed an entirely new version of this tank, the OF-40 'Lyon' (whose engine is manufactured by FIAT), which is particularly suited for desert warfare. The Lyon has been bought by several Middle Eastern countries.

The real development of the postwar Italian defence industry took place during the 1970s. In this period, the value and volume of total sales increased more than ten-fold, as shall be seen in more detail below. In the 1970s, a certain industrial restructuring and concentration also occurred, led mainly by the state-owned sector, which currently dominates defence production, primarily through the shipbuilding and aircraft industries. The private sector is, however, also very well represented, especially in engineering, aircraft parts and engines, and land transportation. Moreover, in 1980-1, with the general economic recession that hit the weaker sectors of state-owned industry especially hard, the private sector took over a sizeable part of defence electronics production. This trend appears, however, to have been short-lived as the state sector was again dominant in 1981-2.

Present Structure of the Defence Sector

The first question to be examined here will be the current size of Italy's defence industry and its relevance to the country's industrial system as a whole. Some estimates have been made in the past by other analysts but the present discussion will rely, especially for data covering 1979 and 1980, on the author's own survey of the 50 largest Italian defence companies with annual sales over 10 billion lire (about $10 million in 1980). The results of this survey are shown in Table 7.1, which indicates that, in 1980, Italian defence industries had total sales of over 4,000 billion lire (about $4 billion). (Preliminary data for 1981 show that sales increased to over 5,000 billion lire.)

The 1980 total increases to 4,650 billion lire ($4.6 billion) if the estimated sales of a network of subcontractors – not necesassarily producing only military items – are also taken into account.[6] The figures cited here are for gross total sales. There is a possibility of duplication, that is, that part of the defence production by some industries manufacturing parts or components of major weapon systems, such as aircraft or ships, that are subsequently supplied to other national defence industries for the final manufacture or assembly have been counted twice (or as many times as transfers of components have occurred). It is true that other sources have tried to establish the extent of such double-counting, notwithstanding the difficulties involved. This leads to reductions of between 20 and 40 per cent on the figures presented here for total sales, although there is agreement among different studies on the order of magnitude of the personnel employed in the defence sector.[7]

Table 7.1: The Development of the Italian Defence Sector, 1977-81

	1977	1978	1979	1980	1981
Major Contractors					
Total sales (billion lire)	1,900	2,500	3,200	4,000	5,000
Number of employees	73,500	76,500	78,500	80,000	–
Subcontractors					
Total sales (billion lire)	200	390	500	650	–
Number of employees	9,000	10,500	11,500	12,000	–

Source: Various authors. The estimates provided here are based on figures available for 50 major defence producers with sales in excess of ten billion lire. Only that proportion of total production which is purely military has been used for civilian and civilian-military firms. The estimates of the number of employees involved in military production provided here are higher than those from other sources but are consistent with the latest figures released by the Minister of Defence.

Sales by Italian defence companies in 1980 represented about 2.5 per cent of the total estimated sales of the entire Italian manufacturing sector for that year. The official statistics on gross sales by Italian industry between 1974 and 1980 also show that while the total sales by industry as a whole have increased about 3.5 times, the total sales of the defence sector have increased some 4.4 times. As will be seen below, this is one of several indicators which show a more dynamic rate of development for the defence sector, even allowing for differential rates of inflation. The rate of growth of the Italian defence sector over the last decade has been very high, averaging 30 per cent per year and never dropping below 25 per cent. This is a rate several points higher than the inflation index (16-20 per cent), which also contributed to the overall upward trend. It must be remembered, however, that the inflation index for high technology products is usually higher than the national average. This difference has been estimated to range, for military production, from over 4 per cent in the engineering subsector to 8 per cent in electronics.

A significant role in the development of defence production has been played by the increasingly large orders placed with domestic industry by the Ministry of Defence (Table 7.2). This is also an indication of the steep rate of growth of the defence sector. According to

Table 7.2: Military Spending and Defence Contracting in the Italian Economy (billion current lire)

Year	Defence Budget	As percentage of GNP	As percentage of State Budget	Government Defence Contracting[a]
1974	2,373	2.9	9.2	
1976	2,956	2.5	7.8	4,500[b]
1978	4,313	2.4	7.1	
1980	5,780	2.4	4.7	1,700
1981	7,511	2.4	4.4	2,300
1982[c]	10,149	2.3	4.3	3,000

Notes: a. Between 70-80 per cent of these sums were contracted in the Italian market.
b.Data for 1975-1979.
c.Preliminary data.
Source: Author's elaboration from official data.

figures released by the Ministry of Defence in January 1982, government orders for 1975-9 amounted to 4,500 bilion lire, of which 85 per cent were placed with domestic industries. In 1980, orders (or defence contracts) were worth 1,700 billion lire, of which 70 per cent were placed in Italy. (Of the domestic orders, 25 per cent went to companies located to southern Italy.) About half of the orders placed abroad went through NATO agencies which means that part of the foreign expenditure has 'returned' to the domestic market through Italy's participation in the NATO agencies. Preliminary, but reliable, estimates based on the 1982 national defence budget put the value of total (domestic and foreign) defence industry orders at over 3,000 billion lire.

The total number of workers directly employed in military production is estimated to have been about 80,000 in 1980. If subcontractors are taken into account, the number is more than 92,000. These figures represent from 1.6 to 1.8 per cent of employment in the manufacturing industry for 1980. They constitute a more accurate attempt at estimating defence-related employment than the Italian Ministry of Defence's official estimate. The 1977 *White Book on Defence* reported that there were 8,000 companies on the official list of defence contractors and suppliers (which includes trading firms and non-military suppliers) and that these had a total of 1,300,000 employees. The 150 companies directly producing weapons or weapons-related components were estimated to have 150,000 employees.

Subsequently, however, in an interview with the Roman daily newspaper *Il Messaggero*, the Minister of Defence Lelio Lagorio* gave more accurate figures: Italian defence employment is about 100,000 or almost 2 per cent of the personnel working in the manufacturing industry. Moreover, according to the Minister, defence production increased from 900 billion lire in 1972 to 3,000 billion in 1977 and to 4,000 billion lire in 1980.

A more precise elaboration and subdivision by main production sectors is not yet possible on the basis of the figures available for 1980. Such a breakdown can be made with more accuracy using the figures elaborated by Battistelli for 1977. Up to this point, these figures have constituted the only published attempt at an in-depth analysis of the structure and evolution of the Italian defence industry over the last ten years.

Table 7.3: Degree of Concentration in the Italian Defence Industry, 1980

Number of Companies	Military Sales (billion lire)	Percentage of Total Italian Defence Sales	Employees in Military Production	Percentage of Total Employees in Italian Defence Production
10	1,700	42.5	50,000	62
15	2,100	52.5	55,000	68
20	2,500	62.5	61,000	76
25	2,800	70.0	65,000	81

Note: The difference in the sales-per-employee index (not calculated here) for the first group of ten companies and those for the other groups is due to the sizeable amount of personnel working in civilian production for at least three major companies.
Source: Author's elaboration.

*Lagorio is the first socialist politician to hold this portfolio.

Table 7.4: Italian Major Defence Contractors[a], 1980

Firm	Production Sites	Products	Total Sales (billion lire)[b]	Number of Employees	Market[c]
Cantieri Navali Riuniti (CNR)	Genova, Riva Trigoso, Ancona, Palermo	Audace cruiser, Lupo and Maestrale frigates, ASM corvettes, Fac (M) P-420, Sparviero-Swordfish.	490	8,700	M/C
Costruzioni Aeronautiche Giovanni AGUSTA	Varese	Helicopters (Agusta-Agusta Bell, Agusta Sikorsky).	340 (400)	4,200	M
Italcantieri	Genova Sestri Monfalcone, Castellamare di Stabia	Toti and Sauro submarines, V. Veneto missile cruiser, missile destroyer Ardito, helicopter cruiser Garibaldi.	250	9,700	M/C
Aeritalia soc.	Torino, Pomigliano d'Arco, Napoli, Milano	F-104-S fighterbomber, G-91-Y tactical reconnaisance, Mero Tornado, multi-role combat aircraft, G-222 transport, avionics, aerospace and satellite systems.	280 (530)	11,500	M
Oto-Melara	La Spezia	Control systems and naval guns, Albatros SAM, Otomat antiship missile systems, Leopard and OF-40 tanks, various armoured troop carriers.	200 (290)	2,500	M
Selenia Industrie Elettriche Ass.	Roma, Pomezia, Napoli	Missile systems (air, sea and surface), radar and navigation systems, air defence systems, electronic countermeasures/ electronic counter-countermeasures.	170 (304)	6,400	M
FIAT Aviazione	Torino	Aircraft engines and components, air and naval turbines, various parts.	150 (210)	3,600	M
SIAI-Marchetti[d]	Varese, Sesto Calende, Borgomanero	Light trainer aircraft (SF-260 m), tactical aircraft (SF-260 w), reconnaissance and light transport aircraft.	140 (200)	3,100	M

Grandi Motori Trieste spa.	Trieste	Diesel engines, especially for ships.	130	3,300	M/C
Aeronautica Macchi spa	Varese	MB-326 and MB 339 advanced trainers, MB-326 K light tactical support, AMX tactical support fighter, parts of G-222, F-104 G/S, Atlantic, and MRCA.	86 (125)	1,000	M
Snia Viscosa (Space and Defence Division)	Colleferro (Roma), Ceccano	Artillery, ammunition, explosives, propellers, rockets.	120	4,300	C
Telettra spa (Division of Military Systems)	Chieti	Telecommunications systems.	120	4,800	C
Borletti F:11. spa (Defence Division)	Milano	Artillery, gun and mortar fuses, rockets, aircraft bombs.	120	3,000	M/C
Elicotteri Meridionali	Frosinone	Helicopters.	90	1,000	M
Elettronica San Giorgio (ELSAG)[e]	Genova, Sestri	Naval gun systems, rocket launchers, submarine control systems.	80 (115)	1,600	M
Oerlikon Italian	Milano	Anti-aircraft gun battery 35 mm., several types of land and naval guns, ammunition	90	1,100	M

Notes: a. Major contractors are firms with annual sales in excess of $100 million. In 1980, $ 1 million was roughly equivalent to 1 billion lire.
b. Figures in brackets are for 1981.
c. M indicates a military market; C indicates a civilian market.
d. SIAI-Marchetti now belongs to the Agusta group.
e. Elettronica San Giorgio (ELSAG) now belongs to the Selenia group.
Source: Author's elaboration.

It is possible, none the less, to use the most recent figures to show the remarkable degree of concentration in the Italian defence sector (Table 7.3). The ten largest of the 50 major defence contractors, with sales of 1,700 billion lire and 50,000 employees directly involved in military production, account for 42.5 per cent of total sales by the Italian defence sector, and for 62 per cent of the personnel employed. These figures increase to 52.5 per cent of sales and 68 per cent of personnel if the fifteen largest companies are considered. (A detailed list of the names and types of production of these companies is found in Table 7.4.) Going further down the list, twenty companies account for 62.5 per cent of total sales and 76 per cent of the personnel, while 25 companies – none of them having sales below 40 billion lire – account for 70 per cent of total sales and 81 per cent of the personnel employed. To put these figures in perspective, it should be noted that each firm in the second group of 25 companies among the 50 major defence contractors has sales ranging from 10 to 30 billion lire and employs between 100 and 300 workers.

As has already been mentioned, a detailed picture of the structure of the Italian defence sector is available only for 1977. Based on a survey of some 294 firms involved to some degree in defence production, the sectorial breakdown for 1977 is shown in Table 7.5. Taking into account only the production devoted to military purposes and the corresponding number of employees, sales in 1977 are estimated at 1,854 billion lire, to which another 200 billion lire of defence subcontractors' sales should be added for a total of more than 2,000 billion lire. There are some 63,600 employees directly involved in defence production plus another 9,000 employed by subcontractors for a total of 72,600.

The principal features of the defence sector revealed by this survey are as follows. The greatest number of companies is to be found in the electronics subsector. Here there are 107 mainly medium-sized firms but of these, only about 40 are engaged solely in the production of electronics. The remainder produce electro-mechanical items and missile systems. The next largest subsector is the engineering industry with 66 firms. The naval subsector follows with 64 firms, of which only two or three are very large. The aircraft subsector is composed of 40 firms. The chemical subsector has only 17 firms.

The two most politically and economically important subsectors are the aircraft industry, with 20,100 employees (31.6 per cent of the total) and the electronic industry with 20,000 employees (31.4 per cent of the total). The engineering subsector has 11,000 employees (17.3

Table 7.5: The Structure of the Italian Defence Industry by Sectors, 1977

Sectors	Companies		Employees		Sales	
	Number	%	Number	%	Billion lire	%
Aircraft	40	13.6	20,100	31.6	650	34.0
Electronics[a]	107	36.4	20,000	31.4	523	28.2
Engineering	66	22.4	11,000	17.3	419	22.6
Shipbuilding	64	21.8	6,700	10.5	130	7.0
Chemicals	17	5.8	5,800	9.2	152	8.2
Total	294	100	63,600	100	1,854	100

Note: a. Included here are electronics, electromechanical and missile systems.
Source: F. Battistelli, *Armi: Nuovo modello di sviluppo? L'industria militare in Italia*, Torino: Einaudi, 1980, p. 184.

per cent of the total). Less relevant are shipbuilding with 6,700 employees (10.5 per cent of the total) and the chemical industry with 5,500 employees (9.2 per cent of the total). The figures for sales by subsector confirm this hierarchy: since 1977 the share of shipbuilding in military production has increased in comparison to the chemical sub-sector. There is unquestionably a high degree of concentration in the defence sector. The employment figures for the largest five firms in every subsector, as a percentage of employment in the entire military sector, are: electronics, 46.6 per cent; engineering, 64.2 per cent; aircraft, 67.0 per cent; shipbuilding, 70.7 per cent; and chemicals, 80.6 per cent.

Ownership in the Defence Sector

State-owned, Semi-public and Private Firms

As is shown in Table 7.6, state-owned industry accounts for the largest percentage of employment in the defence sector (46 per cent). If firms of mixed ownership (public/private) are included, this figure becomes 58.1 per cent. By contrast, the privately owned companies account for only 41.9 per cent of total defence-related employment. Since the fall of 1979, however, when the private corporation Bastogi acquired electronics companies previously owned in part by the public holding-company Montedison, private industry's share increased to at least 45.2 per cent.

If instead total sales are considered, as is done in Table 7.7, the private sector records 45.5 per cent of the total against 43.1 per cent

Table 7.6: Employment in the Defence Industry by Sector and Ownership, 1977

Industrial Sector	State-owned firms		Mixed-ownership firms		Privately owned firms		Total
	Number	%	Number	%	Number	%	
Aircraft	12,900	64.2	–	–	7,200	35.8	20,100
Electronics[a]	7,650	38.3	2,150	10.7	10,200	51.1	20,000
Engineering	3,540	32.2	510	4.6	6,950	63.2	11,000
Shipbuilding	5,150	76.9	–	–	1,550	23.1	6,700
Chemical	–	–	5,070	87.4	730	12.6	5,600
Total	29,240	46.0	7,730	12.1	26,630	41.9	63,600

Note: a. Included are electronics, electromechanical and missile systems.
Source: F. Battistelli, *Armi: Nuovo modello di sviluppo? L'industria militare in Italia*. Torino: Einaudi, 1980, p. 188.

Table 7.7: Sales in the Defence Industry by Sector and Ownership, 1977 (billion lire)

Industrial	State-owned firms		Mixed-ownership firms		Privately owned firms		
Sector	Sales	%	Sales	%	Sales	%	Total
Aircraft	430.6	68.3	–	–	199.4	31.7	630
Electronics[a]	134.0	25.6	59.0	11.3	330.0	63.1	523
Engineering	147.0	35.1	17.0	4.0	255.0	60.9	419
Shipbuilding	87.0	66.9	–	–	43.0	33.1	130
Chemical	–	–	136.0	89.5	16.0	10.5	152
Total	798.6	43.1	212.0	11.4	843.4	45.5	1,854

Note: a. Includes electronics, electromechanics and missile systems.

Source: F. Battistelli, *Armi: Nuovo modello di sviluppo? L'industria militare in Italia*. Torino: Einaudi, 1980, p. 189.

for the state-owned sector. The latter's proportion is increased, however however, to 54.5 per cent if the sales accounted for by the mixed-ownership sector (11.4 per cent of the total) are added in. Here again, the share of the private sector has increased since 1979 due to Bastogi's acquisitions to about 48.7 per cent of total sales. One conclusion to be drawn from the 1977 figures as far as the productivity of the defence sector is concerned is that, generally speaking, the private sector fares better than the public sector, employing fewer people and earning more money. The sales-per-capita ratio is higher for the private sector, with 31 million lire, against 27 million lire for the public sector.

There are also a number of other points that should be made regarding the distribution of ownership by sectors of production. The first is the prevalence of state ownership in the aircraft industry. One explanation for this might be that only the government can sustain the very high investment rates required by the aircraft industry. A second point to note is that there is even greater public predominance in shipbuilding. This industry has been particularly hard hit by the world-wide economic recession and by increasingly strong competition from Japanese companies. Once again, although for different reasons pertaining mainly to social stability, that is, the need to contain unemployment, only the state can sustain losses such as those experienced by the shipbuilding industry in recent years and provide the funds necessary to restructure this industry without losing too many jobs. The current plan for revitalizing the Italian shipbuilding industry relies, to a large extent, on growing military production, especially for navies of developing countries.

In the chemical industry, on the other hand, it is firms of mixed ownership which are predominant. Within this group Snia Viscosa is the dominant firm. Finally, it is the private sector which dominates electronics. (After 1979, the percentages would have increased at least to 61.7 per cent of the employees and 74.4 per cent of the sales following Bastogi's acquisition of some former Montedison companies.) Private firms also control much of the engineering/transport subsector.

There are two public holding-companies that currently dominate almost all types of defence production. The first is IRI (Istituto per la Ricostruzione Industriale) which controls Aeritalia and Alfa Romeo in the aircraft sector, Selenia and Elettronica S. Giorgio (Elsag) in the electronic subsector and Italcantieri, Cantiere Navale Muggiano and Cantiere Navale Breda in shipbuilding. These are only the more important and well-known companies controlled by IRI. The second most important state holding-company is EFIM (Ente Finanziario per l'Industria Mineraria), which controls a host of aircraft companies (the entire Agusta group, which also includes Siai-Marchetti and Elicotteri Meridionali) and firms in the engineering subsector (Oto-Melara, Breda Meccanica Bresciana, Breda Fucine and others).

The mixed-ownership or semi-public sector is dominated by two large companies, Montedison and Snia Viscosa, which primarily produce chemical and explosive products. As mentioned above, until 1979 Montedison also controlled a group of electronic companies (Montedel, Officine Galileo, Sistel) but these now comprise the privately owned Bastogi Systems.

In the private sector, Fiat is the leading industrial group. It owns companies in the aircraft subsector (Fiat Avio, Motoravio Sud), in the engineering and military-transport industries (Iveco, Lancia Special Vehicles, Whitehead Motofides, Grandi Motori Trieste) and also in the electronics and telecommunication subsectors (Telettra and minority shares in both Sistel and Sigme).

The purchase by the Libyan State Bank of a minority share (between 12 and 13 per cent) of the Fiat group in December 1976 did not necessarily mean automatic access by the Libyans to Fiat's more advanced military technology. It is true that Fiat, through Iveco's special-vehicles section, normally supplies the Libyan army with trucks and MC-6616 armoured cars, but this involves no transfer of advanced military technology. Fiat is building a truck factory in Libya but this is for civilian production.

Finally, the private defence sector also includes other Italian companies such as Aermacchi, Piaggio, Microtecnica, Beretta, and foreign companies such as Oerlikon, Marconi, Contraves, Litton. Many of these are simply the military divisions of large industrial groups active in civilian production.

The 'Defence Industrial Area'

No survey of the Italian defence sector should fail to mention, even if only briefly, the sector known as the 'defence industrial area'. This is the complex of 30 plants and arsenals plus several technical, research and maintenance centres directly controlled by the Ministry of Defence. Six of these belong to the Army. These six centres do not undertake repairs but ensure that all contract obligations are fulfilled. Their R&D activities are extremely limited. According to the 1977 White Book, this complex has 3,300 military and 16,000 civilian employees, a total of nearly 20,000 workers. These employees are not included in the figures cited above for defence-industry employment, because they are not, strictly speaking, involved in military production. This is one of several reasons why estimates for total defence-industry employment, direct or indirect, vary so much. The estimate of 80,000 defence-industry employees used here has recently been endorsed by the Secretary-General of the Ministry of Defence, General Giuseppe Piovano.[8]

During the past decade, this particular industrial area has suffered from a lack of funds and also from the inefficient management of those resources that have been available. These problems have commonly been linked to the scarcity of resources allocated through the national defence budget. Following the approval of funding increases in 1981, it is believed that the situation will improve. This could have a positive effect on the maintenance and repair of the different weapon systems belonging to the Italian armed forces and should expand the number of persons employed in the various Ministry of Defence plants and arsenals. According to a long-term plan for the rationalization and restructuring of the whole defence industrial area, approved by the Italian Chiefs of Staff in 1978, the personnel employed will increase within a decade to over 26,000. These workers will be employed, however, in fewer installations (sixteen instead of 36). These changes should allow the armed forces to take direct responsibility for up to 70 per cent of the repair and maintenance work for the Army and up to 80 per cent for the Navy instead of having to rely on the more expensive private contractors. The latter usually have to pay higher salaries and bear heavier labour costs but attract the most skilled and highly trained technicians.

The Growth of the Italian Defence Sector in the 1970s

A comparison of the development of the Italian defence sector and its main trends and characteristics, on the one hand, and the growth of Italian industry in general, on the other hand, is available for the ten-year period from 1968 to 1978. The results of analyzing the main economic indicators during this period for a selected group of 38 defence companies are shown in Table 7.8.

The first consideration is that Italian military sales increased almost ten-fold in these ten years (from 274 to 2,655 billion lire). No comparable performance has been registered by any civilian sector. The average annual rate of growth was greater than 25 per cent. There were peaks of 36 per cent in 1975 and 44.5 per cent in 1976. The second consideration is that defence-sector employment doubled during the same

Table 7.8: The Development of the Italian Defence Industry, 1968-78 (billion lire)

	1968	1969	1970	1971	1972	1973	1974	1975	1976	1977	1978
Sales	274	314	380	468	582	759	917	1,248	1,804	2,180	2,655
Value-added	120	134	183	222	255	357	477	625	844	952	1,071
Profit on equity/assets	15	12	17	11	10	9	34	61	130	152	144
Average invested capital	313	362	448	577	691	809	996	1,211	1,476	1,783	1,993
Net fixed-assets	90	99	119	167	220	260	320	412	489	558	626
Employees (thousands)	36.8	39.9	46.7	48.4	53.3	63.5	66.5	70.0	72.0	72.7	74.8

Source: F. Battistelli, *Armi: Nuovo modello di sviluppo? L'industria militare in Italia*, Torino: Einaudi, 1980, pp. 208-9.

period (from 36,800 to 74,800 according to Battistelli's estimates, which are lower than those presented in Table 7.1). Its rate of growth was highest in the first half of the decade. It then declined somewhat and stabilized in the second half of the period when economic growth declined in general in Italy as in all Western economies, and employment in industry as a whole actually decreased.

A third consideration concerns value-added in the defence sector which has also shown remarkable progress, increasing almost nine-fold from 120 billion lire in 1968 to 1,071 billion in 1978. The rate of growth of defence value-added in this ten-year period (an average of 24.7 per cent annually, with peaks of 31 per cent in 1975 and 35 per cent in 1976) compares rather favourably with the corresponding rates of growth for civilian industrial subsectors, industry as a whole, and gross domestic product (GDP).

The value-added of corresponding civilian industrial subsectors, taken together, grew at an average rate of 17 per cent a year, with peaks of 31 per cent in 1974 and 32 per cent in 1976. The value-added of Italian industry as a whole had an average annual growth of 16 per cent, with peaks of 27 per cent in 1974 and 28.5 per cent in 1976. Finally, Italy's GDP grew on average 15.6 per cent annually, with peaks of 23 per cent in 1974 and almost 25 per cent in 1976. An additional confirmation of the greater dynamism of the defence sector compared to industry as a whole (where civil production is dominant) is portrayed in Tables 7.9 and 7.10. The growth index of value-added for the defence sector between 1970 and 1978 (5.85) is higher than the corresponding indexes both for general industry (3.68) and for the more dynamic civilian sectors, such as the engineering and electrical industry (4.06) or transport (4.29). Gross investments also increased much more in the defence sector: 4.44 against 2.87 for industry as a whole during the 1970-8 period.

Where data are available for several civilian sectors, as for the years 1970-7, the growth index of gross investments in the defence sector (3.97) remains higher than even that of the more dynamic branches of civilian industry (3.25 for the engineering and electrical subsector). It is also important to remember that, according to Battistelli, the increase in productivity (defined as value-added at constant prices) of the defence sector was 40 per cent between 1969 and 1978 while for industry as a whole, it was 34 per cent.

A fourth important consideration is that the rate of growth of value-added in the defence sector has been more stable, especially during the recession years, than that of other industrial subsectors or the GDP.

Table 7.9: Value-added in the Defence Sector and Industry as a Whole, Selected Years, 1970-78 (billion current lire)

Industrial Sectors	1970	1975	1977	1978	Growth Index 1978/1970
Defence	183	625	952	1,071	5.85
Total Industry	23,069	48,177	73,253	85,019	3.68
Engineering and Electrical	4,054	9,421	14,588	16,488	4.06
Transport	1,149	2,726	4,190	4,938	4.29
Chemical	1,498	3,140	4,322	5,060	3.37

Source: Author's elaboration from official data from the Italian Central Office of Statistics for civilian industry and F. Battistelli, *Armi: Nuovo modello di sviluppo? L'industria militare in Italia*. Torino: Einaudi, 1980, for data on the defence sector.

Table 7.10: Growth of Gross Investment in the Defence Sector and Industry as a Whole, Selected Years, 1970-78 (billion current lire)

Sector	1970	1975	1977	1978	Growth index	
					1977/1970	1978/1970
Defence	448	1,211	1,783	1,993	3.97	4.44
Total Industry	3,932	7,680	10,632	11,291	2.70	2.87
Engineering and Electrical	490	977	1,596	–	3.25	–
Transport	276	549	888	–	3.21	–
Chemical	555	1,225	1,523	–	2.74	–

Source: Author's elaboration from official data from the Italian Central Office of Statistics for civilian sectors and F. Battistelli, *Armi: Nuovo modello di sviluppo? L'industria militare in Italia*, Torino: Einaudi, 1980, for the defence sector.

In 1975, for instance, the GDP grew at 13.1 per cent; industry as a whole, 10.3 per cent; the four corresponding industrial subsectors, which include civilian plus defence production (electronics, engineering/chemical, aircraft and shipbuilding) 9.6 per cent; and the defence sector 31 per cent. Finally, taking into account the poor performance of the shipbuilding industry, the available data suggest that since 1975 the profitability, or return-on-investments index (ROI) [which is not shown in Table 7.8], has been more stable and consistently higher in the defence sector than for Italian industry in general. This trend was particularly evident in 1977 and 1978, when the ROI figures were 8.5 (defence) versus 4.8 (total industry) and 7.2 (defence) versus 4.6 (total industry).

Comparing the main development indicators for each of the four main subsectors (aircraft, electronics, engineering/chemical and shipbuilding) with trends in the defence sector as a whole, the following observations can be made. First, the aircraft industry appears to have developed faster between 1968 and 1978 than the defence sector taken as a whole. Sales increased fifteen-fold in the former as against ten-fold in the latter. The value-added in the aircraft subsector has increased ten-fold as compared to nine-fold for the defence sector in general. The aircraft subsector's profitability has tended to be somewhat higher and more stable than the average even during periods of economic recession, which have also been shorter for the aircraft-makers than for the other subsectors of military production. Employment in aircraft companies has doubled, which is in line with the overall defence sector trend.

Aircraft and helicopter production for military use is likely to continue to represent a significant share of production in the Italian aerospace industry and even to increase that share. Available data on sales by the five major aerospace companies (Aeritalia, Agusta, Fiat Avia, SIAI-Marchetti and Aermacchi) for 1980 and 1981, show that these companies now account for 70 per cent of the total aerospace production by the members of AIA (Associazione Industrie Aerospaziali). The combined sales of the 'big five' increased over 50 per cent 1980-1, going from 1,000 billion lire in 1980 to almost 1,550 billion in 1981. Taking into consideration the whole aerospace industry, which had sales of 2,200 billion lire in 1981 and knowing approximately the military share of production for the 'big five', it seems safe to assume that over 50 per cent of the production in the Italian aircraft industry depends on defence contracts, especially from abroad.

The electronic subsector has also displayed a high rate of development. While the increase in sales and employment has been in line with the general trend, especially since 1974-5, profitability has been higher than in the defence sector as a whole. The return on investment (ROI) more than doubled between 1968 and 1978, while in the aircraft industry the same ROI was stable or increased by 19 per cent (if Aeritalia is eliminated from the calculation). The profitability of the military and civil sectors of the electronic industry has recently been compared for a sample of 32 companies operating in the subsector designated as 'professional or advanced electronics'. This subsector accounts for most of the defence-related electronics production.[9] Six companies involved exclusively in military production, each employing more than 500 persons in 1978, had sales of 212 billion lire versus 353 billion for the whole subsector and accounted for 5,975 of the subsector's 9,433 employees. The sales-per-employee index (35.4 million lire) and labour-cost per employee (12.6 million lire) were in line with those for the subsector as a whole. The military sector was, however, superior in terms of value-added as a percent of production (55 per cent, against 45 per cent in electronic processing systems, and 29 per cent in mechanical automation), value-added per employee (23.3 million lire, against 21.7 for biomedical electronics, 11.9 for electronic processing systems and 11.6 for mechanical automation) and the return-on-investment or ROI index (14.39 per cent for military electronics against 9.32 for biomedical electronics and 3.30 for power transmission electronics).

The degree of reliance on military production by the electronic sector* can be estimated fairly accurately for 1979-81. According to ANIE data and estimates produced by the specialized weekly *Mondo Economico,*[10] military electronics sales in 1979 totalled 500 billion lire. That amounted to 11.5 per cent of the 4,353 billion lire figure for total sales by the Italian electronic industry in 1979, which at that time employed about 118,000 persons. According to *Mondo Economico*, approximately 30,000 employees are now working in defence production. Moreover, while total employment in the electronics sector decreased from 115,000 in 1977 to 113,000 in 1981 (with peaks of 120,000 in 1978 and 1980), employment in the main defence electronic companies remained stable or increased. The data from *Mondo*

*The electronics sector as defined by ANIE (Associazione Nazionale Industrie Elettroniche) includes consumer electronics (colour television sets, washing machines and so on), professional electronics, computers, information and telecommunication systems.

Economico cannot be compared with Battistelli's estimates of defence-electronics employment for 1977 as presented in Table 7.6 since different criteria have been used by these two sources in preparing their estimates. The latest data for 1981 put military electronics sales at about 750-800 billion lire, that is, between 12.2 and 13 per cent of total sales in the electronic sector, which totalled 6,146 billion lire. The importance of military production has therefore increased for electronics, both in terms of sales and employment, especially since 1976/7. Military-related production can be expected to grow in the future by about 5 per cent a year for the major companies. Finally, while exports as a percentage of total electronics sales oscillated between 33 and 40 per cent between 1977 and 1981, the export percentage of military electronics sales oscillated between 60 and 80 per cent during this period.

The highest return on investment during the 1968-78 period was recorded by the engineering-chemical subsector. The value of the ROI index almost tripled while employment increased rather more slowly (35 per cent) than for the defence sector as a whole (100 per cent). As sales increased tenfold, which is in line with the overall trend, this subsector clearly raised its production and the value-added of its products more than the other defence subsectors.

The dependence of the engineering-chemical subsector on defence contracting is much more difficult to assess accurately, even using the latest data for 1981. This is because of the larger number of companies and the correspondingly greater difficulty in obtaining data on military versus civil production. However, taking into account only engineering (excluding the iron and steel industry and chemicals), the degree of reliance on defence contracting is probably no greater than 5-6 per cent. Considerably lower percentages, perhaps less than 1 per cent, could be applied to chemicals.

Finally, as expected, shipbuilding showed the least impressive performance of the four defence-related subsectors between 1968 and 1978. Sales increased seven-fold but one must take into account the acquisition of Cantieri Navali Riuniti by IRI in 1973. In reality, sales in the shipbuiding sector only doubled between 1974 and 1978, while they increased, for instance, five-fold for engineering during the same period. Employment also increased very slowly and the ROI index has consistently been very low; in some years it has even been negative. The degree of reliance of Italian shipbuilding on defence contracting is clearly shown for 1980 and 1981 in the latest yearly report to the shareholders of the leading shipbuilding group, Fincantieri. After having acquired control (99.9 per cent) of Cantiere Navale Breda in 1979, this state group now accounts for well over 80 per cent of the entire shipbuilding industry in Italy. It owns shares in 24 shipbuilding companies, including Italcantieri and Cantieri Navali Riuniti. Official data released on 25 June 1982 show that military construction increased its share of hours worked from 39.3 per cent in 1980 to 43.6 per cent in 1981. This increased reliance on the military sector has occurred because orders for civil shipbuilding have plummeted. More than 1 million of working hours were lost between 1980 and 1981 and this trend is likely to continue in the foreseeable future.

Moreover, in terms of economic results, it is significant that the only company of the Fincantieri Group that posted a profit in 1981 (after several years of losses) is Cantieri Navali Riuniti, which is heavily involved in military shipbuilding, especially for foreign navies. Cantieri Navale Breda, equally involved in defence production, also improved its performance, reducing previous losses by 30 per cent. Simultaneously, the other civil shipyards, like Italcantiere, recorded additional losses. Finally, in terms of employment, the Fincantieri report disclosed that during the last few years, the Italian shipbuilding-engineering industry has lost more than 9,000 employees from civil production. Nearly 5,000 of these 9,000 have been transferred to military production.

Regional Distribution and Concentration of Employment in the Defence Sector

An interesting aspect, and one which is particularly relevant to the conversion question considered in the United Nations study on the relation between defence production and employment and the economic system as a whole, is the present regional distribution and concentration of employment in the defence sector. As no statistics or data of any kind are available concerning this question for Italy, it has been necessary to extract the data shown in Table 7.11 from a brief survey of over 500 Italian defence companies. It is probable that the margin of error is no more than 10 per cent. The estimates presented here take into account the difficulties, especially for the major industrial groups and companies, of separating the personnel working mainly in civilian production from the personnel employed primarily in military-related production. This task becomes even more complicated when the production of weapons systems is divided among plants located in different regions.

The first consideration is that over 80 per cent of defence industry employment is absorbed by only five of Italy's twenty regions. These are Campania (with 16,000 defence-related employees), Lombardy (15,000), Piedmont (13,000), Liguria (13,000) and Lazio (9,000). Within these five regions, however, the importance of defence industry employment varies dramatically. Although the absolute number of persons employed is large, the percentage of total employment accounted for by defence industries is relatively low in the two most industrialized Italian regions, Lombardy and Piedmont. Only 0.9 and 1.6 per cent, respectively, of the industrial employees in these two regions work in defence-related industries. This increases to 1.1 and 1.8 per cent, respectively, if only maufacturing industry is considered. Even if one takes into account only employment in regional defence-related subsectors of industry (engineering, electronics, chemicals, transport),* the percentage of defence employees increases only to 2.3 per cent for Lombardy and 3.3 per cent for Piedmont.

The predominant subsectors of production in both regions are aircraft (Aeritalia, Fiat Aviation in Piedmont; Agusta, Aermacchi in Lombardy) followed by engineering (Fiat and Lancia military vehicles in Piedmont; Breda, Beretta, Oerlikon guns, light weapons and weapons-systems components in Lombardy). In Lombardy there is also a sizeable group of electronics firms, such as Montedel and Galileo (although Galileo's main plant is in Tuscany). This can be seen from the percentages for production subsectors in the last column of Table 7.11, although it must be emphasized that these are only gross estimates and should not be taken to indicate anything other than rank ordering.

The situation is rather different for the remaining three regions in this group of five. The highly developed Liguria, which along with Piedmont and Lombardy forms the Northern Industrial Triangle, displays the highest degree of dependence on defence-industrial employment. This dependence amounts to 3.4 per cent of regional

***The statistics on regional employment used in this section list employment only according to very broad sectors of manufacturing and thus differ from the breakdowns used in earlier sections. The sectors used here are chemicals, metal products and machinery (engineering), electrical products (including electronics), transportation. Aircraft production and shipbuilding are included in transportation. Naval engines, however, come under machinery. It was possible to distinguish the traditional regional subsectors by using the author's own list of the 50 largest defence corporations with their location and employment figures.**

Table 7.11: Regional Distribution and Concentration of Employment in the Italian Defence Industry, 1980

Region (Towns)	Employment in Defence Sector	As percentage of National Defence Sector	Total Employment in Regional Industry	As percentage of Regional Employment in Industry	As percentage of Regional Employment in Manufacturing Industry	As percentage of Regional Employment in the Main Defence-related Industrial Sectors[a]	Main Sectors of Production[b]
Liguria (Genova-La Spezia)	13,000	15.8	382,000	3.4	10.2	21.0	60% Shipbuilding 25% Electronics 10% Engineering
Campania (Naples)	16,000	19.4	453,000	3.5	5.3	16.3	60% Aircraft 30% Electronics 10% Shipbuilding
Latium (Rome)	9,000	10.9	403,000	2.2	3.7	10.1	80% Electronics 10% Aircraft 5% Chemicals
Friuli-Venezia Giulia (Trieste)	4,000	4.8	155,000	2.6	3.3	8.0	90% Shipbuilding, naval engines
Marche (Ancona)	3,000	3.6	170,000	1.8	2.4	10.8	90% Shipbuilding
Tuscany (Florence, Pisa, Livorno)	4,500	5.5	455,000	1.0	1.2	5.0	80% Electronics
Piedmont (Turin, Novara)	13,000	15.8	820,000	1.6	1.8	3.3	55% Aircraft 40% Engineering-vehicles

Lombardy (Milan, Varese, Brescia)	15,000	18.2	1,580,000	0.9	1.1	2.3	60% Aircraft 30% Engineering 10% Electronics
Veneto (Venice, Treviso)	3,000	3.6	555,000	0.5	0.7	2.1	50% Shipbuilding 40% Engineering
Sicily (Palermo)	1,000	1.2	318,000	0.3	0.6	2.2	95% Shipbuilding
Puglia (Bari, Taranto)	1,000	1.2	300,000	0.3	0.6	2.8	95% Ship maintenance and repair
Total (Nationwide)	82,500	100	6,537,000	1.2	1.7	4.2	

Notes: a. The main defence-related industrial sectors are: engineering (all types of construction), electronics, chemicals, and transport.
b. The sectoral percentages are gross estimates and are only to be used to indicate rank order.
Source: Author's elaboration.

employment in industry but reaches 10.2 per cent of regional employment in manufacturing industry and 21 per cent – that is, over one-fifth – of regional employment in the four main defence-related industrial subsectors. The dominant production subsector in Liguria, where the civilian harbour of Genova and the military harbour of La Spezia are located, is of course shipbuilding (Cantieri Navali Riuniti, Cantiere Navale Muggiano). This is followed by electronics (Elsag, Marconi) and by engineering (Oto-Melara).

Just after Liguria in terms of dependence on defence production comes Campania, a developing region of southern Italy, where a number of military and civilian industries were established in the late 1960s (Alfa Romeo for aircraft and aircraft engines, Selenia for electronics) in addition to the more traditional, but less important, shipbuilding industry. The approximately 16,000 persons employed in the defence sector (19.4 per cent of the national total) represent 3.5 per cent of Campania's employment in industry, 5.3 per cent of its manufacturing employment, and 16.3 per cent of employment in the main defence-related industrial subsectors.

These three employment ratios decrease somewhat for Latium and the Rome industrial area (22 per cent of regional industrial employment, 3.7 per cent of manufacturing employment, and 10.1 per cent of employment in the main defence-related subsectors). Here the large majority of defence output is produced by several medium-sized firms such as Contraves, Elettronica, Montedel, Selenia, Sistel. The aircraft industry is represented by Elicotteri Meridionali.

Between Liguria, Campania and Latium, which can be termed regions with a *high* dependence on defence-related employment (from 10 to 20 per cent of employment in the general engineering, chemical, electronic and transport subsectors), and Piedmont and Lombardy which can be characterized as regions with a relatively *low* level of dependence on defence-industry employment (3.3 per cent or less), there is another group of three regions that can be said to have a *medium* dependence on defence-related employment (from 5 to 10 per cent of employment in the four key subsectors). These are Friuli-Venezia Giulia, Marche, and Tuscany. In two of these, Friuli and Marche, the shipbuilding or naval industry is the predominant subsector (Italcantieri and Grandi Motori Trieste in Friuli; Cantieri Navali Riuniti in Marche). In Tuscany the major defence-related subsectors are, first of all, electronics (Galileo, Whitehead-Motofides) and, then, chemicals (Sipe-Nobel). For certain indicators, however, such as defence employment as a percentage of regional industrial employment (1.0 per cent) and as a percentage of regional manufacturing industry (1.2 per cent), Tuscany is on the low side. The one exception is its defence employment as a percentage of key industrial subsectors (5 per cent).

Finally, at the very end of the scale come Veneto, Sicily and Puglia. These are regions with a *rather low* dependence on defence-industrial employment. This ranges from 0.3 to 0.5 per cent of regional industrial employment, from 0.6 to 0.7 per cent of regional manufacturing employment, and from 2.1 to 2.8 per cent of employment in the four key industrial subsectors. It must be pointed out, however, that only the Veneto region has a sizeable employment in defence production (about 3,000). These employees are found in shipbuilding (Cantieri Navali Breda in Veneto) and in chemical-engineering (Simmel in Castelfranco Veneto). In Sicily there are an estimated 1,000 military employees of Navaltecnica and Cantieri Navali Riuniti in Palermo. As for the Puglia region, there are also some 1,000 defence-related workers employed in various capacities, mainly ship maintenance and repairs. The navy has facilities at Taranto and some of the dockyard facilities are under the direct control of the Ministry of Defence.

Exports by Italy's Defence Industry

On 15 May 1981, the national financial daily *Il Sole 24 Ore* of Milan, announced on its front page that the president of Agusta, one of the leading Italian aerospace firms, had signed a preliminary agreement in Peking for the transfer of know-how to the People's Republic of China. The agreement, the value of which has not been disclosed, provides for the construction of a plant to manufacture helicopters and parts of aircraft engines in China. Also during the first half of 1981, a second agreement involving Brazil which had been initialled in March 1980 was confirmed. This deal will enable the co-production of the new AMX aircraft by Aermacchi of Italy and Embraer of Brazil. The AMX is a tactical support fighter which will be bought in the 1980s by both the Italian and the Brazilian air forces (200 and 100 units respectively).

These are two significant examples of the growing vitality of the Italian defence industry and its presence in the sharply expanding international market of arms exports. The 1981 volume of *World Armaments and Disarmament*, the yearbook edited by the Stockholm International Peace Research Institute (SIPRI), confirmed that since the late 1970s Italy has been the fourth largest exporter of major conventional weapon systems (warships, tanks, missiles, guns, aircraft and helicopters). Only the United States, the Soviet Union and France have sold more than Italy. The Italian share of the international arms market is now estimated at 4 per cent. This can be compared with 3.7 per cent for Great Britain, 10.8 per cent for France, 27.4 per cent for the Soviet Union and 43.3 per cent for the United States.

SIPRI figures, however, are for major weapons only and are compiled mainly on the basis of unclassified data or open sources. Well-informed Italian defence sources say that data for UK exports are underestimated, and that Italy ranks behind Britain in overall defence exports. This view seems to have been confirmed by the *Financial Times* of London according to which British arms exports provide jobs for 14,000 persons and account for 2.5 per cent of all UK exports.[11]

As far as the total value of Italian arms exports during the last five years are concerned, the conservative estimates presented in Table 7.12

Table 7.12: Italian Military Exports 1976-80

Year	Exports (billion lire)	As percentage of sales	As percentage of total exports
1976	660	44	2.1
1977	850	46	2.1
1978	1,200	48	2.5
1979	1,600	50	2.6
1980	1,900-2,100	50-52	2.8-3.1

Source: Author's estimates and elaboration from official data from the Italian Central Office of Statistics.

indicate that in 1980 arms exports were somewhere between 1,900 and 2,100 billion lire ($1.9 to 2.1 billion) and accounted for over 50 per cent of total military industry sales. They also represented between 2.8 and 3.1 per cent of total Italian exports for the same year. This is a percentage which has increased considerably since 1976, when it was about 2.1 per cent. Higher estimates for 'export orders' were released during an interview with the Minister of Defence in January 1982: 350 billion lire in 1972, 1,800 in 1977 and 2,500 billion lire in 1980, representing 60 per cent of defence production and almost 4 per cent of all Italian exports.

Defence industry exports accounted in 1980 for over 4 per cent of the total exports of complex (as opposed to semi-processed) products by Italian industry. Furthermore, they represented over 8 per cent of Italian engineering exports. The Italian Central Office of Statistics

(ISTAT), however, includes in this latter category not only the normal engineering and metal industries and the manufacture of cars, ships, and aircraft, but also the electromechanical, electronic and telecommunication industries. In other words, almost every main subsector of production in which the defence industry operates is included in this category. The one exception is chemicals. To give a more precise idea of the growing importance of Italian military exports compared with other leading civilian industrial subsectors, it should be recalled that in 1980 Italy exported over 5,000 billion lire worth of textile products, 3,000 billion lire worth of automobiles, and 1,200 billion lire worth of

Figure 7.1: Compared Growth of Exports in the Military and Civilian Sectors (1976-80)

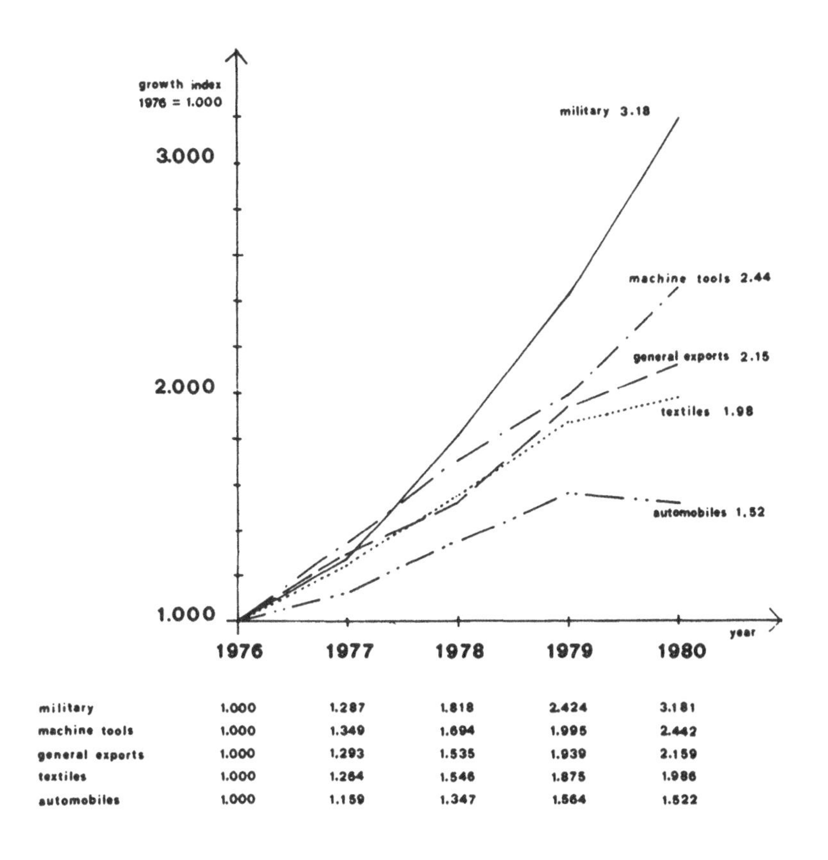

	1976	1977	1978	1979	1980
military	1.000	1.287	1.818	2.424	3.181
machine tools	1.000	1.349	1.694	1.995	2.442
general exports	1.000	1.293	1.535	1.939	2.159
textiles	1.000	1.264	1.546	1.875	1.986
automobiles	1.000	1.159	1.347	1.564	1.522

Source: Author's estimates and elaboration from official ISTAT data

machine-tools. The latter subsector is one in which Italy ranks fourth world-wide in terms of exports. The dynamism of Italian defence exports can easily be seen in Table 7.12. Military-sector exports grew from 660 billion lire in 1976 to 1,900-1,200 billion in 1980. This is a more than three-fold increase and an average annual rate of growth of over 30 per cent. Moreover, as Figure 7.1 clearly shows, the growth index of military exports for the same five-year period (3.18) is considerably higher than the corresponding growth index of total Italian exports (2.15). It is also higher than that of the most dynamic and export-oriented civilian subsectors, such as machine-tools (2.44) or textiles (1.98). In the automobile industry, the export growth index was rather low (1.52). Preliminary data for 1981 confirm this upward trend of defence exports compared with civilian exports.

Table 7.13 provides a more detailed analysis of the trends in defence industry exports by main subsectors of production. This table shows the evolution of export sales as a percentage of total sales for a selected

Table 7.13: Exports as Percentage of Total Sales for 16 Leading Italian Defence Firms, 1972-78

Sector	1972	1973	1974	1975	1976	1977	1978
Aircraft	50.3	59.4	67.9	67.1	70.7	60.1	58.2
Electronics	45.2	44.2	44.3	41.6	47.6	47.5	46.0
Engineering-Chemical	63.0	59.6	59.0	67.5	61.4	60.9	61.3
Shipbuilding	28.4	19.8	8.8	12.4	13.5	25.9	37.1
Total	40.8	36.0	35.4	39.6	44.0	44.3	48.9

Source: F. Battistelli, *Armi: Nuovo modello di sviluppo? L'industria militare in Italia*, Torino: Einaudi, 1980, p. 283.

group of sixteen leading defence companies between 1972 and 1978. Included here are the aircraft producers Aeritalia, Aermacchi, Agusta and Elicotteri Meridionali; the electronics firms Elettronica, Elettronica S. Giorgio, Montedel and Selenia; the engineering-chemical companies Breda Meccanica Bresciana, Oerlikon Italiana, Oto-Melara and Simmel; and the shipbuilders, Cantieri Navali Riuniti, Grandi Motori Trieste, Italcantieri and Navaltecnica.

Table 7.13 shows, first of all, that in the seven years between 1972 and 1978, the average of exports as a percentage of total sales increased from 40.8 per cent to almost 49 per cent. This occurred, however, in two separate stages. Between 1972 and 1974/5, the overall export percentage actually decreased to 35 per cent of sales. This was due mainly to the effects of the energy price increases of 1973/4 and to the subsequent economic decline which was felt especially severely by the weaker subsectors of production such as shipbuilding. Since 1975/6, however, there has been a remarkable recovery, led by the aircraft and electronics subsectors. After the peak of 1976 (when the growth rate of the whole defence-industrial sector was 45 per cent), the defence-sector growth rate stabilized in the late 1970s. The positive effects of this trend even extended to the shipbuilding subsector, aided by the Promotional Law for the Navy passed in 1975. This law authorized the state to spend over 1,000 billion lire on a ten-year (1975-1985) military shipbuilding plan. Subsequently, however, the heads of the Italian military services complained that the law had failed to take into account the high rates of inflation in Italy. Therefore, the implementation of the Navy plan (and of similar plans for the Army and the Air Force) had to be delayed until 1990 for lack of funds. Additional funds are now being sought through the ordinary defence budget, and for a new ten-year plan (1982-91).

The relative decrease for 1977-8 compared to 1974-6 in the percentage of the output of the aircraft subsector exported can be ascribed to the large financial requirements of the development of projects such a as the MRCA-Tornado which are mainly oriented towards domestic military use and are therefore not yet ready to enter the export market. The fact that nearly 50 per cent of the total output of the Italian defence sector was exported in 1978 and that an even greater percentage was exported in 1979-80 can be seen as a confirmation of the competitiveness of Italian defence products in the international market. The relatively low prices and high quality of Italian weapons make them rather attractive to foreign countries with limited financial means.

There are analysts, for example Thomas Ohlson of SIPRI, who believe that the boom in Italian defence exports is not entirely due to the quality of the weaponry, rather that '. . . the export surveillance scheme enables firms to export to virtually any country in the world'.[12] It would be more correct to say that the Italian government has the legal means to control arms exports (for instance by withholding the required export licences) but that it has sometimes chosen not to enforce it. Complete data on Italian arms sales have been collected by the Ministries of Foreign Trade and Industry, particularly in the last two years, but these remain classified.

Two other factors, however, remain to be considered in the assessment of the international competitiveness of the Italian defence industry. The first is that the primary destination of Italian arms sales are the developing countries, largely those countries in Africa, the Middle East, and Latin America where Italy has traditional or historical trade ties, but also to some areas in Asia. According to SIPRI computer data, 76 per cent of Italian exports of major weapons go to less developed countries, primarily Libya, Egypt, South Africa, Peru, Argentina and Brazil.[13] Of the total value of Italian exports, in four categories of major weapons, aircraft account for 36 per cent, missiles for 34 per cent, armoured vehicles for 17 per cent and ships for 13 per cent. One of the largest and most well known orders received by the Italian arms industry was finalized at the end of 1980. It involves the purchase by Iraq of four Lupo-class frigates, six corvettes, an 8,700 ton replenishment tanker, a 6,000 ton floating dock, a contract for the construction of an entire naval base, and a training package.[14] The first available data on exports for 1981 are primarily for military aircraft sales. Table 7.14 shows the major contracts recorded, worth almost 580 billion lire.

Table 7.14: Major Exports of Military Aircraft and Helicopters, 1981

Type	Number of units	Value (billion lire)	Purchaser
Helicopters (under two tons)	18	19.3	FRG, UK, Spain, Libya, USA, Mexico, Venezuela
Helicopters (over two tons)	73	332.8	Egypt, Iran, USA, Spain, Austria, Greece, France
Aircraft (under two tons)	18	2.4	Belgium, FRG, Denmark, Burundi, USA, Venezuela, Argentina, Australia
Aircraft (2-15 tons)	31	64.2	France, Morocco, Peru, Argentina
Aircraft (over 15 tons)	36	160.5	Libya, Zaire, USA, Argentina
Totals	176	579.2	

Source: *Mondo Economico* no. 23 (16 June 1982): 51.

The fact that Italian arms exports are generally not directed towards other NATO countries can be explained by the difficulties encountered in attempting to penetrate a highly competive market, which is already full of highly sophisticated products from the more advanced defence industries of France, West Germany and the UK.

The second factor that must be considered when assessing the international competitiveness of Italian arms is that the Italian defence industry still depends to a very large degree on foreign licences, or on the export of foreign parts of weapon components, which are then used to assemble the finished products. In 1979, according to the 1980 SIPRI Yearbook, Italian defence produce held about twenty production licences for major weapons systems.[15] Italy thus ranks first among those countries dependent on licence production, with Japan a close second. Recent estimates place the degree of dependence on foreign technology for military aircraft production (particularly aero-engines) at 25 per cent of the value of output. This kind of dependence may in some cases act as a constraint on the Italian policy of exporting armaments to third-world markets. Often these markets are quite promising, but their exploitation involves a high degree of political and economic risk, including competition and conflicts of interest with other European arms exporters and, of course, the United States. The sale of Italian warships to Iraq, which was partially blocked by the US, the supplier of part of the engines used in these ships, is but one recent example. On the other hand, it is true that licensing agreements sometimes allow Italy to sell arms to countries to which the licence-holder cannot or does not want to sell directly, often for some political reason.

Finally, this degree of dependence on foreign weapons technology is reflected in the Italian arms imports. Informed sources in the Ministry of Defence estimate that in 1977 the value of these imports was about 730 billion lire, that is, not much less than the value of arms exports for 1977, about 800 billion lire.

Since then, however, Italy's weapons trade balance has improved considerably. According to a public statement by the Minister of Defence, the value of arms imports during 1980 was less than 1 per cent of total Italian imports (85,390 billion lire) that is, no more than 850 billion lire. This would mean that since 1977/8, arms imports have increased only moderately in nominal terms and even decreased in real terms in view of inflation. Arms exports, however, have grown considerably. This means that the trend towards a gradual emancipation of the Italian defence industry from foreign technology has been strengthened.

Promotion and Control of the Defence Industry: The R&D Conversion Question

Italian political and social forces have only recently become aware of the fast evolution and growth of the defence sector and of all its attendant implications, for example, its economic effects. A debate has begun on how to exert more control over the development of this sector, especially military exports. A number of legislative bills have been presented, especially by the Socialist and the Communist Parties, which are designed to give Parliament real control over decisions on major defence export contracts, particularly when politically sensitive areas or countries are involved. As things now stand, Parliament has not played an important role where the Italian government's military export policy is concerned. Furthermore, this policy has been more *ad hoc* than coherent and well-formulated. Such an approach has been encouraged by the fragmentation of the bureaucratic-administrative decision-making processes. The Ministries of Defence, Foreign Affairs, Foreign Trade and Industry, can all have a say and have sometimes failed to reach agreement. At the same time, this question has become part of a general need for a rationalization of and improvement in the complex relations between Parliament, the government, and the defence sector.

In the past, Italian political authorities have tended to ignore, officially at least, the reality and problems of the defence sector as well as its role in the economic system. Quite recently, responding in part to pressure from the defence sector for more financial and political support from the government, and in part to pressure from some political parties (mainly those of the centre-left) for more control over defence developments, the Minister of Defence has inaugurated a new policy of official support for Italy's military exports. This export policy, will seek to support the priorities of Italian defence, foreign, and economic policy. Its guiding principles will be defined and controlled both by the government and by the Parliament.

The lack of a unified approach to defence contracting and inadequate funding for military-related R&D have caused those funds allocated to defence in the Italian budget to be used less efficiently than they could be. The lack of a unified approach to defence contracting primarily affects the domestic market but also extends to the international one. In practice it means that the different branches and services of the Armed Forces and the Ministry of Defence offer separate bids and make procurement decisions independently of each other. The defence industry suffers from this lack of coordination. Other Western countries have agencies to coordinate weapons procurement. The post of National Director for Armaments has only recently been created within the Italian Ministry of Defence, and it still has very little to say about the formulation of an efficient procurement policy.

The failure of the State to allocate adequate resources to military research and development (R&D) has been stressed by several managers of the most advanced Italian defence industries, aerospace and electronics. They claim that although the more dynamic and economically viable defence companies have up to now managed to maintain a minimal level of R&D primarily with their own resources, the situation will have to change in the future. The R&D problem is partly due to the skyrocketing costs of personnel and materials not directly engaged in production. It is also due to strong international competition, both in terms of increased technological sophistication of weapons and in terms of economies of scale that can be exploited by larger arms producers.

To demonstrate the extent of the military R&D problem in Italy (which the OECD has shown to be a problem affecting all Italian industry not only the defence sector), the official defence budget allocated 75.4 billion lire to military R&D in 1977. This amount, according to the Defence White Book of that year, represented only 2.1 per cent of the Italian defence budget, while comparable figures for other leading industrialized countries were: United States, 10.5 per cent; Federal Republic of Germany, 4.9 per cent, France, 4 per cent and Britain, 2.5 per cent. The low priority accorded to R&D in the defence budget is also related to the fact that a sizeable amount of R&D is included in ordinary weapons procurement programmes. This preference for unified 'packages' (weapon construction, plus R&D) is furthermore encouraged by the excessive red tape which R&D contracts must force their way through. The most glaring example of this problem is the delay experienced in refinancing the special R&D fund of IMI (Istituto Mobiliare Italiano), an important State banking institution which provides special credit and loans for business enterprises.

Only a few data are available to indicate the magnitude of actual military R&D expenditures in Italy. For military electronics for example, some 8 per cent of sales are devoted to R&D. Some major companies, however, such as Elettronica, Contraves, and Marconi allot as much as 10 per cent. The biggest spender is probably the recently formed Selenia-Elsag group which has assigned 15 per cent of its personnel to a new research unit of 1,200 people, the largest in this sector. The major engineering company Oto-Melara, allocates 10 per cent of the value of its sales to R&D. At present, it is quite difficult to compare these figures with those from the civilian sector. The only indication available, for whatever it is worth, is that the Italian manufacturing subsector devotes between 0.5 and somewhat over 1 per cent of its total sales to R&D. Those estimates vary according to whether or not one includes funds allocated to 'scientific or theoretical research' and to 'applied or operational research' which are distinct from the 'R&D' category proper.

With regard to the issue of converting military industry to civilian production a recent debate among businessmen, politicians, the military, trade unionists and various experts leads one to the conclusion that Italy is only at the beginning of a serious discussion and analysis of this problem.[16] On one side, trade unionists (particularly those on the political left) such as Alberto Tridente, national secretary of FLM (Federazione Lavoratori Metalmeccanici) are trying to reach a compromise between their moral and political goals in favour of conversion and their desire to avoid exacerbating the already difficult Italian economic and social situation. A gradual approach has therefore been advocated, involving the identification of new and promising civilian sectors to which it would be possible to transfer human, financial and technical resources from military production.

On the other side, spokesmen for leading defence sector companies, such as Gustavo Stefanini, President of Oto-Melara, claim that any political decision to start a real conversion process, which would result in the disappearance of the Italian defence sector must cope with the problem of compensating for a decline of about 3 per cent in the national standard of living. The only civilian sector believed to have a technological level and industrial-commercial perspectives comparable to the defence sector is nuclear energy. It, however, is not immune to other social and technical problems.

Conclusions

On the basis of the data provided and the trends analyzed in this chapter, it is possible to make some general statements about the role of the defence sector in the Italian economy:

– Italy's defence industry, in its present form, is still relatively young compared with those of the other main industrialized countries. It really only emerged as an important sector in the second half of the 1970s. Since Italy received very large injections of US aid during the first post-World War II decade, it was possible to delay the reconstruction of a national defence industry until the 1960s. (Some estimates place cumulative US military aid at 13,000-14,000 billion 1981 lire, about $11 billion in 1981 values.)

– In the 1970s, the defence sector proved itself to be one of the most dynamic sectors of production in Italy. In terms of employment, it grew by 100 per cent as against 5 per cent for manufacturing as a whole and 11 per cent for the oil industry. The growth index for sales was 9.7 in the military sector as against 5.7-6.0 for manufacturing as a whole. The growth index of value-added for military producers was 5.85 against 3.68 for industry as a whole. The growth index of gross military investments was 4.44; for industry as a whole, it was 2.87. The productivity of the military sector grew by 40 per cent while the productivity of industry as a whole expanded by only 34 per cent.

– The net result of this sustained development is that in 1980 the defence sector accounted for between 80,000 and 92,000 employees (1.6-1.8 per cent of the personnel in manufacturing) and for over 4,000 billion lire of sales (about 2.5 per cent of total manufacturing sales). Of these sales, at least 2,100 billion lire worth (over 53 per cent) were exported. This sum represented about 3.2 per cent of total industrial exports for 1980. This is of the same order of magnitude as the exports of other major arms producing countries such as the UK and France.*

– The profitability of the Italian defence sector expressed in terms of the return-on-investment index (ROI) was lower than the average value for industry as a whole up to 1974-5. Since 1975, it has been consistently higher. This trend is clearly evident in those individual subsectors, such as military electronics, where detailed anlayses are available.

– Despite the dynamism of the Italian defence sector, particularly when compared with the performance of the economy as a whole, it must be borne in mind that this sector still accounts for a relatively small proportion of total economic activity in Italy. In 1980, total defence-related sales were only 2.5 per cent of total manufacturing sales and the defence sector employed no more than 1.8 per cent of Italy's manufacturing labour-force.

– The degree of concentration in the defence sector is rather high. The fifteen largest companies account for 53 per cent of sales and 68 per cent of all employees. Concentration is particularly marked in certain subsectors such as aerospace where the five largest companies account

*Editors' Note: This would seem to underestimate the economic importance of French arms exports. Chapter 3 indicates that arms exports accounted in 1977 for 4.6 per cent of all French exports and a more important share of French capital goods exports. See pages 91-96.

for over 70 per cent of sales and electronics where the six largest companies account for over 80 per cent of sales. Consequently, there is not much competition among the largest Italian companies for major defence contracts. One partial exception is found in the aircraft subsector where Aermacchi and SIAI-Marchetti compete with each other. There is more also competition among the smaller companies, for instance defence subcontractors, although it is somewhat limited by the high degree of product specialization.

– State intervention in the defence industry is fairly high, both in terms of contracting and ownership. There is almost 50 per cent state ownership; the figure reaches 60 per cent if firms of mixed ownership are counted. The Italian Defence Ministry now purchases between 45 and 50 per cent of annual domestic military production. Both trends are on the rise as state-owned groups such as CNR, Aeritalia, and Selenia-Elsag have increased their participation in other naval, aircraft and electronic companies during 1981/2.

– In terms of military exports, Italy has emerged as one of the major world suppliers of weapons.[17] It ranks fourth or, more probably, fifth (depending on criteria of classification used) among industrialized countries. Over 75 per cent of arms exports go to less developed countries which prefer weapon-systems that are relatively cheap, not overly sophisticated but quite effective.

– Defence exports grew faster in the 1976-80 period than any civilian export: the growth index was 3.2 for military as against at most 2.5 for civilian. The 1980 figure of 2,100-2,500 billion lire worth of exports might even be an underestimation. What is certain is the heavy reliance on exports exhibited by several military subsectors. For aircraft it is about 60 per cent and for electronics, about 70 per cent.

– The Italian defence sector seems to be characterized by higher R&D expenditures than civilian industry. From the little data available, it can be ascertained that defence electronics firms devote an average of 8 per cent of the value of their sales to R&D. For some of the major companies this figure rises to 10-15 per cent. Some leading engineering companies allocate 10 per cent of their sales to R&D. For manufacturing as a whole this figure averages 0.5-1 per cent. Nonetheless, R&D expenditure of the Italian defence sector still seems somewhat low when compared with the R&D outlays of other European and the American defence industries.

– In terms of regional distribution, 80 per cent of defence-sector employment seems to be concentrated in only five of Italy's twenty regions. However, while (with one exception) the more developed regions show a low dependence on defence-related employment, the medium- and less-developed regions display a medium-high dependence on defence employment. This is especially true for certain subsectors such as shipbuilding. In part the regional concentration is due to the special state subsidies given to companies that build new plants in the less-developed regions of southern Italy.

– The degree of reliance of some industrial sectors on defence contracting appears to be fairly high, especially aerospace (over 50 per cent) and shipbuilding (over 43 per cent). For electronics it is considerably less (13 per cent), but it is increasing; the dependence on military-related production is lower for engineering (5-6 per cent) and chemicals (less than 1 per cent). The trend seems to be towards an increase in the Italian economy's reliance on defence contracting. It has been accentuated, at least in the short-medium term, by the sluggish growth of the civilian sector.

– With respect to the conversion question, informed opinion in Italy holds that the conversion of military production facilities to civilian uses is not likely to be given serious consideration in the near future. This assessment arises primarily out of the general economic crisis and the potential social consequences of a quick conversion of the military sector, that is, still higher unemployment. Only significant changes in the current stagnant state of the world economy and a movement toward a more stable politico-strategic system internationally, particularly in those areas which directly affect Italy's national security, could alter or even reverse this conclusion.

Notes

1. Institute of International Affairs, *L'Italia nella Politica Internazionale 1979-80*, Rome: Ed. di Comunità, 1981. See Section IV, 'La Politica strategica e militare', pp. 190-218.

2. F. Battistelli, *Armi: Nuovo modello di sviluppo? L'industria militare in Italia*, Torino: Einaudi, 1980, p. 412.

3. For instance, see S. Silvestri and M. Cremasco, *Il fianco sud della NATO*, Milano: Feltrinelli, 1980, p. 206, and *La politica di sicurezza dell'Italia. Problemi e Prospettive*, Doc. no. 51 bis. Rome: Dipartimento Affari Internazionali, Documentazione per le commissioni parlamentari, Servizio Studi, Camera dei Deputati, February 1980, p. 322.

4. They are ISTRID (Istituto Studi e Ricerche sulla Difesa) in Rome and CESDI (Centro Studi e Documentazione Internazionali) in Turin. Moreover, in two universities, Genoa and Rome, special seminars or courses in strategy or defence problems have been started.

5. This is clearly shown in a recent useful essay, P.V. and V.I., 'Il campo di

studio della politica militare e il suo sviluppo in Italia', *Politica militare* no. 8 (June 1981): 25-34.

6. For this particular estimate, data provided by Battistelli, *Armi: Nuovo modello di sviluppo?* have been used.

7. The Federation of Metalmechanic Workers' (FLM) trade union working group on military industry and Gianluca Devoto estimate the total net defence sales in 1979 to be 2,000-2,500 billion lire with 65-75,000 employees. See Institute of International Affairs, *La Politica Strategica e Militare*, p. 198.

8. G. Piovano, 'La promozione dell'industria bellica nazionale da parte della difesa', *Città e Regione* no. 4 (August 1981): 57.

9. S. Rolfo, 'Alcuni aspetti strutturali dell'industria italiana dell'automazione e strutturazione elettronica', *Bollettino Ceris* no. 8 (1982).

10. 'Business delle armi: Esplosivo (Inchiesta)', *Mondo Economico* no. 23 (16 June 1982): 51.

11. *Financial Times*, 22 June 1981.

12. Thomas Ohlson, 'The Trade in Major Conventional Weapons', p. 74, in *The Arms Race and Arms Control*, Stockholm International Peace Research Institute, London: Taylor & Francis, 1982.

13. Stockholm International Peace Research Institute, *World Armament and Disarmament, SIPRI Yearbook 1981*, London: Taylor & Francis, 1981.

14. For more technical details, see R. Dicker, 'Shake up in the Italian Naval Industry', *International Defense Review* no. 4 (1982): 437-48.

15. Stockholm International Peace Research Institute, *World Armament and Disarmament, SIPRI Yearbook 1980*, London: Taylor & Francis, 1980.

16. See 'Difesa e Industria in Italia' (Special Survey), *Città e Regione* no. 7 (August 1981): 5-136.

17. It is surprising, in this context, to see how the recent, supposedly comprehensive, book by Andrew Pierre, *The Global Politics of Arms Sales*, Princeton, NJ: Princeton University Press, 1982, has almost entirely neglected Italy, both as a weapons producer and as an arms supplier.

8 CHINA

Sydney Jammes

Over the decade of the 1970s, the defence industry has changed from the premier claimant on China's capital and labour resources to a lower priority sector that must bow to Beijing's overriding goal to build China into a major economic power by the end of the century. Moreover, the defence industry has been charged with unprecedented research and development and production responsibilities in support of the civilian economy. This does not mean that the Chinese have abandoned the goal of modernizing their armed forces, but rather reflects the present leaders' recognition that they must correct fundamental weaknesses in the pattern and rate of national economic development before they can undertake any dramatic upgrading of defence capabilities.

China is a large, slowly developing country, with at least three-quarters of its labour force engaged in agriculture and with a low level of output per capita in industry. Nevertheless, it has advanced further towards military self-sufficiency than any other Third-World state. In terms of output, the Chinese defence industry remains the third largest in the world – exceeded only by those of the superpowers, the USSR and USA – producing a wide range of weapon systems, including nuclear weapons and delivery systems (long-range missiles, jet aircraft, and submarines). The sophistication of the military equipment produced and the technology used in its production, however, lag far behind that of the industrially-developed countries. Almost all of the output consists of copies or modifications of Soviet weapon designs of the 1950s, and a number of constraints – political and economic – prevent any major remedies to the industry's backwardness in the near future.

Problems of Assessment

The security precautions surrounding the Chinese arms industry are effective and have long hindered accurate assessment. The scope and the organization of the industry have remained largely a mystery to the Western world, while information on the location of factories, their products, and their productivity has been scant and difficult to analyze. The whole problem of assessment has been further complicated by

China's periodic political upheavals – the Great Leap Forward, the Cultural Revolution, and the activities of the 'Gang of Four'. The presentation in this chapter, therefore, must provide a description of an industry that only can be seen from the West as through a glass darkly. The lack of firm data prevents an exposition as thorough as those in the other chapters and permits only the barest outline of the character of the Chinese defence industry.

The Setting

China has a long history of arms production, but it was not until the Communist Party came to power in 1949 that attempts were made to develop a national defence industry. Assistance on a broad scale was provided by the USSR in the 1950s, when a number of large Soviet-designed factories were built to make duplicates of then-current Soviet weapons. By the late 1950s, these plants were producing a wide variety of military equipment, including jet aircraft. The aid, which left a characteristic and lasting imprint on Chinese design and production practices, ended abruptly in 1960, and the Chinese were left to make the best they could of the situation.

The Soviet departure, together with the Great Leap Forward, virtually stopped arms production in the early 1960s; by the middle of that decade, however, output of all types of arms had reached new peaks. In the mid-1960s, the Chinese also launched an ambitious construction programme in the military machine-building industry. Under the general slogan of 'war preparation', the PRC constructed hundreds – possibly thousands – of small, medium, and large-scale industrial projects in every region of the country, including the remote interior. The size of that effort apparently caused severe dislocations in the economy.

In 1966, just when the Chinese armament industry seemed to have recovered completely from the difficulties attendant with the Great Leap, Mao Zedong launched his Great Proleterian Cultural Revolution. This was not basically economic in nature, as the Great Leap had been; nevertheless, it affected the Chinese defence industry in a variety of ways. Although the central authorities sought to insulate the industry from the troubles of the Cultural Revolution, political activity in the factories frequently caused disorders. Disruptions in the transportation and communications systems created bottlenecks in the delivery of raw materials, parts and sub-assemblies. During 1967, the first full year of the Cultural Revolution, the value of new military equipment produced fell by about 20 per cent.[1]

During the Cultural Revolution, the disruptions in military production were shorter and less severe than those of the Great Leap. By late 1968 there were signs that the worst effects on the weapons industries were over, and heightened tensions with the Soviet Union then began to spur another period of growth. Overall, growth in annual defence production for the period 1965-71 averaged 10 per cent.[2]

In 1972, defence output and the construction of new production facilities were again cut severely. Hardest hit was the aircraft industry, where output dropped about 70 per cent, but production of naval ships and land arms declined also.[3] Three factors apparently were responsible: the government's new emphasis on agriculture; the military's reduced influence in policymaking as the shattered party and government apparatus recocovered from the Cultural Revolution and the fall of Lin Biao*; and a realization by the military that continued large-scale output of older weapons was taking resources away from its long-term efforts at defence modernization. In early 1975, however, defence began to return to favour. As part of a major reassessment of economic policy, Premier Zhou Enlai presented to the Fourth National People's Congress a broad outline for revitalizing the economy and raising the level of technology in China. In what is now billed the 'four modernizations', Zhou's long-range economic plan called for the modernization of agriculture, industry, science and technology, and national defence with the aim of achieving developed-nation economic status for China by the year 2000.[4] Within this framework, the Military Commission of the party Central Committee proposed a comprehensive plan in mid-1975 for modernizing China's military forces and defence industries.[5] In keeping with their opposition to the general modernization programme, leftists in the leadership attacked the military plan for disregarding 'revolutionary' principles.

By early 1977 the leftists had been defeated, Hua Guofeng and his moderate allies were in control, and the military planners were moving quickly to modernize the military machine-building industry. In the brief period between mid-1975 and the end of 1977, Beijing started to expand a large number of military industrial facilities, made sweeping changes in the miltary and scientific/technological institutional

* Before his fall in 1971, Lin Biao was Defence Minister and Mao's constitutionally designated successor. The growth in defence production between 1965 and 1971 coincided with his rise in power.

structures, resumed testing of a variety of weapon systems (after a hiatus in some cases of as much as six years), and launched a massive study of foreign military technology and equipment.

The whole period was one of strong debate, however, over the appropriate level of defence spending, priorities within the military, and the role of defence in deciding what science and technology to acquire. The leaders of the People's Liberation Army (PLA), led by Defence Minister Ye Jianying, wanted larger allocations of resources for modernizing the armed forces. These demands apparently conflicted with the development plans for the civilian sector, which were supported by a large part of the party leadership. The civilians did not deny the importance of a defence build-up but felt that China could best achieve it by first encouraging growth in non-defence investment. Moreover, because overall economic planning was inadequate, the leaders did not immediately understand the difficulties and costs of military modernization. After Vice Premier Deng Xiaoping's return to power in August 1977, a compromise plan apparently was reached. The extent of the compromises and their impact on China's military machine-building industries can only be inferred. It is likely, however, that the military modernization plan as it was envisaged in 1981 is considerably less ambitious than the one the military commission originally proposed in mid-1975 and that military production and the development of the military-industrial sector of the Chinese economy now have a relatively low priority in recognition of competing economic goals. According to present policy, a full-blown effort to modernize the military must await further development of agriculture and light industry and the development of greater skills in machine-building, metallurgy, electronics, and chemistry.

Even though definitive information on the compromises is lacking, a scaling back became apparent in February 1978, when Premier Hua Guofeng in his report to the Fifth National People's Congress called for a return to Zhou's 'four modernizations'.[6] Subsequently, Beijing selectively curtailed the expansion that had begun in military-related industries, established stronger control over the military-industrial bureaucracy, and increased the transfer of existing weapons manufacturing capacity to non-military production.

Overall, defence production increased 1 or 2 per cent per year between 1972 and 1979, but production in 1979 was still below the 1971 peak.[7]

The Defence Industry and the Economy

Defence absorbs a substantial portion of China's economic resources, particularly its output of high-technology machinery. Despite its relative low priority, more than 10 per cent of the PRC's industrial output is believed to be in the form of military goods. (It should be pointed out that data are very rough, both on overall industrial output and on defence industry production, so that the ratio of the two is subject to rather wide margins of error.[8]) The military's share of production, however, is currently much smaller than it was in the past, particularly during 1965-71. Figure 8.1 shows how general industrial production has outstripped defence procurement since that period. A comparison of their respective trends shows that defence procurement growth conformed closely to the growth of industrial output through 1971. Since 1972, however, industrial output has continued its upward trend, while defence procurement has increased only slightly.

No estimates are available on the size of the defence industry's labour force. If employment is assumed to be proportional to output, and if defence output is assumed to be roughly 10 per cent of Chinese industrial production, the defence industry can be expected to include about one-tenth of China's industrial workers, or 4 million people.[9] The defence industry's production of military equipment, however, does not fully utilize its productive capability. Much of this excess capacity is devoted to producing civilian goods. The magnitude of this civilian production is not known with certainty, but some sources claim it to be as much as 30 per cent of the total output of the industry.[10] Using this figure to obtain total output – and assuming employment proportional to output – the number of defence industry workers could be almost 6 million. A lower figure of 2 to 3 million workers can be obtained by using the labour productivity figures estimated by Rawski for industrial-worker productivity in Shanghai and Liaoning.[11] These estimates are of necessity very rough and differ from each other by sizeable margins. They range from one-seventh to one-twentieth of the total industrial labour force.

Similarly, no estimates are available on Chinese capital investment in the defence industry. It is clear, however, that military production consumes many of those resources needed for sustaining economic growth. A large share of the capacity to produce machinery and equipment is diverted away from civilian programmes. Furthermore, much of China's industrial sector devoted to the production of military equipment represents the nation's most modern production capacity.

Figure 8.1: China: Defence Procurement and Industrial Production (1965 = 100)

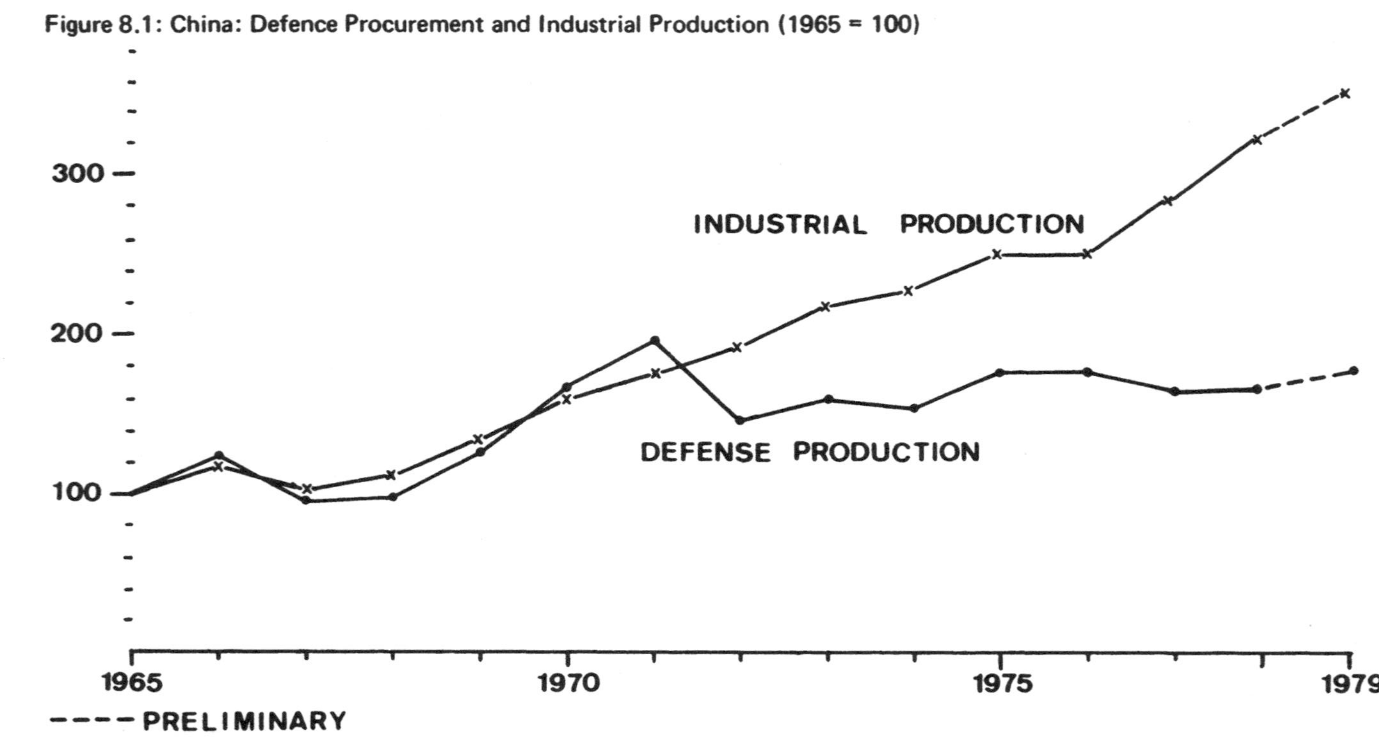

Source: Central Intelligence Agency, *Chinese Defence Spending, 1965-79*, SR 80-10091, Washington, DC: CIA, 1980, p. 6.

More important from the standpoint of economic growth, however, is the fact that the defence effort preempts a large share of the finest scientific, engineering, and managerial talents of the economy – assets that might otherwise improve productivity in the civilian sector. Indeed, American government sources believe that the proportion of China's advanced industrial sector committed to defence production is 'far larger . . . than is the case in the US or the USSR'.[12]

Economic Constraints on the Defence Industry

The relatively low priority currently being given to military modernization reflects the present Chinese leaders' recognition that they must correct fundamental weaknesses in the pattern and rate of economic development before they can undertake any dramatic upgrading of defence capabilities. The 6 per cent average annual rate of growth of the Chinese gross national product between 1957 and 1979 has masked a host of economic problems that constitute an immense impediment to the upgrading of the Chinese defence industry. The impetus for growth has come primarily from industry (9 per cent per year), while agricultural growth has lagged behind, growing at about 2 per cent annually during the same period.[13] If population also grew at approximately a 2 per cent per-year-rate between 1957 and 1979 – as is generally accepted – per capita food availabilities have not changed significantly during the past 21 years.

Industrial performance, though creditable in terms of overall output, also has been flawed. Electric power is in chronically short supply, and the production of such basic commodities as building materials, cement and finished steel is insufficient for the economy's needs and has regularly had to be supplemented by imports in recent years. For years, volume of production has been emphasized over quality; inappropriate pricing policies and inadequate attention to matching the output of both light and heavy industrial products to the needs of consumers has meant that much of what China produces is of low quality or must be stored because it is of little use. In general, China's industry employs levels of technology that, with few exceptions, are from 10 to 30 years behind those in the developed countries.

Moreover, the resource allocation policies adopted in December 1978 (with the establishment of the three-year programme of 'readjustment' by the Third Party Plenum) do not, at least for the near term, support defence modernization. Investment in heavy industry – particularly the iron and steel industry – was cut back, and allocations to agriculture, light industry, and the building materials industry were increased.[14] Additionally, while maintaining their interest in acquiring foreign equipment and technology, and continuing to solicit (and receive) long-term credits to pay for such imports, the leadership has suspended or postponed a number of planned purchases from abroad.

Technology Problems in the Defence Industry

Technological weaknesses in the Chinese defence industry are apparent throughout the entire system, from basic research to the maintenance of finished products. The most critical shortcomings are in the design of technology and manufacturing know-how. These shortcomings are exacerbated by an array of other problems, including lack of qualified personnel, poor production and quality control methods, limited standardization procedures, inadequate machine tools, limited instrumentation and equipment, and shortages of alloy and special steels, certain nonferrous metals, and a variety of plastics and fibres.

Human

A key weakness at all levels of the military industry establishment is the rapidly growing shortage of well-trained scientists, engineers, and technicians. The core of China's technical professionals is composed of a small cadre of foreign-trained scientists and engineers who returned to build a 'new China' in the 1950s. They are now in their seventies and eighties and are having difficulty carrying the burden of research. These are followed, in decreasing levels of ability, by two other distinct groups of professionals: those trained in China prior to 1966 and those trained since 1970. Because many military scientific personnel were chosen for their political reliability rather than their technical qualifications, they are often less competent than their counterparts in non-military institutes and production facilities. Additionally, because of rigid security compartmentation throughout the military industrial sector, researchers and technicians do not have easy access to information from inside their organization or from the outside.

Progress in higher education and in the development of military technology was severely disrupted by the anti-science policies of the Cultural Revolution era and the 'Gang of Four' period. Long-term military research programmes were curtailed, military industrial and technological institutes closed, laboratories ransacked and destroyed, and faculty and scientists harassed. Memories of the anti-science period still influence the attitudes of many military researchers despite Beijing's new support of science and technology. Veteran professors and scientists, now in their early sixties, could be doing productive work, but those who were subjected to humiliating criticism during the Cultural Revolution still have little enthusiasm to begin again. By the same token, middle-level management cadres, who owe their current positions to political performance during the Cultural Revolution rather than to expertise in managing a defence research facility or operating an industrial enterprise, consider the modernization effort a threat to their careers.

The government is fully aware of the crisis it faces in training large numbers of new technicians. Until China has sufficient engineers and technicians trained in modern methods and conversant in modern technology, staffing for most major weapons programmes will include many engineers lacking needed design and production skills. Even with this weak cadre of trained engineers and technicians, China can nevertheless mobilize enough technical talent for selected high-priority projects in the strategic weapons, aircraft, and naval fields. Growing numbers of young students will begin to enter the defence industries in the next few years as the current emphasis on science and technology in the colleges and universities throughout China bears fruit. Additionally, several thousand engineering and technical students are abroad for advanced training.

Research and Development

In establishing China's defence industry during the 1950s, the Soviet Union seems to have purposely withheld the expertise and means to develop new weapon systems. Consequently, the Chinese have had to limit their research and development efforts to a few major projects. These have generally included one or two models in each major type of weapon system (such as aircraft, missiles, and ships). The systems being developed show a substantial technological improvement over those currently being produced, but they still only represent weapons technology levels achieved by the Soviet Union in the early 1960s. Progress in general has been slow, and many projects that were begun in the late 1960s are still under development.

Materials

The capabilities of the industrial infrastructure to support modern weapons development and manufacture are uneven. While China produces most of the materials and basic types of machinery required to support its current weapons production effort, any attempt to improve its military manufacturing processes will require the import of a variety of modern industrial technologies.

One serious deficiency is in the capacity to produce alloy and special steels and some nonferrous metals. China can produce small quantities of superalloys, electrical steels, and stainless steels, but will have to make considerable investments in new capacity before it can substantially increase its high-grade steel production.[15]

A key weakness in the nonferrous metals industry – a weakness that has particularly affected aircraft and missile development – is China's limited ability to produce and fabricate such quality nonferrous metals as aluminium, magnesium, titanium, cobalt, and nickel. The technology currently in use is based on older Soviet equipment and production methods.

Machine Tools

Beijing has developed a substantial machine-tool industry made up of several thousand plants, which range in size from backyard shops to large, modern factories. This industry can meet the country's needs for low- and medium-grade machine tools and can produce some good-quality general purpose machine tools for export. It is far less capable, however, of making the precision tools needed in the production of sophisticated weapons. In numerically controlled machine tools and computer-aided manufacture, China is still in the early stages of development. It can be expected to continue to buy precision machinery and equipment from Japanese and Western suppliers.

Electronics

China has developed a strong and rapidly expanding electronics industry. This industry has held a priority claim on the nation's resources because of the importance of its products to both military development and industrial production. An estimated one-half to three-quarters of its total output is procured by the military, with most of the remainder going to civilian industry. In terms of volume of production, China's electronics industry compares favourably with industries in some of the developed countries of Western Europe, such as the UK, France and West Germany. In technology, however, the industry still lags substantially behind world levels.[16]

Civilian Versus Military Control

Expecting (perhaps with reason) a degree of military objection to their economic policies, the current leadership in Beijing has found it necessary to strengthen control over the defence industry. In a series of moves that underscores the determination to re-establish strong civilian control over the military, the bureaucracy that manages military production has been reorganized. Beginning in late 1977, the ministerial heads of the machine-building industries – the industries responsible for military production – were replaced. Five of the new appointees were civilians supplanting men formerly associated primarily with the PLA. By September 1978, all eight of the ministers of machine-building were civilians (Table 8.1).

In May 1982, the Chinese announced more far-reaching organizational changes in the defence-industry sector (Table 8.2). Under the new arrangement, several defence machine-building ministries have been consolidated into related civilian sector ministries in what appears to be an attempt to enhance civilian control and to foster greater integration of military and civilian production activities. The full details of the new organizational structure are not yet known. The implications of some of the changes will not be completely understood for some time to come.

The organizational changes over the past several years have consolidated China's military-industrial and scientific functions into a more manageable framework. Administrative control bodies that were abolished during the mid-1960s have been re-established under the State Council and given broad authority over the operations and planning of industrial and scientific activities. One military industrial element, the National Defence Industry Office (NDIO), originally appeared in the Chinese media in the mid-1960s, but then disappeared. It re-emerged in 1976 as the principal coordinating unit between the State Council and the military machine-building industries. The NDIO operates in concert with the State Planning Commission on matters of production and allocation of funds – a function which in the early 1970s was handled by the Ministry of National Defence.[17]

The State Scientific and Technological Commission (SSTC) has also reappeared, with seemingly broad powers over the planning, funding and supervising of all scientific and technical work. It has assumed from the military some of the functions which had been performed by the National Defence Science and Technology Commission (NDSTC). The NDSTC is probably still responsible for defence-related scientific and

Table 8.1: Ministries of Machine-Building Prior to May, 1982

Ministry	Responsibility	Date of Initiation	Minister	Appointed
First	Civilian machinery		Zhou Zijian	October 1977
Second	Nuclear	September 1960	Liu Wei	March 1978
Third	Aircraft	1963-65	Lu Dong	March 1978
Fourth	Electronics	May 1963	Qian Min	August 1978
Fifth	Munitions	September 1963	Zhang Zhen	March 1978
Sixth	Shipbuilding	September 1963	Chai Shufan	March 1978
Seventh	Missiles	January 1965	Zheng Tianxiang	February 1978
Eighth	Space	September 1979	Jiao Ruoyu	September 1978

Sources: Ministry, Responsibility, Minister and Date of Ministerial Appointment from: International Institute of Strategic Studies, *Strategic Survey*, London: IISS, 1979, p. 69. Date of Initiation of Ministry's Responsibility: Harlan W. Jencks, *From Muskets to Missiles: Politics and Professionalism in the Chinese Army, 1945-1981*, Boulder, Colo.: Westview Press, 1982, p. 195.

Table 8.2: Chinese Defence Industry Ministries After May, 1982

Ministry	Responsibility	Minister	Comments
Machine-building Industry	Civilian Machinery	Zhou Jiannan	Merges First Ministry of Machine-building, Ministry of Agricultural Machinery, State Bureau of Instruments and Meters Industry, and the National Bureau for Complete Industrial Plants.
Nuclear Industry	Nuclear	Zhang Chen	Formerly Second Ministry of Machine-building
Aviation Industry	Aircraft	Lu Dong	Formerly Third Ministry of Machine-building
Electronics Industry	Electronics	Zhang Ting	Merges Fourth Ministry of Machine building, National Bureau of Radio and Television Industry, and State Administration of Computer Industry
Ordnance	Munitions	Yu Yi	Formerly Fifth Ministry of Machine-building
Space Industry	Missiles and Space	Zhang Jun	Formerly the Seventh and Eighth Ministries of Machine-building

Note: The functions of the former Sixth Ministry of Machine-building (Shipbuilding) have been subsumed by the China State Shipbuilding Corporation (*East Asia Journal*, 26 May 1982, p. 5.).
Source: *Daily Report, People's Republic of China*, Volume 1, Number 87 (Washington, DC: Foreign Broadcast Information Service, 5 May 1982), pp. K1-K5.

technical projects, but direct control over the military research academies and institutes appears to have reverted from the NDSTC to individual ministries.[18]

Beijing's aim in these moves appears to be to erase past special treatment of the military in the allotment of scarce resources, to eliminate overlapping redundancies between civilian and military efforts, and – within the defence industry – to concentrate resources on the most important projects.

Defence Industry Contributions to the Civilian Economy

There is growing evidence that China's military machine-building base is much larger than is needed to support its present relatively modest rate of output. To better use this excess manufacturing capacity, Beijing has adopted a policy under which a sizeable and growing proportion of capacity at military plants is used for non-military production. The harbinger of this policy appears to have been Chairman Hua Guofeng's report to the Congress in February 1978, when he called for greater integration of military and civilian enterprises. He indicated that coordination of production would reduce the need for investment capital, but his speech did not provide any specifics on how this integration was to be accomplished.[19]

Under this policy, an increasing number of military factories are sending specialists to civilian organizations to familiarize themselves with the market for non-military items. The programme has led to new lines of production in military plants, ranging from cameras to mining equipment. The Xiangtan Tank plant, for example, has started to produce sewing machines, electric fans, bulldozers, and tower cranes. And an ordnance factory in Wuxi reportedly has begun manufacturing equipment for use in ear surgery. The Chinese also claim that about 80 per cent of the defence industry enterprises in Liaoning Province have begun to produce daily necessities for local consumption as well as items for export.[20] Overall, production of civilian goods may preempt as much as 30 per cent of the defence industry's output.[21]

Also emphasized under the integration policy are joint defence and civilian research and development projects. The most notable project of this sort to be publicized involves a defence research institute engaged in the development of electronic circuits for the Seventh Ministry of Machinebuilding (the ministry responsible for missile development). As part of its cooperative efforts, the institute reportedly participates in the selection of civilian projects to be undertaken and consequently is able to influence the focus and the progress of civilian research and development in areas vital to military research. The prototype electronic products are first put to use in civilian applications and then are improved over time to meet the generally more stringent military specifications. The publicity about this project has underlined the benefits that have accrued to the defence institute from its cooperative relationship with the civilian sector – an emphasis possibly designed to persuade reluctant elements in the defence establishment to comply with the new policy.[22]

Foreign Technical Assistance

In recent years, China has engaged in a massive effort to modernize its military machine-building industry by studying Western experience in order to achieve higher levels of development. As a result of its heavy reliance since the 1950s on Soviet design and manufacturing practices of that decade, the military machine-building industry now has an enormous need for Western design technology and manufacturing expertise. China's interest in and need for modernizing its defence plants appears to cover the entire technology spectrum, from basic materials to complete weapon systems.

To justify this important policy shift – bitterly attacked by the 'Gang of Four' – the leadership has extensively quoted Mao Zedong, Lenin, and even Karl Marx as favouring the importation of foreign technology. There has been, however, careful recognition and appropriate rewards for domestic inventiveness and publicity of instances in which Chinese workers have improved upon foreign technology. In this way the leadership represents its policy of importing technology as 'making foreign things serve China'.

Beijing has focused much of its efforts on soliciting foreign technical help in areas where the lines between civilian and military technologies are often blurred. Since 1978 the Chinese have opened many defence industries to civilian production and have purposefully been using excess defence industry capacity for civilian purposes. This is different, of course, from the 1971-2 *cutback* in defence output where the resultant excess capacity remained idle. With the 'civilianization', acquisition of technologies with both military and civilian applications has become politically more acceptable within China than purchase of purely military technology. The most notable examples of this have been in the areas of computers, electronics, and aircraft.

Perhaps the foremost indicator of the new policy on receiving foreign technical assistance is a new willingness to send thousands of students to the West to study. In addition, several thousand Chinese technicians and specialists have visited Western factories since 1976, shopping for the latest in industrial technologies. Beijing has also encouraged visits by Western businessmen, who are flocking to China to promote their products and services. These exchanges increased markedly in late 1977, after Deng Xiaoping's return to power, and reached a peak in late 1978. Since early 1979, however, there has been a noticeable decline in the number of delegations travelling to and from China to discuss military equipment, as well as an apparent slowdown in negotiations for foreign military goods. This seems to have resulted from the economic reassessment and consequent readjustment of China's overall modernization plans that occurred in late 1978. In addition to the apparent decline in interest in foreign military goods, China postponed or cancelled several major projects in defence-related industries in 1980 (including the Bao Shan steel production complex) that involved substantial inputs of foreign technology.[23]

Although imports of special high-quality raw materials, machinery, and precision instruments from Japan and the West still play an important role in China's production of its more advanced military equipment, the Chinese appear to be less interested in importing large quantities of finished weapon systems than in obtaining production technology and licensing agreements. With the notable exception of the 1975 purchase of Rolls-Royce Spey engines, no major purchases in the defence area have been concluded; and even in that case, the arrangement was primarily for the purchase of technology and expertise and not of finished items. The purchase of completed weaponry generally has been perceived by the Chinese as unwise because of the possible complications and dependency such purchases might well engender both politically and technologically, especially if they are all from a single supplier as was their experience with the Soviet Union in the 1950s. China also has come to appreciate the difficulty of absorbing advanced foreign technologies into their weak technical base and thus decided that exhaustive study is required before making major purchases.[24] They may also be concerned with the high cost of modern weapon systems and weighing their cost effectiveness, given that no single new foreign or domestically produced conventional weapon system will markedly change the Sino-Soviet military balance or significantly increase China's ability to defeat a Soviet attack.

Arms Exports

China is a relatively minor exporter of arms with less than 3 per cent of the world military export market. Nevetheless, the People's Republic is the fifth largest weapons supplier to the world, ranking behind the USA, USSR, Britain, and France.[25] During the past five years the principal recipients of Chinese arms were Pakistan and a number of African states. The main exports have been aircraft, especially the Shenyang F-6 (the Chinese version of the Soviet MIG-19 fighter) and light coastal-defence craft. Tanks have also been exported, but in minute numbers compared with the exports of the USSR. The recipients have virtually all been 'uncommitted' countries'[26]

It should be noted that China is not a world leader in *sales* of arms, despite its position as fifth largest *supplier*. Over the past 30 years, the Chinese generally have refused to sell military goods. They have, none the less, traded or given arms in an effort to establish or enhance influence in Third-World nations. Beijing objected to selling military equipment because it did not wish to be branded by such countries as an arms vendor. Chinese leaders have probably thought they could encourage better relations through gifts rather than sales.

Arms transfers account for a very small proportion of defence industrial output and are small relative to the nation's total exports and to its heavy industrial exports. Moreover, deliveries have fluctuated widely in response to political upheavals at home and to vacillating attitudes toward their costs and benefits. Overall, arms transfers have declined over the past fifteen years – reflecting the ready availability of better and more modern Western and Soviet weapon systems.* The size of recent arms agreements between China and several other countries suggests a slight reversal in this trend.[27] Although Chinese weapon systems are generally technologically inferior to those of the other major military exporters, Chinese arms – especially infantry weapons do offer some advantages for Third-World countries: they are simple and rugged, and getting arms from China is an alternative to tilting either to the West or to the Soviet Union.

Given the Chinese need and desire for foreign exchange to support

***Part of the decline is also attributable to the schism between China and Vietnam – formerly one of the prime recipients of Chinese arms.**

the economic modernization programme, the temptation to sell arms instead of giving them away must be very strong in Beijing, and a change in policy would be consistent with the new pragmatism now associated with most Chinese economic policies. Indeed, weapons that most likely could have only come from China are openly advertised by international arms dealers, suggesting that a change in policy has already occurred.[28]

Outlook

The Chinese leaders probably will continue in the 1980s to view agricultural development as the foundation for overall economic modernization and growth. They apparently hope that sustained agricultural growth will eventually allow them to pursue import-substitution policies. The Chinese would aim first at decreasing their imports of food products and natural fibres, which now account for over 20 per cent of the import bill. Declining import costs for food and fibres, coupled with aggressive promotion of Chinese exports, should enable them to import more complete plants, equipment, and technology for both civilian and military purposes. Imports of new plant capacity and the expansion of present domestic capacity will in turn begin to diminish the importance of imports in meeting the country's needs for finished and semi-finished industrial materials and for transportation equipment. These three categories now account for over one-third of the import bill.[29]

Fundamental to this policy is an important implicit assumption – that China will not run into balance-of-payments problems. Beijing badly wants to pay off its present foreign debt of some US $3.4 billion.[30] It already can draw, however, on US $20 to $30 billion in firm credits and loans through 1985, and is actively seeking more for the second half of the decade.[31] The Chinese have been adept managers of payments problems and there is no reason why they should not continue to be. The Chinese leadership, however is wary of undue dependence on foreign banks and other creditors in view of China's experience with Western loans in the 1920s and 1930s.

Military modernization under such a scenario is likely to remain slow and gradual, primarily dependent upon China's progress in lifting the levels of technology in its industrial base and in creating conditions for indigenous scientific and technological development. Some improvement in Chinese defence production capabilities can be expected on a selective basis; overall, however, the Chinese defence industry is likely to remain circumscribed by continued deficiencies in technology, personnel, and resources.

Even if China were to acquire foreign defence-related technology, the impact on production would not become apparent until the late 1980s. Time is needed to assimilate new technology into the existing production processes. The Chinese might participate in joint ventures or co-production programmes with other nations, but assimilation would still take time.

In any case, two fundamental imponderables affect all projections of Chinese economic and military development. The first is the ability of the present leaders to consolidate their hold on power and to implement their economic policies while at the same time placating a restless military. Their success in this endeavour is by no means assured, and it is further complicated by the factor of age – the likelihood that most of the present leadership will pass from the scene during the next five to ten years. The second is the government's ability to continue the gradualist approach to Chinese military modernization by linking it to the eventual development of a modern industrial base. A serious deterioration in relations with the Soviet Union or renewed war in Vietnam could lead to increased allocations of resources to the military establishment at the expense of economic modernization.

Notes

1. Central Intelligence Agency, *Chinese Defence Spending 1965-79*, SR-80-10091, Washington, DC: Central Intelligence Agency, 1980, p. 2.
2. Ibid.
3. Ibid.
4. *Daily Report, People's Republic of China*, vol. 1, no. 13, Washington, DC: Foreign Broadcast Information Service, 20 January 1975, pp. D20-D27.
5. *Peking Cieh-Fang-Chun Pao*, 27 August 1977, pp, 1, 4.
6. *Daily Report, People's Republic of China*, vol. 1, no. 39, Washington, DC: Foreign Broadcast Information Service, 27 February 1978, pp. D7-D10.
7. Central Intelligence Agency, *Chinese Defence Spending*, p. 6.
8. Because of a lack of data, all estimates of the defence burden on the Chinese economy are necessarily tenuous, and the one presented here is no exception. The only data generally available are the single-figure entries for 'defence and war preparation' in the Chinese state budgets for 1979, 1980 and 1981 (the first budgets revealed after a hiatus of almost twenty years) and the estimates presented publicly by the United States Central Intelligence Agency. The single figure presented in the Chinese state budget provides little insight into the problem. The figure is low, which suggests that it may include only defence operating expenditures – with military procurement, investment in defence

industries, and research and development subsumed in such other budget accounts as 'capital construction and 'science' in a manner similar to the Soviet practice. Other Chinese economic data seldom include any defence-related statistics. The Central Intelligence Agency data for China's defence expenditures are estimated on the basis of a 'building-block' methodology similar to that which the Agency has used for many years to assess the defence costs of the Soviet Union. It is based on a detailed list of the activities and physical components of the defence programme for each year. This list includes estimates of order of battle, manpower, production of equipment, construction of facilities, and the operating practices of the military forces. These estimates are then converted into monetary estimates. This costing is in yuan terms for some components and for others in US dollar terms. Estimates in dollar terms are converted to yuan using suitable yuan-dollar ratios constructed to reflect differences in the US and Chinese price structures for different goods and services. The published Central Intelligence Agency data are more revealing than the official Chinese data, but they are still rather cryptic and provide only a limited view of the Chinese defence establishment. In the study presented here, the ratio of military production to overall industrial production presented in this chapter was derived by comparing estimated net Chinese military output (15 billion yuan) with the estimated value of overall net industrial output (143 billion yuan) for the year 1979. The estimated net military output was obtained by extrapolating from the figure on page 4 of Central Intelligence Agency, *Chinese Defence Spending*. The estimated net industrial output was obtained by multiplying the official announced gross value of industrial output by an estimated ratio of net value to gross value calculated by Dr Robert Michael Field in an unpublished paper based on data presented in *Jingjiyanjiu* [Economic Research] no. 4 (1975): 51, and *Jingjiyanjiu* [Economic Research] no. 12 (1979): 9. Possible definitional incompatibilities between the CIA's figures and the official Chinese data, as well as a problem of comparing current yuan with 1970 yuan, add to the uncertainty of the estimate. Nevertheless, despite the mental gymnastics used to arrive at it, the estimate appears reasonable in light of other official PRC statistics and the estimate of the percentage of a Chinese gross national product pre-empted by defence (8.5 per cent) presented in US Arms Control and Disarmament Agency, *World Military Expenditures and Arms Transfers 1969-78*, Publication 108, Washington, DC: December 1980.

9. Thomas G. Rawski, *Economic Growth and Employment in China*, New York: Oxford University Press, 1979, p. 163, estimates the number of Chinese industrial workers to be 40 million.

10. *Daily Report, People's Republic of China*, vol. 1, no. 159, Washington, DC: Foreign Broadcast Information Service, 14 August 1980, p. L1.

11. Rawski, *Economic Growth*.

12. Statement by George Bush (then Director of Central Intelligence), 26 May 1976, p. 31, in US Congress, Joint Economic Committee, Subcommittee on Priorities and Economy in Government, Hearings: *Allocation of Resources in the Soviet Union and China, 1976*, Part 2, Washington, DC: US Govt. Printing Office, 1976.

13. Central Intelligence Agency, *China: A Statistical Compendium*, ER 79-10374, Washington, DC: CIA, 1979, p. 3.

14. *Daily Report, People's Republic of China*, vol. 1, no. 248, Washington, DC: Foreign Broadcast Information Service, 26 December 1978, pp. E4-E13.

15. China's drive to lessen dependence on foreign sources of steel is noted in an article on tank production in *Liberation Army Pictorial*, March 1980:

> China's successful development of armour plate and structural steel to replace various lines of chrome-nickel steel from abroad has been an important contribution and has resulted in the granting of first-class awards in national science and technology.

16. 'China's Defence Industries', *Strategic Survey*, London: International Institute for Strategic Studies, 1979, pp. 70-1.
17. Ibid., p. 60.
18. Ibid., p. 69.
19. *Daily Report, People's Republic of China*, vol. 1, no. 39, Foreign Broadcast Service, pp. D7-D10.
20. 'China's Defence Industries', *Strategic Survey*, p. 72.
21. *Daily Report, People's Republic of China*, vol. 1, no. 159, p. L1.
22. *Daily Report, People's Republic of China*, vol. 1, no. 89, Washington, DC: Foreign Broadcast Information Service, 8 May 1978, p. E6. See also, *Daily Report, People's Republic of China*, vol. 1, no. 63, Washington, DC: Foreign Broadcast Information Service, 31 March 1978, p. E11.
23. 'Another Turnaround in China', *Far Eastern Economic Review*, 12 December 1980, pp. 60-2.
24. The Chinese are well aware of the disadvantages of importing manufacturing capabilities without also acquiring design and production technology. The following observation on 2 December 1978 by the leading Chinese science-oriented newspaper, *Guangming Ribao*, may reflect their experience with the Soviets in the 1950s and possibly with the British regarding the Spey transfer:

> However, there are many drawbacks to importing complete sets of equipment, not the least being the great expense incurred, and it is not likely to help raise the levels of domestic research and production. Although in this process some technology may be brought in, such technology is mainly confined to data concerning production capabilities and not related to the technological processes essential to manufacture. This is because the vendors alone have access to basic designs and vital aspects of production technology, whereas the buyers have no way to acquire technological and production know-how beyond certain engineering particulars on the manufacture of required components and accessories for replacement purposes. This is an important reason why for a long time China was unable to manufacture sets of equipment identical to those it had imported in the past.

25. Philip J. Farley, Stephen S. Kaplan and William H. Lewis, *Arms Across the Sea*, Washington, DC: The Brookings Institution, 1978, p. 13.
26. *The Chinese War Machine*, London: Salamander Books, 1979, p. 50.
27. Central Intelligence Agency, *Communist Aid Activities in Non-Communist Less Developed Countries, 1979 and 1954-79*, ER 81-10318U, Washington, DC: CIA, 1980, p. 13.
28. For example, see *International Defense Review* 13 (July 1980): 1172-3.
29. Central Intelligence Agency, *China: International Trade, Second Quarter, 1980*, ER CIT 81-001, Washington, DC: CIA, 1981, p. 8.
30. 'Another Turnaround in China', *Far Eastern Economic Review*, p. 62.
31. Central Intelligence Agency, *China: International Trade, Second Quarter, 1980*, p. 3.

9 ISRAEL

Gerald Steinberg

Introduction

The Israeli defence industry is one of the largest and most sophisticated outside the advanced industrial states and the Soviet bloc. In addition to supplying the relatively extensive requirements of the Israeli Defence Forces (IDF), annual exports in this sector have, according to reports, surpassed $1 billion.[1] Products include small arms, advanced aircraft, tanks, and missile boats, as well as computers, radar, complex electronic equipment, and a variety of missiles. In the span of three decades, the Israeli defence industry has changed from a consumer of war surplus materials, purchased by the pound, to a supplier of advanced technology, spare parts and components to the US and some 60 other nations.

This rapid growth, both in size and scope, is not matched by any other third-world state or newly independent industrializing country. Since 1948, Israel has been involved in four major wars and numerous skirmishes. Until the Camp David talks and the Israeli-Egyptian Peace Treaty in 1979, the presence of the Jewish State in the region was not accepted by any of its Arab neighbours. As a result, successive Israeli governments have placed primary emphasis on national security, in the belief that a strong military is necessary to insure survival.

Domestic defence production plays an increasingly important role in the maintenance of Israeli national security. With a population of 3.9 million, Israel is at a distinct demographic disadvantage when compared with a combined Arab population of over 100 million. To compensate for this imbalance, Israel has attempted to maintain a qualitative advantage, based largely on more sophisticated weapons and military equipment. While this equipment has generally been sought externally, Israel has had difficulty in assuring uninterrupted supplies. Throughout Israeli history, weapons embargoes of varying scope and duration have been imposed. At the same time, as the economic and political power of the Arab states has grown, they have had increasing access to sophisticated weapons while threatening to limit Israel's access. Although the US now supplies Israel with billions of dollars worth of military assistance (over $1.4 billion annually), Israelis fear that by controlling oil supplies and using oil revenues, the Arab states will be able to pressure the US to limit sales to Israel. Thus, domestic defence production has been seen as a major means of assuring access to advanced technology and weapons systems necessary to the survival of the state, and a large proportion of available resources have been devoted to the development of defence industries.

The rapid growth of the Israeli defence industry can also be attributed to the influence of this sector on the growth of the national economy. Defence production has contributed to the development of the economic infrastructure, providing jobs, training and new technology. Furthermore, the large number of skilled technicians and professionals among the immigrants from Europe and the US, which form a large segment of the Israeli population, provided the state with a relatively unique base from which to build.

In this chapter, the development, structure, and impact of the Israeli defence industries on the national economy will be analyzed. In the first section, the evolution of domestic defence production will be traced through four stages: maintenance and repair, licensed production and upgrading, independent production of major subsystems, and, finally, independent design and production. In studying this process, the catalytic role of externally imposed arms embargoes will be discussed. The second section consists of an analysis of the structure of the defence industries, examining the ownership, management and primary products of the relevant firms and government agencies. The third section is devoted to an analysis of the economic effects of defence production in Israel. In particular, the effects on employment, research and development, the 'spillover' of products and skills into civilian industries, and the balance of trade will be considered in detail.

The Evolution of Defence Production in Israel

The origins of the arms industry can be traced to 1929 and the first wave of Arab anti-Jewish riots in Palestine. While the Arab population of mandatory Palestine was able to obtain both guns and ammunition, the Jews had access to neither. In response, a clandestine 'home' arms industry, known as 'Ta'as' was created. This net of workshops, dispersed on agricultural settlements and in the cities, made small arms and ammunition. These workshops played an important role during the War of Independence, providing not only small arms and ammunition, but also assembling components and rehabilitating surplus aircraft, tanks and armoured cars shipped piecemeal from abroad.[2]

Following the cease-fire and Armistice in 1949, the major powers imposed an embargo on weapons deliveries to the region. Britain, however, continued to supply Iraq and Egypt with weapons.[3] From the Israeli perspective, the embargo was a continuation of the pre-1948 policy, and the theme of self-reliance gained new impetus. The weakest link in Israeli defences was perceived to be dependence on external suppliers, which rendered Israel vulnerable to 'outside political pressures'. Thus, emphasis was placed on the 'home production of arms'. In the early 1950s, production in facilities which had existed previously was expanded and surplus machinery and tools were purchased as scrap from the US at a cost of 19¢ per pound.[4] During this period, the manufacture of small arms and ammunition continued to receive the greatest emphasis. In 1952, the Uzi sub-machine-gun was designed. More significantly, a group of American aircraft engineers who had served in the War of Independence moved to Israel and provided the core of the Israeli Aircraft Industries (IAI).

IAI, which is wholly owned by the Israeli government, was founded in 1953 for the purpose of maintaining and repairing military aircraft, although according to Shimon Peres, 'Among the initial aims was nothing less than the eventual construction of jets.'[5] By 1955, IAI was refurbishing and repairing used aircraft for the Israeli Air Force (IAF), and was servicing jet engines which had formerly been shipped to France for repair. This decreased the maintenance costs and substantially increased the operational readiness of the IAF.[6] The maintenance and repair of other equipment was also critical to the Israeli Defence Forces during this period. To keep its surplus Sherman tanks in working order, surplus components and military scrap were purchased all over the world, refurbished and installed by the Army Ordnance Depot. Similarly, naval guns salvaged from Italian World War I equipment were refurbished and installed.[7]

The second phase, which encompassed the years following the Sinai campaign and Suez War in 1956, marked the transition to licensed production. The Israeli defence industry expanded significantly, largely in cooperation with French firms. Prior to the Sinai campaign, a number of Arab states, including Iraq and Egypt, had received advanced weapons from Britain and the Soviet Union, respectively.[8] To offset this potential change in the balance of power, Israel relied largely on the French, who were engaged in the Algerian conflict, and who, like the Israelis, had been excluded from alliances in the region, such as the Baghdad

Pact. In addition to purchasing French weapons, Israel sought to increase local defence production with French assistance. Common interests were apparent as both states stood to gain technically and economically from cooperative defence production. The French arms industry served as a model for Israeli leaders. In addition to providing the basis for military and political independence, the Israelis believed that this industry had been a major factor in turning France from a basically agricultural country into an advanced industrial state. According to this model, arms production provided both the technology and infrastructure for industrial growth in other sectors.[9]

Much of the cooperative effort between Israel and France took place in aircraft production. IAI expanded into licensed production of aircraft and the first flights of Fouga Magister jet trainers manufactured in Israel took place in 1960. During this period, Israel also began to produce artillery and rifles under licence from a Belgian firm.[10] Israeli engineers also began to make design changes and modifications to products produced under licence or purchased outright. Most externally developed weapons had not been designed for optimal performance in the environment of the Middle East. In general, the short ranges of the Arab-Israeli theatre and the temperatures in the region allowed for the removal of long-range radar, fuel tanks, and heaters. Israeli engineers modified second-hand Sherman tanks for desert fighting and added a 105-mm cannon to counter Soviet T-54/55 tanks. Guns and rockets were added to the Fouga Magister aircraft to turn these former trainers into combat aircraft. In the case of the Mirage III aircraft, the wings were strengthened, the ordnance capability increased, and an Israeli electronics package was designed to replace the original system.[11] These changes were adapted by the French and incorporated in the Mirage V.*

In the wake of the 1967 Six Day War, Israel lost its largest arms supplier and partner in cooperative ventures when Charles de Gaulle declared an arms embargo. Further orders of weapons were barred and undelivered orders were withheld, including gunboats and 50 Mirage V combat aircraft. (The gunboats, which were moored in Cherbourg, were eventually reclaimed by Israel.) The embargo led to the third phase of development, in which emphasis shifted from licensed production to

*These aircraft were sold but not delivered to Israel after the French embargo. These aircraft were, however, commercially successful and sold to many Arab States. Similarly, Israeli design changes in the US F-4 aircraft were included in aircraft which were sold to Kuwait.

independent design and manufacture of major subsystems. While the Arab states had access to Soviet, French, and increasingly, American aircraft and missiles, Israel became totally dependent on the US for advanced weapons. As Israel's sole supplier, the US was in a position in which it could either directly or indirectly link arms sales to political pressures and conditions or deny equipment for a variety of reasons (see below).

As a result of its experience with previous embargoes and its dependence on the US, the Israeli government decided that 'it would be virtual suicide to leave production of vital equipment in non-Israeli hands and moved full force towards setting up an ultra-sophisticated manufacturing capability . . .'[12] Investments in the defence industry grew quickly and were distributed among a variety of sectors. The largest projects undertaken in this period were in the development and production of major components for combat aircraft. An Israeli version of the Mirage V, known as the Nesher, was produced, with the assistance of French parts which became available in 1969, when the embargo was partially lifted. However, IAI, which manufactured this system, produced the necessary machine tools, jigs and fixtures.[13]

This experience led quickly to the fourth phase of the evolution of the Israeli defence industry, the independent design and production of sophisticated complete weapons systems. After demonstrating its capability with the Nesher, IAI proceeded to the higher performance Kfir combat aircraft, based on the more powerful American General Electric J79-17 engines (which power the US F-4 Phantom). The first production models of the Kfir appeared in 1973, and these aircraft were used extensively during the war. An improved version, the Kfir C2, is currently in production.[14]

In the 1970s, despite production of the Kfir, the IDF continued to rely heavily on the US high performance fighters such as Phantoms and F-15s. During this period, the continued unrestricted availability of such aircraft was brought into question by a number of actions and events. In 1969, for example, in the midst of the War of Attrition, the Nixon administration pressured Israel into concessions by withholding approval for the sale of aircraft.[15] During the Yom Kippur War, the supply of US weapons was linked to Israeli willingness to accept a cease-fire in place.* More recently, the availability of arms from the US

*Early in the war, after Israeli troops had been pushed back, the US began to airlift weapons only after Israel agreed to a cease-fire. The US effort failed in the face of Arab resistance. At the end of the war, during successful Israeli counterattacks, the US threatened an embargo unless Israel again agreed to an immediate cease fire.[16]

was affected by the Carter administration's broader effort to limit US arms exports,[17] and in response to Israel's raid on an Iraqi nuclear facility in 1981, as well as in reaction to subsequent Israeli action in Lebanon, the Reagan administration delayed the transfer of a number of aircraft for a short period of time. The flow of arms to Israel has also been affected by a variety of other factors. Within the past few years, sales to Israel have been linked to similar sales to Egypt and Saudi Arabia.[18] In addition, some components and technical data packages have been withheld in an attempt to prevent the transfer of technology which would allow Israel to compete with US defence firms.

On the basis of these developments and the successful production of the Kfir, the government has embarked on the development and production of an advanced light 'workhorse' aircraft designed to Israeli specifications and requirements. Designated the Lavi (Lion), this aircraft is to be powered by a US Pratt and Whitney 1120 engine. In order to limit dependence on the US as much as possible, this engine will be manufactured under licence in Israel. Production is scheduled to begin in 1986 and the first squadron to be delivered in 1988.[19]

Domestic defence design and production has also expanded into a number of other weapons systems, including tanks. The decision to proceed with tank production was taken after the British government suddenly backed out of an almost concluded agreement for the sale of of Chieftains.[20] After successfully modifying and extending the lifetime of its Sherman tanks, the design of the Merkava, or Chariot, tank was completed after the 1973 war. In battles during that conflict, the Armoured Corps suffered two-thirds of Israeli casualties and therefore the Merkava was designed for increased mobility and crew protection, incorporating a 900 horsepower frontal engine.[21] In addition to the Kfir and the Merkava, the Israeli defence industry has designed and produced other sophisticated weapons, including the Shafrir air-to-air missile (based on the US Sidewinder), the Saar-, Reshef-, and Dvora-class patrol boats, and the Gabriel sea-to-sea missile. To replace the Uzi sub-machine-gun, a new assault rifle known as the Galil has been developed and is currently a very popular export item.

From a few small firms, defence production has grown to the point where this sector is of central importance to the national economy. It consumes a substantial portion of the national budget, employs a large number of skilled workers, and accounts for a major and growing portion of Israeli industrial exports. In the following sections, these aspects of the Israeli defence industry will be examined.

The Structure of the Israeli Defence Industry

Decisions affecting investment in the defence industry and the development of new systems are made by the Ministry of Defence (MOD), the Israeli Defence Force, and the managements of the large firms such as IAI and IMI.

The broader structure of the 'military-industrial complex' encompasses over 150 defence firms and production units in Israel.[22] Given the historical evolution of the Israeli defence industry, the fact that the state was the primary source of capital, and the socialist ideology of the Labour Party which led the government until 1977, it is not surprising to find that many firms are partially or totally owned by the Israeli government. The four largest and most important firms, in terms of technological innovation and production, fit into the former category. These are IAI, IMI (Israeli Military Industries), Raphael (National Weapons Development Authority) and the Army's Main Ordnance Factory.

Since its founding in 1953, IAI has grown to be Israel's largest employer and consists of five divisions and fifteen subsidiaries. While primarily still in the aircraft industry, IAI and its subsidiaries also produce electronic equipment and a variety of other components, systems and services. Formally a state-owned corporation, the manager of IAI is appointed by the Minister of Defence, and the Board of Directors includes representatives from the Air Force, Defence Ministry, other government ministries, and large Israeli industrial concerns.* Although its activities are closely linked to and ultimately controlled by the Defence Ministry, IAI has some degree of independence in its research and development activities.

IMI and Raphael are more closely linked with and managed by the Defence Ministry. IMI, which has evolved from the Ta'as Underground workshops described above into a multi-million dollar firm, produces munitions, artillery and various guns, projectiles and components for military vehicles. Raphael is primarily engaged in R&D, but also produces the Shafrir air-to-air missile and various other systems. The

*While the ambiguity of the distinction between military and civilian spheres in Israel makes discussion of a 'military-industrial' complex somewhat overstated, the role of former military commanders in the management of defence firms should be noted. For example, Benyamin Peled, who has served as the Chief of Staff of the IAF is the Chief Executive Officer at Elbit. Other former military personnel play major roles at IAI, Koor, and other firms.

Main Ordnance Factory is managed by the Army and is largely responsible for tank development and production, including the Merkava.

The second largest sector of the Israeli defence industry consists of joint ventures between Israeli groups and foreign partners. In many cases, these partners are American firms, such as Control Data Corporation (CDC), which has a major interest in Elbit (computers), and G.T.&E., which owns 50 per cent of Tadiran (electronics). Some European firms also participate in joint ventures in Israel: for example, Bet Shemesh Engines, Ltd involves a French firm, Turbomeca, and Soltam includes a Finnish firm, Tampela.[23] Israeli partners in these ventures include banks, IAI, and firms owned by the major labour organization (the Histadrut).

Such joint ventures provide Israel with capital, marketing, advanced facilities, and access to new technology from abroad, the US in particular. Foreign partners are able to take advantage of the relatively low wage-scale for engineers, scientists and trained technicians, and, in an increasing number of cases, gain access to Israeli products, design, and R&D. Thus, the benefits are distributed to both sides.

The third group of defence industries includes those which are wholly owned by Israeli firms but do not include large government participation. These include Telkoor, a subsidiary of Koor Industries, which is an industrial conglomerate, owned by the Histadrut, and Israel Shipyards. Telkoor is active in the field of military electronics and radar systems. Israel Shipyards, which is jointly owned by Koor and Clal Industries (another conglomerate owned by Israeli banks), produces missile boats and related equipment. The Urdan armour plant is also a subsidiary of the Industrial Division of Clal.[24]

Finally, the fourth category of defence industries is those which are wholly owned subsidiaries of foreign (generally US) firms. These include Astronautics CA, which was formed in 1971 and manufactures cockpit displays and other components, and Intel. This group is significantly smaller than the other categories and plays a relatively limited role.

The Impact of Domestic Production on the Israeli Economy

In discussing the origins of the Israeli defence industry, Shimon Peres has noted that, 'The setting up of local industrial and research organizations . . . is not only a political and strategic concern, but it is also of far-reaching economic significance.'[25] As noted above, defence production was seen by Israeli leaders as the foundation for the industrialization of the economy, the source of technology, facilities, and skills, and, potentially, of foreign exchange. As a country with few natural resources to exploit and limited area and water for agriculture, economic development in Israel depended on industrialization. Given the reservoir of skills and training in the Western-educated segment of the Israeli population, as noted above, emphasis on technology was logical. While investment in defence production may be far from the optimal path for industrial development, as this sector produces no 'consumables' and diverts other resources from their production, given Israel's security concerns, the central role played by defence production in economic development may have been unavoidable. In this section, the impact of defence production on the economy will be examined.

Inputs

Since the mid-1960s, the Israeli defence budget has grown from less than 10 per cent to over 30 per cent of the GNP and rising defence costs have consumed most, if not all, of the increase in the GNP since 1967.[26] Approximately one-third of annual government expenditures go to defence, and, in 1978, military expenditures totalled over $4 billion out of $11 billion in government allocations (see Table 9.1 and Figure 9.1). An additional one-third of the national budget goes to servicing of the debt, much of which is the result of past defence expenditure.

The annual defence budget is composed of allocations for imports, including weapons and related equipment, military pay, and other local expenditures. The cost of imports varies from year to year and is highly dependent on delivery schedules, but in general, such imports account for one-third to one-half of the annual defence budget.[27] Another 20 per cent goes to military pay,[28] leaving from one-third to one-half of the defence budget for other local defence expenditures (see Table 9.2). Some fraction of this remaining total is allocated for food and clothing and military construction. With the exception of those years in which large new bases are constructed (with local funds), perhaps 50 per cent to 75 per cent of local defence spending, excluding pay, can be considered to be allocated to defence production. According to this accounting, the domestic defence industry receives from 20 per cent to 33 per cent of the total defence budget, or over $1 billion in recent years.[29]

In a general sense, it is clear that allocations for the domestic defence industries have been rising recently, in both relative and

Table 9.1: Annual Fiscal Data (1977 US dollars[a])

Year	GNP (1)	Government Budget (2)	Defence Expenditures (3)	Defence Imports (4)	Total Imports (Current $) (5)
1981		17.0[b]	5.0[b]	1.5	
1980				1.6[b]	9.9[b]
1979			3.1[c]	1.2	8.0[b]
1978	15.0	11.1	3.6	1.6	6.9
1977	14.3	12.4	4.3	1.1	5.8
1976	14.3	11.4	4.7	1.6	6.0
1975	14.4	10.6	4.6	1.8	6.7
1974	14.1	9.3	4.0		6.6
1973	13.2	9.6	5.0		5.7
1972	12.7	6.7	2.4		3.5

Notes: a. Except where otherwise indicated.
b. *Israeli Economist*, March 1981, pp. 8, 10. Projected estimate in current dollars.
c. Based on a reported 14 per cent decrease. *Israel Government Yearbook, 1979*, Jerusalem: Israeli Government Printing Office, 1980.
Sources: Columns 1-3, 5: US Arms Control and Disarmament Agency, *World Military Expenditures and Arms Transfers, 1969-1978*, Publication 108, Washington, DC: December 1980, unless otherwise stated.
Column 4: *Bank of Israel Annual Report, 1979* [D'in V'Heshbon], Jerusalem: Israel Government Press, 1980, p. 54. Note: This total is in current US dollars and includes only actual disbursements. New orders not included unless they were received in the same year.

Table 9.2: Defence Budget

Expenditure Category	Total Expenditure million current Israeli pounds				Percentage Annual Change constant prices		
	1976	1977	1978	1979	1977	1978	1979
Military Pay	4,645	7,405	11,408	22,991	1.4	3.3	4.9
Local Purchases	10,815	13,805	21,106	40,960	- 6.6	- 4.4	2.9
Imports	14,357	13,262	30,500	31,838	-36.7	40.0	-35.4
Total	29,817	34,472	63,014	95,789	-19.1	13.7	-14.2

Source: *Bank of Israel Annual Report, 1979* [D'in V'Heshbon], Jerusalem: Israel Government Press, 1980, p. 167.

absolute terms. In 1979, amidst pressure from the Minister of Finance to decrease the defence budget, the Government attempted to limit weapons imports. In order to gain the support of the military and the Defence Ministry, these cuts were balanced by an increase in allocations for local defence production. In the 1979 national budget, local defence allocations grew by 3 per cent (*not* including construction of

Figure 9.1: Relative Defence and Non-defence Portions of Israeli Government Budgets, 1960-79 (percentages, current prices)

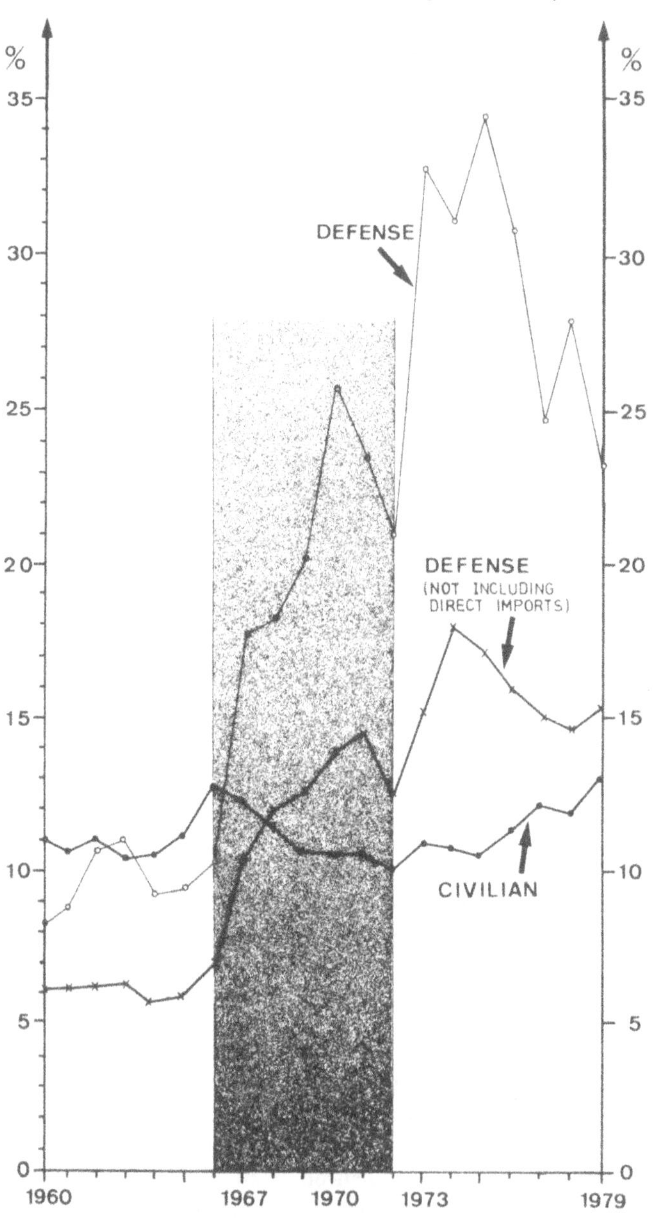

Source: *Bank of Israel Yearbook*, Jerusalem: Government Press, 1979, p. 166.

the new air bases in the Negev) while import allocations decreased by 35 per cent and the overall defence budget was reduced in real terms by 14 per cent[30] (see Table 9.2). The net increase in local allocations can be largely attributed to increased investments in the domestic defence industry.

The scale of investment can be expected to increase significantly over the next decade as R&D for the Lavi high-performance aircraft increases. According to press reports, over $1 billion has been budgeted for this programme.[31]

In order to understand the relative importance of the over $1 billion that is allocated annually to defence production, it is useful to compare this total to investments in other economic sectors. In 1981, the Israeli government allocated approximately $2 billion for investment and development in all the civilian branches of the economy, including agriculture, transportation and communication, housing and construction, and industry.[32] Thus, investment in defence industries accounts for perhaps one-third of the total annual governmental investment. Capital investment, both public and private, in non-military industry averaged some $350 million per year in the 1970s, rising from $235 million in 1970 to $555 million out of a national total investment of over $3 billion in 1978 (the latest available figure).[33] While investments in defence production were approximately equal to non-defence industrial investments through the mid-1970s, recent expansion of the defence industries has left the non-military sector far behind in terms of capital investment. At over $1 billion, allocations for the Lavi aircraft alone are likely to account for a large percentage of total available resources in the mid-1980s.

Impact on Inflation

The large investment in defence production contributes in no small measure to Israel's high rate of inflation, now measured in three digits and currently one of the highest in the world (135 per cent in 1980). As noted above, defence spending, both direct and in the form of debt servicing, accounts for a major proportion of GNP. Government spending, in general, exceeds the resources which are available in the form of taxes (the highest income tax rates in the world) and revenues, and the difference is financed by deficit spending. To meet debt obligations including both principal and interest payments, the government must increase the money supply, leading to inflation. In addition, while a great deal of money is added to the economy through the defence industry, this sector does not add to the availability of consumer goods and services nor contribute to the standard of living. As the national product stagnates or even decreases while the money supply increases, the rate of inflation increases. In the terms of economists, 'more money. is chasing less goods and services'.

Employment

In contrast to most states in the Middle East and other 'new', post-colonial nations, the Israeli population possesses a large cadre of technically skilled and trained workers. Immigrants to Israel from Europe brought these skills, and additional sources of training were provided by the British Ordnance Corps which was stationed in Palestine during World War II. In addition, a number of advanced technical and educational institutes have been created in Israel. The Israelis who served in the Ordnance Corps were able to maintain armoured and other weapons, and passed these skills on to trainees. The Technion (Israeli Institute of Technology), the Hebrew University in Jerusalem, and the Weizman Institute have provided students with advanced engineering and scientific training. These institutions are also a source of research and development facilities and personnel for the defence industry (see page 292).

Additional technical training is provided by secondary vocational and technical schools, including some established by the IDF. These schools turn out technicians for aircraft maintenance, the operation of production lines, and the construction of electronic equipment. According to one source, approximately half of all Israeli high school students attend such vocational schools.[34]

As in the case of output and production data, detailed information on employment in the defence sector is not published. A general picture in this area can, however, be drawn by examining the available information. According to the Israeli Central Bureau of Statistics, the number of people employed in the manufacture and repair of aircraft, ships and boats has tripled in the past decade.[35] Since 1967, the number of employees at IAI has grown from 4,000 to 22,000, making this firm the largest employer in Israel. Approximately an equal number are employed by IMI, Raphael and the Army's Main Ordnance Factory combined. The total can be estimated at approximately 60,000,[36] out of an industrial work force of some 300,000.[37] Thus, with 20 per cent of the work force, it is clear that industrial employment in Israel is heavily concentrated in the defence industries and in the larger firms in particular.

There are some costs associated with this concentration of employment in the government-owned defence industries. In particular, if military exports are reduced (see pages 298-9), and reductions in the defence budget continue, reductions in domestic defence production could result in large unemployment. One reason cited for proceeding with the Lavi is the need to maintain employment in this industry.

The concentration of employment in the defence sector is particularly strong among scientists, engineers, and skilled industrial workers. There are over 275 scientists and engineers per 10,000 people in Israel.[38] This ratio is one of the highest in the world and is double that of developed industrialized economies. It is growing with the addition of new graduates from Israeli and foreign academic institutions, the return of Israeli professionals from abroad, and the immigration of professionals from the US and the Soviet Union. In 1977, of 9,200 new immigrants, over 2,000 were scientific and academic workers.[39]

The defence industry provides a major source of employment for scientists and engineers, and in order to prevent emigration of trained engineers and craftsmen, the government and the larger firms will often support redundant production lines or uneconomic ventures. For example, in the 1960s, during a major Israeli economic recession, IAI's management decided to proceed with the development of the Arava STOL (Short Take-off and Landing aircraft) to avoid losing aeronautical engineers.

Regional Distribution and National Development

Israel is a relatively small country. It has an area of 20,000 square kilometres, and extends 400 kilometres from north to south and, at its widest point, 100 kilometres from east to west. Within these boundaries, the country is divided into three economic zones: the coastal plain, which includes Tel Aviv and Haifa, the north (Galilee), and the south (Negev). While the coastal plain is highly developed and heavily populated (70 per cent of the population), the northern and southern regions are underdeveloped and sparsely populated. Since the state was founded, successive governments have sought to promote settlements and industrialization in the underdeveloped areas, and the defence industries have played a role in this effort.

Most of the large defence industry installations and plants are located in the Tel Aviv area, and, to a lesser degree, in Haifa. IAI's main aircraft plants are in this area, as is the Army's Main Ordnance Depot, where the Merkava tank is manufactured. Israel Shipyards and Elbit are located in Haifa. Some small arms and ammunition plants, however, are located in northern outposts, such as Kiryat Shmona, and Elbit has established a plant in the development town of Karmiel in the Galilee.[40] In the south, IAI and Beta (a small electronic control firm) have built plants in Beersheva, and other firms have facilities in other plants of the Negev area. The government has encouraged the movement to remote areas, and has sought to double or perhaps triple the number of employees in government-owned plants in the development towns. In the next few years, additional plants are scheduled to be moved from the Tel Aviv region to the 'development' areas in the north and south.[41]

Research and Development

Research and development in the defence sector has received a great deal of attention in Israel, and this attention can be attributed to a variety of factors. In the first place, as noted above, in order to offset the demographic imbalance in the region, Israel has sought to maintain qualitative weapon superiority and technological innovation. New technology is imported as rapidly as it can be obtained and partly for this reason, co-production of weapons systems has become a major goal for the Israeli government (see pages 303-4). Second, as the Israeli economy has become increasingly industrialized, its industry has, in turn, become 'technology intensive'. In this context, the government has looked to the defence sector for the development of new industrial technologies, facilities, techniques, and, most importantly, products. On an annual basis, research and development accounts for approximately 2.3 per cent of the Israeli GNP, or $450 million in 1980.[42] According to reports, from one-third to one-half of this total is defence related.[43]

Defence-related R&D in Israel, as in the US, takes place at four levels: the Ministry of Defence (MOD), defence production firms, military forces, and academic institutions. Allocations are made for basic and applied research in general areas of interest as well as for specific projects and specialized areas. Recipients and contractors include the large firms, such as IAI, and small research groups in or affiliated with academic institutions, such as the Technion, Weizman Institute, and the three major universities. While the quality of output and the availability of expertise is a major factor in the distribution of contracts, concern for the maintenance and continued support of these groups is also considered in the granting of research contracts.

New products and weapons systems resulting from these processes reach production via a variety of paths. Individual technologies are often identified by Raphael, while the IDF will define operational requirements for a specific weapons system. Contacts between these groups are often informal, but once technologies and requirements are defined, a formal development proposal is written. Various firms may compete in the case of relatively small projects, while large projects involving aircraft or tanks will be assigned to the obvious firms. In the former case, competition is generally not in terms of price, but rather design approach, and the government will often fund different firms to try multiple designs and to maintain full employment.

In cases in which the Israeli government decides to produce and test systems, a process resembling the DSARC* system in the US is used. The planning staff of the IDF is heavily involved in this process and officers from interested branches of the IDF are assigned to such weapons development programmes. If a project loses military support at some stage, the individual firm may proceed with testing and development in the hope that government assistance or foreign interest will offset investments. As noted above, IAI's management decided to proceed with the development of the Arava STOL in the 1960s to avoid losing aeronautical engineers. While this effort was opposed by the military, who saw it as a waste of capital, this aircraft became a successful export product and at a later stage, was purchased by the IAF. Similarly, IMI began development of a short-range (40 km) 290-mm rocket as an export project.[44]

Output

As in the case of investment, the output of the Israeli defence industry, and the percentage of total and sectoral industrial output and profitability are not reported systematically. While contracts between the defence ministry and government-owned firms such as IAI and IMI are negotiated, the value of the contracts and the total annual revenues of these firms are not made public. As a component of the Israeli industrial base, however, defence production constitutes a major portion of net industrial production. Detailed articles in the military and aviation press, which appear to rely on semi-official data, reported a total output of approximately $1.5 billion in 1979 (not including

*In the US, a Defense Systems Acquisition Review Council (DSARC) is established for each major new weapons system development programme. This council consists of representatives from the military service and defence agencies with an interest in the programme. It meets periodically to examine the progress of the programme, and to decide whether to continue funding, accelerate or slow development, or end the project.

Israel Shipyards and a few other firms).[45] Of this total, IAI and IMI accounted for $560 and $400 million respectively. In the case of IAI, however, a significant portion of this total, including business jets and maintenance for commercial aircraft, should not be considered as defence production (see the following section).

The validity of this total is supported by available export statistics. There is a great deal of evidence, both official and unofficial that approximately half of Israeli defence production, including the output of large, government-owned industries, such as IAI and IMI, as well as the joint ventures such as Elbit and Tadiran, is exported (see page 301).[46] According to a variety of estimates, the volume of exports for 1980 exceeded $1 billion,[47] which, according to the above assumption, places the total production at approximately $2 billion. This compares with a GNP of $19.6 billion (estimated) in 1980[48] and an output of $4 billion in the manufacturing and mining sectors of the economy, including defence.[49] Thus, defence production would appear to account for as much as one-half of the total Israeli industrial output and one-tenth of the GNP.

Spinoffs

In addition to its direct impact on the Israeli economy, defence production also contributes indirectly to the growth of the civilian economy. The Israeli government has sought to make some of the results of military research and development available to the civilian economy. The Ministries of Commerce and Industry have explicitly commissioned studies aimed at facilitating this transfer.[50] The transmission of technology from the military to civilian sectors takes the form of spinoffs of both personnel and products. Many, if not most of the individuals employed in the civilian technology-intensive sector received their training in military research, development and production activities, and it is clear that this sector is a source of personnel for civilian industries. In addition, technologically oriented defence production has contributed to a number of high-technology civilian manufacturing facilities, most notably in the areas of aircraft production and electronics.

Most of the technologically advanced non-military production in Israel is aimed at the export market. Technology intensive non-military exports, including electronics, transport and machinery, grew by 200 per cent between 1967 and 1977, increasing from $29 million to $580 million, and from 7 per cent to 22 per cent of total industrial exports[51] (Table 9.3). Currently the volume of high-technology non-military industrial exports constitutes approximately one-quarter of total Israeli industrial exports.

Table 9.3: Israeli Defence Exports and Imports (million current $)

Year	Total[a] Exports (1)	Industrial[b] Exports (2)	Electronics[c] Exports (3)	Defence[d] Exports (4)	Defence Imports (5)
1980	4,917[e]			1,000[f]	1,600
1979	4,301	2,501	967	600[g]	1,158
1978	3,716	1,923	795	450	1,624
1977	2,963	1,551	615	380	1,099
1976	2,306	1,227	455	230	1,561
1975	1,835	971	297	110	1,846
1974				80	
1973				60	

Notes: a. Excludes export of services.
b. Total excludes diamonds.
c. Total includes defence-related products.
d. Official defence-related export data is not published and any totals which are cited must be regarded as guesses. Furthermore, different assessments often include different assumptions. For example, the SIPRI data, which is far lower than that listed here, includes only major weapon systems. (See, Stockholm International Peace Research Institute, ***World Armaments and Disarmament, SIPRI Yearbook 1981***, London: Taylor and Francis, 1981.) Higher figures may include re-exported equipment which is not manufactured in Israel.
e. *Israel Economist*, February 1981, p. 7.
f. Estimate reported in 'Israel's Expanding Capabilities', *Defence Attaché* no. 1 (1981).
g. *Military Technology and Economics* 4:20 (1980).
Sources: Columns 1-3: ***Bank of Israel Annual Report, 1979*** [D'in V'Heshbon] Jerusalem: Israel Government Press, 1980, p. 76.
Column 4: Center for Policy Alternatives and Center for International Studies, ***Disarmament and Development: The Case of Relatively Advanced Developing Countries (Possible Economic Pay-offs from Military Production: The Case of Aircraft Industries in Brazil, Israel and India)***, Cambridge: Massachusetts Institute of Technology, 1980, unless otherwise specified.
Column 5: US Arms Control and Disarmament Agency, ***World Military Expenditures and Arms Transfers, 1969-1978***, Publication 108, Washington, DC: December 1980.

The most important civilian spinoffs of defence production are the civilian aircraft produced by IAI. In 1959, IAI investigated the possibility of producing a light STOL aircraft primarily aimed at the civilian market, particularly in the Third World. After an unfavourable market analysis and in recognition of IAI's then limited capabilities, this project was cancelled. In the mid-1960s, however, the project was revived in part to maintain skilled employees (see page 291), and the first aircraft

Table 9.4: Major Reported Israeli Defence Exports

Recipient	Weapons System	Year[b]	Source
Argentina	4 Dabur Patrol Boats		(1)
	18 Gabriel II	1975	(2)
	42 Nesher (Mirage Derivatives)[a]	1979	(1)
Bolivia	6 Arava STOL Aircraft	1975	(1)
Chile	Shafrir Air-to-Air Missiles	1976	(1)
	6 Reshef	1979-81	(3)
Colombia	3 Arava	1980-81	(1)
Ecuador	9 Arava	1975	(1)
	12 Mystère B2 Fighters[a]	1977	
	Gabriel Mssiles		(1)
El Salvador	24 Magisters and Ouragan[a]	1974-75	
	5 Arava		(1)
West Germany	4 Westwind	1981	(1)
Guatemala	8 Arava	3 in 1976	(1)
Honduras	12 Mystère B2 Fighters[a]	1976	(1)
	3 Arava		(1)
	5 Fast Patrol Boats	1980	(2)
	1 Westwind Business Jet		(1)
Indonesia	16 Skyhawks	1980	(4)
Kenya	48 Gabriel II	1979	(2)
	Patrol Boats[a]	1978	(3)
Mexico	10 Arava		(1)
Nicaragua	2 Arava		(1)
Singapore	50 AMX-13 (light tanks)	1969	(1)
	155-mm Howitzers	1977	(1)
	Gabriel		(1)
South Africa	6 Dvora		(1)
	6 Reshef	1976-78	(5)
	108 Gabriel II	Ordered 1977	(2)
Swaziland	1 Arava	1979	(2)
Taiwan	Gabriel	1976	(1)
	Shafrir		(1)
Thailand	Gabriel		(1)
Uganda	10 Sherman Tanks	1970	(1)
United States	600 M-48 Tank Cupolas	1976-	(1)
Venezuela	3 Arava	1980	(2)
Zimbabwe	11 Bell Helicopters[a]	11/78	

Notes: a. Supplied from IDF Inventory.
b. Date received, unless otherwise specified.
Sources: (1) International Institute for Strategic Studies (IISS), *The Military Balance*, London: IISS, annual publication. (2) Stockholm International Peace Research Institute, *World Armaments and Disarmaent, SIPRI Yearbook*, London: Taylor and Francis, annual publication. (3) Unconfirmed SIPRI report. (4) According to SIPRI, these were transferred in a triangular transaction in which the US acted as middleman. (5) According to SIPRI, three were produced under licence in South Africa.

flew in November 1969. The Arava is powered by a Pratt and Whitney engine and is capable of carrying 24 passengers at a speed of 200 mph to a distance of about 500 miles.[52] Licensed by the US Federal Aviation Administration in 1972, over 60 have been sold to date to a large number of South and Central American states, and ten have been ordered by a US commuter aircraft company at a cost of $2 million per plane.[53] (see page 301 and Table 9.4). In an example of reverse spinoff, this aircraft has also been adapted for use as a military transport and for counter-insurgency.

In addition to the Arava, IAI has also produced the Westwind Executive Jet. Formerly known as the Jet Commander, the production rights for this aircraft were acquired from Rockwell Standard for $25 million plus royalties in 1966, and some design changes have been made subsequently.[54] Although the lucrative Arab market is closed to IAI, Westwind sales are second only to Gates Learjet in the medium-sized business jets market. By the end of 1980, 223 aircraft had been sold, 70 in 1980 (for a total of $200 million in 1980), and an additional 60 were on order.[55] (In 1981, production rates were scheduled to increase to five aircraft per month.) This aircraft has also been sold for military uses, largely for training, towing and naval support missions. In addition, an Israeli-designed, advanced long-range business jet, the Astra, is scheduled for introduction in 1985.[56]

Civilian spinoffs have also been developed in other product areas. In 1977, 350 Israeli firms exported $580 million in products derived from local research and development.[57] The outputs of many firms which began as military suppliers have become increasingly oriented towards civilian production. For example, although Israel Electro-Optics Ltd. was founded as a defence firm, its production is now evenly divided between civilian and military products. Similarly, Elron was originally formed to produce military computers and weapons delivery systems for the A-4 aircraft, but now has an equal non-defence output. In addition, an Elron subsidiary, Elscint, has established itself in the area of medical electronics and has now grown to be a major exporter of sophisticated equipment, including CAT scanners. Elta, which began by producing military radar units, now also produces radar for civilian air-traffic control systems.[58]

As the general emphasis on technology and high-technology products in the Israeli economy has grown, the diffusion of trained personnel, technological knowledge, and products can be expected to increase in a similar manner. Given the limited domestic market for such products, however, this trend will become increasingly important in the export sector.

Defence Exports

Since 1948, Israel has relied heavily on weapons imports, and as the cost of these weapons has increased, so has the import bill. In the 1960s, defence-related imports totalled approximately $1.5 billion, and during the following decade, this total grew by an order of magnitude, to $15 billion. Defence imports totalled over $1 billion annually between 1976 and 1980. After a decline in real terms in the volume of defence imports in 1979, the volume has begun to rise again with the delivery of F-15 and F-16 aircraft from the US, and $1.8 billion was allocated for defence imports for 1981.[59] (Note that these amounts are in current dollars and do not correct for inflation.) Despite the growth of domestic production, arms imports have not decreased significantly to date as more sophisticated and expensive technologies continue to be introduced into the region. Weapons imports currently account for approximately 17 per cent of all Israeli imports (military and civilian).

This heavy import burden has not been balanced by a comparable volume of exports and arms imports have contributed to a major imbalance of trade (between $2.5 and $4 billion from 1975 to 1979).[60] This imbalance has resulted in deficit financing which, as noted above, has contributed to the rate of inflation. In addition to local defence production, an increasing proportion of the national budget is devoted to the servicing of loans for weapons imports. According to one estimate, 'repayment and interest charges may soon consume all of the current foreign military sales financing available to Israel, thus precluding any significant new Israeli military purchases from the United States.'[61]

To offset this imbalance, Israel has looked to its own domestic defence industries not only for import substitution but also to increase the volume of exports, both directly and indirectly. Direct assistance is provided through the export of specific products, such as aircraft and defence electronics (see page 300). Indirect assistance to civilian exports is provided by the defence industry through the infrastructure which the industry provides, the research and development and 'spin-offs'. As noted above, the defence sector makes 'state of the art' technology and facilities available to the rest of the industrial sector. The 'value-added' to the raw materials and labour in the area of military systems and non-military high technology magnifies the contribution of these sectors to the economy.

In addition to decreasing the national trade deficit and providing support for the growth of exports in non-defence sectors of the economy, defence exports serve to reduce the cost of development and procurement of weapons for the Israeli Defence Forces. As in other technology and capital-intensive industries, large production runs lower the cost per unit and allow the amortization of R&D and tooling costs over a greater number of units. As long as the IDF is the only customer for the Kfir combat fighter, the Israeli economy will bear the entire cost of R&D and capital investment. Increasing foreign sales would allow for greater recovery of these costs. (In contrast to the practices of other arms-exporting states, the Israeli government provides no financing and in fact claims a 3 per cent royalty on the sales of products which were developed or produced with the assistance of government funds.[62])

While the reported value of Israeli arms exports varies with the source of data, a number of sources report that sales have gone from a few tens of millions in 1973, to over $750 million* in 1979 and over $1 billion in 1980. Of this total, IAI alone accounted for one-third with $336 million in 1979.[64] In the years immediately after the 1973 war, the output of IMI went largely to rebuild depleted stockpiles in the IDF, but between 1978 and 1979, IDF orders decreased by 8 per cent while the volume of exports rose by 77 per cent to more than half of IMI's total output. According to government figures, during 1979 IMI accounted for approximately 5 per cent of total Israeli exports and 10 per cent of industrial exports, which would give IMI over $200 million in exports. Unofficial sources report that defence exports now account for approximately one-third of total Israeli exports.[65]

The success of the Israeli defence industry in the international market, which now includes over 60 countries, can be attributed to a variety of factors. In the first place, Israeli equipment has been combat tested and has demonstrated its capabilities repeatedly. In addition, Israeli R&D skills and the advanced technology which is available has placed the Israeli industry at level comparable to that of advanced industrial countries. At the same time, due to the low wage levels relative to the industrialized world (40 per cent lower than the US),[66] Israeli products are significantly less expensive than comparable equipment.†

*Other sources report a total of $600 million for defence exports in 1979.[63] One factor in this large discrepancy may be a large volume of unannounced sales which may not be included in the lower figures. The higher figure may include re-exports of outmoded IDF equipment, including aircraft, and non-military items, such as aircraft that are produced within the nominal defence industry.

†According to some reports, the competition presented by Israeli firms is among the factors responsible for the restrictions which are placed by the US on the sale of some advanced systems and data packages to Israel.

Defence exports began during the first phase of Israeli defence production and consisted of small arms and ammunition, most notably the Uzi sub-machine-gun which itself has been purchased by at least 60 nations, including the United States. By the 1960s, the sale of spare parts for aircraft and the overhaul and maintenance activities of the Bedek division of IAI became increasingly important. Maintenance stations were opened in Germany, Thailand, Britain, and Switzerland.[67]

In the last decade, Israeli defence exports have included increasingly sophisticated products, including aircraft, electronics, patrol boats, air-to-air missiles (the Shafrir), and sea-to-sea missiles (Gabriel). In addition, an extensive but so far unsuccessful effort to market the Kfir combat aircraft has been undertaken. As the sophistication of these products has grown, the markets for the Israeli defence industries have changed. In the early 1960s, the major customers for Israeli products were African states. Beginning with small arms and ammunition, including the Uzi, these states also purchased larger systems. Uganda, for example, purchased a number of Israeli-made Magisters. During this period, Israeli arms were also provided to strategic allies, such as Ethiopia and the Kurdish rebels in Iraq. Assistance to Ethiopia, largely in the form of aircraft maintenance and spare parts, has continued despite the revolution and establishment of a Marxist government supplied with Soviet weapons.[68]

In the 1970s, however, the African market declined while new markets developed in Western Europe, the US, South America, and, to a limited degree, in some of the industrializing states such as Iran, Thailand, and Singapore. Exports to the US are increasingly taking the form of 'offsets' or commodities which are transferred in exchange for Israeli purchases, and many Israeli weapons imports include provisions for offsets. Sales to the US and Western Europe consist largely of weapons, components, and electronics in which Israeli firms specialize and which are either more expensive or unavailable elsewhere.*

In addition to exporting weapons systems originally designed for the IDF, many firms now place major emphasis on the development of products explicitly for export. In fact, within the past few years, IDF

*The volume of sales to the US was expected to increase in the wake of the 1979 Memorandum of Understanding between the US and Israel under which Israeli firms are exempt from the 'Buy American' provisions of defence contract competitions. According to US sources, however, only one Israeli firm has taken advantage of this provision. This had led to complaints within the US defence establishment which feels that Israel has failed to appreciate an important concession.

orders have decreased while exports have increased, so that major firms such as IMI and IAI now export more than 50 per cent of their total output.[69] In 1979, Tadiran exported over half of its output ($120 of $225 million) largely to Western Europe and the US.[70]

South American and Third-World customers have expressed interest mainly in Israeli aircraft and small arms. While discussions concerning the sale of Kfir fighters have been held with a number of states, the US had, until recently, refused to approve such sales (see below). Many states have purchased the Arava STOL in its military configuration, however, and sales of Israeli equipment to South American states totalled over $1 billion between 1973 and 1979.[71]

During this period, sales to Iran had become increasingly important. Prior to the revolution, Iran had been at least a tacit ally of Israel and the two states had a number of strategic interests in common. Although Israeli arms sales and assistance to the Shah's forces may have begun as a result of such common interests, they quickly assumed great economic importance for Israel, and were largely used to offset oil imports from Iran. (Recent Israeli efforts to increase arms sales to Mexico have a similar purpose.) A number of products, including avionics, electronics, patrol boats, missiles, and spare parts for American-supplied aircraft and other weapons were sold to Iran.

With the fall of the Shah, Israel lost a potentially large customer. Other political factors have further limited Israeli arms sales, most notably the economic war which the Arab states have waged against Israel. (Sales of Israeli produced, US-designed radio equipment to Arab states have reportedly been channelled through West German and Brazilian intermediaries. Since the beginning of the Iraq-Iran war, there have also been reports of Iranian attempts to purchase Israeli equipment and spare parts through third parties.) The sudden loss of African markets can be attributed to the pressure exerted by the Arab states after the 1973 Yom Kippur War. The cancellation of a projected sale of Kfirs to Austria is also attributed to Arab pressure. The Malaysian government's recent decision against purchasing Kfirs or refurbished Skyhawks from the IAF inventory is explained similarly.

A second major source of obstacles to Israeli arms exports derives from the fact that most Israeli weapons systems include parts that are either produced in the United States or are manufactured under licence. The Kfir, for example, is powered by engines manufactured under licence in Israel. As a result, as noted above, Israeli efforts to sell aircraft and other systems to South and Central America and Taiwan have been subject to American vetoes. In addition, the sale of engine components produced under licence for US-designed engines has been blocked by the State Department.[72] During 1981, however, US policy in this area shifted and Israeli sales to South America and other states are likely to gain approval.[73] In addition to the Reagan Administration's general receptivity to arms transfers and sales, three explanations specific to the Israeli case have been offered: (1) As a recognition that the efforts to keep sophisticated weapons out of particular regions – the original justification for the policy – have failed; (2) As a means of compensating Israel for increased military aid to Egypt and Saudi Arabia; and (3) As a less costly and indirect form of aid to Israel.[74] There have also been unconfirmed reports of arms purchases by the Chinese government, with the approval or even encouragement of the US.[75]

To offset these obstacles to exports, Israel has increased its arms sales to so-called international 'pariah' states, whose access to other sources of weapons is limited. South Africa, Taiwan, Somoza's Nicaragua and Chile have purchased and integrated Israeli weapons systems into their arsenals. Although these sales have been subject to controversy and domestic debate and foreign criticism, the Israeli government has argued that it has few options. An indigenous defence industry is deemed essential to national survival, and the cost of defence production requires arms exports. Since sales to a number of more 'legitimate' states such as Austria and Ecuador are blocked, Israel is left with a limited range of customers.

Summary and Conclusions

The primary objective of the development of a large and sophisticated defence industry in Israel was to gain independence from foreign weapons suppliers. Independent production would limit vulnerability to embargoes and 'political strings' attached to weapons purchases and would limit the impact of weapons imports on the national economy. The success of this strategy appears to be limited. Despite the development of production of aircraft, tanks, missiles, and so on, Israel still remains dependant on external suppliers, and the US in particular, for major systems such as air defence, early warning, helicopters and advanced combat aircraft. In the latter category the Kfir and, if it is produced, the Lavi, cannot by themselves meet all of Israel's needs, and there is little prospect for a domestic arms industry which would be able to meet even a large fraction of the IDF's diverse needs. Furthermore, even these 'Israeli' systems rely on US engines and other key components. The prospects for complete independence in such advanced weapons systems in the near future are limited. Israel has succeeded, however, in increasing its ability to produce and stockpile ammunition and other consumables, and this has decreased requirements for external re-supply in the course of a conflict.

The economic benefit of import substitution is also yet to be attained. There are groups within the Ministry of Defence and IDF which question whether the investment in local production matches the benefits. The cost of arms imports is still very high, and is now accompanied by the major investments in local production. Critics also argue that the billions which have been spent locally could have gone towards the purchase of more advanced and (taken as a whole) less-expensive weapons systems. For example, the US M-60 tank is cheaper than the Merkava. Given the limitations on the size and versatility of the defence industry, imports are also inherently more flexible. New systems and technological improvements can be purchased rapidly, while local production requires a commitment to a particular design or system which can only be altered at great cost. In summary, the twin goals of independence and import substitution have not been nor are likely to be achieved in the foreseeable future.

Co-production

Partly as a result of these shortcomings, Israel is currently attempting to increase the level of interdependance in its relationship with the US and, to a lesser degree, European manufacturers. For the past few years, the Israeli government has sought to offset its purchases of US hardware by participating in the production of the weapons systems involved. For example, Israel has sought to participate in the co-production arrangements that have been created between the US and NATO countries in the F-16 project. Co-production would not only have gone a long way toward offsetting the purchase price of this aircraft, but would also have increased Israeli access to Western markets. In addition, the interdependant relationship that results would provide Israel with some leverage in its relationship with the US. In case of a shortage or conflict involving the US, Israel would be in a position to increase the flow of components and spare parts.

Israeli co-production requests in the F-16 programme were denied on the grounds that the 'pie' was already sliced too thinly among US and NATO firms.[76] In March 1979, an agreement between the US and Israel was signed which was aimed, in part, at 'affording Israeli sources improved opportunities to compete for agreed upon procurement of the US Department of Defense'. Israeli firms were authorized to receive Invitations for Bids (IFBs) or Requests for Proposals (RFPs) from the DOD on a long list of programmes, including, the F-18 aircraft (airplane structural and accessory components) and XM-1 battle tanks.[77] In 1982, as part of an agreement on strategic cooperation, this agreement was reiterated, but to date, nothing significant has resulted.

Achievement of Secondary Objectives

The 'secondary' economic impacts of the defence industry are also quite significant. This industry has provided for the employment of a large and advanced segment of the work force, which until recently, had few other potential employers. The establishment and support of defence firms prevented a substantial 'brain drain' and has also been used as a tool for regional development. The establishment of defence industry factories in underdeveloped sectors of the north and south promote the industrialization of these regions and provide steady employment which, given Israel's current economic condition, few if any other sectors could offer. Furthermore defence exports, which now exceed $1 billion annually, and account for one-third of all Israeli exports, are a major factor in redressing the trade imbalance. In this sense, defence production has served as an important instrument of national economic planning and development.

The defence industry was also designed to provide an industrial and technological base from which civilian industries could develop. There is some evidence that this objective has been successful, at least to some degree. A number of firms originally created for defence production now show large outputs of non-military products. Skills developed in the context of defence production have been applied to the design and production of civilian goods. For example, Israeli computer, laser and electronics industries have contributed to the growth of industrial exports with a high value-added component. On the other hand, the profitability of civilian aircraft and production is uncertain and has been subject to some dispute within Israel. In 1976 the Government Comptroller reported that the Westwind and Arava had lost $10 and $11 million, respectively.[78] Since that time, however, a number of aircraft in each series have been sold and these programmes may now be profitable.

However, even if these projects are now profitable, it is clear that the economic effects are not all positive. 'Non-consumable' defence production contributes heavily to Israeli inflation and the subsidies given to the defence industry are replacing weapons imports as a perceived source of economic difficulty. With the conclusion of the Peace Treaty with Egypt, and while Syria and Iraq are bogged down in Lebanon and Iran, respectively, attempts have been made in Israel to decrease the resources devoted to defence production.[79]

More significant, perhaps, is the point that the technologically advanced civilian facilities and products can be developed more readily by direct investment than by relying on spill-over from the military sector. Direct investment and subsidies for electronics, computers and other industries would build the necessary infrastructure and skills but at a much lower cost. In these areas, the same results could have been achieved with direct application of a small fraction of the billions of dollars spent on defence industries. In Israel as in other countries, however, it is not clear that the spending of public funds, even in moderate amounts for such industrial development would have been politically feasible or acceptable in the absence of defence requirements.

Applicability of the Israeli Model

The Israeli experience in establishing a local defence industry can, to a limited degree, be generalized to other states. In particular, the progression from small-arms production and assembly to licensed production and upgrading, independant production of major systems, (unlicensed production?), and finally independant production and design is a path which is followed by other states in developing an independent defence production capability.

The pace and scope of this development, however, is particular to Israeli strategic requirements and its economic foundation. In the first place, the geopolitical position of Israel and its overriding concern with survival has provided a major and continuing impetus for the development of this industry, and the continued allocation of national resources to local defence production. Externally imposed embargoes and the determination to avoid dependence on a limited number of suppliers have spurred the Israeli defence industry and led to the development and production of advanced weapons systems. The size of the Israel Defence Forces itself, and in particular, the fact that the Israeli Air Force is the third largest (in terms of combat aircraft) in the world, contributed to the decisions to invest heavily in local manufacture of weapons.

The impact of this production on the national economy is also somewhat unique to the Israeli case. In contrast to other industrializing states, such as Brazil, or India, for example, Israel has few natural resources to exploit and inadequate land and water to expand agricultural production significantly. Furthermore, its relatively small population limits the internal market for industrial goods and manufactures. Defence production was also able to provide the large segment of technologically trained Israeli population employment commensurate with its skills. Finally, the manufacture of military systems and products provided a source of technology and infrastructure for the development of civilian products.

Thus, the development of an industrial and technological base in Israel through defence production is the result of the confluence of three factors; security requirements, the absence of natural resources, and the availability of skilled technicians, scientists, and engineers. In the absence of the first factor, this emphasis and investment priority would not have been rational.

Conversion Questions

Given current political realities in the Middle East, the problems of converting from military to civilian production in Israel would appear to be distant and of little current significance. With the exception of Egypt, Arab acceptance of Israeli legitimacy in the region and a solution to the Palestinian problem seem unlikely in the forseeable future. After 50 years of conflict, Israel's perception of the need to maintain a strong military force for national security is not likely to change radically, and as a result, the emphasis on domestic defence production is likely to continue for some time, even if some diplomatic progress is made.

If, however, the situation changes, it would appear that the Israeli industrial structure has laid a strong foundation for rapid conversion to civilian production, as long as an export market exists. Civilian production in most defence firms already is significant, often exceeding 50 per cent of total output. This sector is highly profitable, takes advantage of similar skills and facilities and could be readily expanded. In addition, the large segment of Israeli society which finds arms production and sales incompatible with Zionist ideology except when necessary for national survival, would welcome such conversion.

Notes

1. Official figures are not published, but this figure is reported by a number of unofficial sources, including *The Israel Economist*, January 1979, p. 6; and 'Israel's Expanding Capabilities', *Defence Attaché* no. 1 (1981):35. As with other

numbers cited in this study, however, this amount should be considered a rough approximation.

2. See Yigal Allon, *Shield of David*, New York: Random House, 1970, p. 161, and Shimon Peres, *David's Sling*, London: Wiedenfield and Nicolson, 1970, p. 109.

3. Gunther Rothenberg, *The Anatomy of the Israeli Army*, New York: Hippocrene Books, 1979, p. 86.

4. Peres, *David's Sling*, p. 110.

5. Ibid. p. 126.

6. Ibid.

7. Rothenberg, *The Anatomy of the Israeli Army*, p. 87.

8. A major arms agreement was signed between Egypt and Czechoslovakia in 1955, although the Soviet Union was the primary source of these weapons. See Jon D. Glassman, *Arms for the Arabs*, Baltimore, Md.: Johns Hopkins University Press, 1975. See also Chapter 6 in this volume.

9. Peres, *David's Sling*, pp. 115-18.

10. Stockholm International Peace Research Institute (SIPRI), *Arms Trade with the Third World*, Stockholm: Almqvist and Wiksell, 1971, p. 771.

11. See Rothenberg, *The Anatomy of the Israeli Army*, p. 112, and SIPRI, *Arms Trade*, p. 769.

12. *Aviation Week and Space Technology*, Special Israel Advertising Section, 8 October 1979.

13. Center for Policy Alternatives and Center for International Studies, *Disarmament and Development: The Case of Relatively Advanced Developing Countries (Possible Economic Payoffs from Military Production: The Case of Aircraft Industries in Brazil, Israel, and India)*, Cambridge, Mass.; Massachusetts Institute of Technology, 1980, p. III-27.

14. Ibid.

15. William Quandt, 'Influence through Arms Supply: The US Experience in the Middle East', pp. 124-5, in *Arms Transfers to the Third World: The Military Buildup in Less Industrialized Countries*, eds. Uri Ra'anan *et al.*, Boulder, Colo.: Westview Press, 1978.

16. Ibid.

17. *Wall Street Journal*, 13 May 1977.

18. US Congress, Senate Committee on Foreign Relations, Hearings: *Middle East Arms Sales Proposals*, 95th Cong., 2nd Sess., Washington, DC: US Govt. Printing Office, May 1978.

19. *Aviation Week and Space Technology*, 22 June 1981.

20. 'Israel in Peace, but . . . ', *Military Technology and Economics* 4:20 (1980): 34.

21. Rothenberg, *The Anatomy of the Israeli Army*, p. 218.

22. *Israel: A Country Study*, The American University Foreign Office Area Series, Washington, DC: US Govt. Printing Office, 1979, p. 213. See also *Aviation Week and Space Technology* 102 (31 March 1975): 17.

23. *Aviation Week and Space Technology*, 8 October 1979.

24. Ibid.

25. Peres, *David's Sling*, p. 113.

26. *Christian Science Monitor*, 14 January 1981, p. 6.

27. *Bank of Israel Annual Report 1979* (Hebrew – D'in V'Heshbon) Jerusalem: Israel Government Press, 1980, p. 67. See also *Israel: A Country Study*, The American University Foreign Office Area Series, Table 24, Appendix A.

28. *Bank of Israel Annual Report, 1979*, p. 167.

29. Ibid.

30. Ibid, p. 169.

31. David Shipler, 'Israel to Build Jet Fighters to Ease Reliance on US', *New York Times*, 8 February 1982, p. A11. In addition, over $300 million has already been invested in the development of the Merkava tank. ('Monthly Report on Industrial R&D and Science Based Industry in Israel', August 1981, AG Publications, Jerusalem.)

32. *Israel Economist*, March 1981, p. 3.

33. *Statistical Abstract of Israel 1979*, Jerusalem: Central Bureau of Statistics, 1979, Table VI/8.

34. *Aviation Week and Space Technology*, 10 April 1978, p. 33, and Center for Policy Alternatives and Center for International Studies, *Disarmament and Development*, p. 5.

35. *Statistical Abstract of Israel 1977*, p. 417.

36. Based on data in *Military Technology and Economics* 4:20 (1980).

37. *Bank of Israel Annual Report 1979* (Hebrew – D 'in V'Heshbon), p. 202.

38. Leah Shinan and K. Nagaraja Rao, *Supply and Demand for Professional and Technical Manpower in Israel*, Cambridge, Mass.: Center for Policy Alternatives, Massachusetts Institute of Technology, 1980, p. 14.

39. Ibid., p. 24.

40. *Israel Government Yearbook 1979*, Jerusalem: Israel Government Printing Office, 1980, p. 98.

41. Ibid.

42. *Israel Economist*, April 1981, p. 23.

43. Ibid.

44. *Military Technology and Economics* 4:20 (1980):27

45. Ibid.

46. *Israel Government Year book 1979*, p. 97.

47. See note 1.

48. Based on data in the *Israel Economist*, April 1981, p. 23.

49. *Statistical Abstract of Israel 1977*, Jerusalem: Central Bureau of Statistics, 1977, p. 165.

50. Samuel N. Bar-Zakay, 'Technology Transfer from the Defence to the Civilian Sector in Israel – Methodology and Findings', *Technological Forecasting and Social Change* vol. 10 (1977):144.

51. *Statistical Abstract of Israel 1977*, Table viii/7.

52. Centre for Policy Alternatives and Centre for International Studies, pp. III-19/20.

53. *Aviation Week and Space Technology*, 15 June 1981.

54. Ibid.

55. *Jerusalem Post*, February 1981.

56. *Aviation Week and Space Technology*, 20 April 1981, p. 55.

57. Centre for Policy Alternatives and Centre for International Studies, p. III-8.

58. *Military Technology and Economics* 4:20 (1980).

59. *Christian Science Monitor*, 14 January 1981, p. 6.

60 *Bank of Israel Annual Report 1979* (Hebrew – D'in V'Heshbon), p. 6.

61. US Congress, House of Representatives, Committee on Foreign Affairs, Report: *US Security Interests in the Persian Gulf*; Washington, DC: US Govt. Printing Office, 16 March 1981, p. 80.

62. *Military Technology and Economics* 4:20 (1980): 36.

63. *Israel: A Country Study*, The American University Foreign Office Area Series, p. 214.

64. *The Israel Economist*, September 1980, p. 9.

65. Ibid.

66. *Israel: A Country Study*, The American University Foreign Office

Area Series, p. 214.

67. *Aviation Week and Space Technology*, 29 April 1974, pp. 32-3.

68. *Quarterly Economic Review*, Second Quarter 1978, p. 6.

69. *Israel Government Yearbook 1979*, pp. 97-9.

70. *Aviation Week and Space Technology*, 29 April 1974, pp. 32-3.

71. *Quarterly Economic Review*, Fourth Quarter 1980, p. 14. This figure probably includes refurbished aircraft from the IAF inventory which are then resold.

72. *Aviation Week and Space Technology*, 13 December 1976, pp. 14-17.

72. Planes have reportedly been sold to Ecuador and negotiations have begun with Columbia. See *Aviation Week and Space Technology*, 30 March 1981, p. 21, and 25 January 1982, p. 15.

74. *Washington Post*, 21 March 1981, p. 18.

75. *The Economist*, 22 November 1980.

76. *New York Times*, 9 March 1977; *New York Times*, 23 June 1977.

77. *Memorandum of Agreement between the Government of Israel and the Government of the United States of America Concerning the Principles Governing Mutual Cooperation in Research and Development, Scientist and Engineer Exchange, and Procurement and Logistics Support of Selected Defence Equipment*, March 1979.

78. *London Times*, 25 June 1976.

79. *Defense Electronics*, March 1981.

10 DEVELOPING COUNTRIES

Herbert Wulf

Introduction

The domestic production of arms in developing countries is – with remarkably few exceptions – a fairly recent phenomenon. This new trend, usually based on the importation of sophisticated technology from industrial countries, has profound implications for disarmament and development, yet its analysis has been surprisingly neglected.[2]

The creation of arms production programmes in developing countries has been associated with a variety of conflicting motives and expectations. The first group of these are military and political in nature. In a great number of developing countries, the drive to be self-sufficient and to reduce the dependence on decision-making in highly industrialized countries has led to domestic arms production. The possession of strong armed forces supplied with locally produced arms is considered an attribute of political independence. Yet the acquisition of weapons or the import of production technology from outside powers may in reality perpetuate the state of dependence.

A perception of threat – actual or potential – from neighbouring countries can also precipitate the move towards domestic arms production. An example is the arms industry in Israel, on the one side, and in Egypt, on the other. The availability of domestically produced arms is considered a safeguard in crisis situations. Occasionally it is claimed that the arms industry can promote regional dominance. For example, Brazil's arms industry 'is seen to complement that nation's long-term policy goal of maintaining dominance in Latin America and exerting greater influence in sub-Saharan Africa and the third world in general.'[3] Finally, in addition to the desire to provide one's armed forces with the most sophisticated arms imported from industrialized countries, the prestige attached to domestic arms production might serve to accelerate the growth of the arms industry.* Spectacular

*It should be mentioned, however, that observers from industrialized countries quite frequently try to downgrade self-reliance strategies in developing countries as prestigious, i.e., too ambitious, unworkable, impractical or irrelevant. Paternalistic attitudes on the part of industrialized countries are as deceptive as false assessments on the part of developing countries are disappointing in the long run.

weapons developments have been seen as a means of emphasizing the strategy of national independence.*

A second set of arguments in favour of domestic, third-world arms production is economic in nature. The world-wide transfer of arms has accelerated during the last decade, but simultaneously fewer arms have been supplied under 'military assistance' programmes while more arms have been delivered on commercial terms. As a result, more countries have been importing arms at increased cost. Developing countries believe that they can reduce these costs – and particularly save foreign exchange – by substituting domestic arms production for arms imports. It is also argued that arms production programmes might contribute to the civilian economy indirectly by improving the skills of the manpower engaged in arms production and by increasing the productivity of the work force. Finally, the inflow of sophisticated arms production technology is expected to help an economy keep abreast of modern technology in general. As with the manpower argument, arms production technology is claimed to have spin-off effects for other industrial branches, thus stimulating overall industrial development. Whether these economic expectations can be fulfilled will be discussed in this chapter.

It is important to note here that strong military, political and economic stimuli exist for further expansion of domestic arms production and increasing transfers of arms production technology. The incentives on the demand side are only one part of the picture. There has also been a general willingness on the part of governments and companies in industrialized countries to meet the demand for technology imports, which are not only conducive to the process of expanding domestic arms production in developing countries but are necessary if such programmes are to be carried out. A major political motive behind the supply of arms and arms production technology has been and still is to exert political influence; it has rested on the maintenance of friendly regimes. As recipient states have increasingly attempted to diversify their supply source in order to maximize independence, suppliers (governments as well as companies) have, occasionally rather reluctantly, agreed to export production technology in order to maintain political influence and expand their markets.†

*Such ambitions are attributed to the early arms programmes under Perón in Argentina.

†The Soviet Union is – according to the standard sources on arms transfers (SIPRI and the US Arms Control and Disarmament Agency) – the largest supplier of arms. In terms of supply of production technology for arms, however, only a limited number of projects have been assisted, as is seen in the next section.

A second rationale for the arms producers' willingness to transfer production know-how and technology to developing countries is the reduction of costs for certain components produced with cheap labour in less-industrialized countries. Occasionally, such components are produced in developing countries on a subcontracting basis, and then exported to the licencer. The extent to which the production of arms in developing countries is possible will depend on the answers to the following broad questions:

a) How diversified is the country's industry; is there already a competent industrial base with skilled manpower and a minimum level of research and development (R&D) facilities available?

b) Can production costs (given the small production runs) be kept low; are there possibilities for exports of arms produced; to what extent will the financial resources of the developing country be allocated to install these production facilities?

c) Who controls production technology; can access to know-how, licenses, patents, etc., be secured by developing countries or are arms producers in industrialized countries or their governments reluctant to cooperate; if suppliers are willing to provide technology, at what financial and political cost to the recipient; does the capacity for indigenous development and production exist?

MORPHOLOGY OF ARMS PRODUCTION IN DEVELOPING COUNTRIES

Projects and Countries at a Glance

The register of arms production (Table 10.1) lists 32 developing countries engaged in arms production or planning the manufacture of arms.* In addition, there are several countries not included in this list

*The group of countries described as 'developing' is not homogenous. It is therefore not surprising that the number of countries included in this category varies. While the World Bank, for example, classifies Greece, Portugal, Turkey, Yugoslavia, and Israel as 'middle-income countries' (along with such countries as Brazil and South Korea), the United Nations calls them 'developed-market economies'. For the purpose of this chapter it was considered appropriate to include all the aforementioned countries as well as South Africa and North Korea. If indicators such as per capita income and the level of industrialization are taken as a measure, it would not make sense to include countries like Singapore, Brazil, or South Korea but to exclude Portugal, Turkey, Greece, Yugoslavia, Israel or South Africa. The People's Republic of China is excluded since it is, first, considered as a class in itself and, second, because such limited information on its arms industry is available. It should be stressed that the classification chosen here is different from the often used UN-classification. (In this volume, China is discussed in Chapter 8 and Israel in Chapter 9.)

where only ammunition is reportedly produced (Algeria, Congo, Ghana, Guinea, Morocco, Nepal and Zaire) or where in the past a small number of simple ship hulls have been produced locally (Bangladesh, Gabon, Ivory Coast, Malagasy Republic and Sri Lanka).* The register of arms-producing countries includes four countries in Europe, eight in Latin America, four in Africa, five in the Near/Middle East and eleven in Asia. The level of domestic arms production attained so far in the respective countries differs substantially. Many industrial undertakings are restricted to the production of small arms (30 countries), ammunition in relatively small quantities and the construction of small naval craft (22 countries).

Several countries, however, have attained a high level of arms production and a relatively high degree of diversification. Modern fighter aircraft are built in eight countries, generally under licence (four additional countries are planning such programmes), while light aircraft, trainers and transport aircraft are manufactured in eleven countries. Eleven aircraft manufacturers throughout the developing world produce or assemble helicopters and another twelve produce guided missiles. For military electronics or avionics, the figure is ten. Major fighting ships like destroyers, frigates, corvettes and/or fast patrol boats are constructed in fourteen countries, but often only the hulls are produced locally while the engines, weapons and electronic equipment are imported. The production of submarines is limited to four countries and the manufacture of tanks to five. Artillery and cannons are manufactured in six of the 32 countries and armoured personnel carriers and other military vehicles are built in nine countries.

Of the four European countries, Yugoslavia possesses the most extensive experience and produces arms in all the twelve categories of weapons covered in Table 10.1. It is interesting to note that Yugoslavia's arms industry has developed several indigenous projects and was also able to secure production licences, know-how and components from both East and West. Sizeable quantities of weapon systems, especially patrol boats, have been exported to developing countries by Yugoslavia. Both Greece and Turkey pursue diversified arms production programmes, but economic constraints have limited these ambitions considerably, whereas Portugal has had to reduce its incipient production for economic reasons.

The foremost arms producers of Latin America are Brazil and

*These are such isolated efforts that they are not considered relevant to this examination of arms production.

Table 10.1: Arms Production in Developing Countries

COUNTRY	(1) fighters jet trainers engines	(2) light planes transporters	(3) helicopters	(4) guided missiles rockets	(5) major fighting ships fast patrol boats	(6) small fighting ships	(7) submarines	(8) tanks	(9) artillery cannons	(10) light tanks APCs trucks jeeps	(11) electronics avionics optronics	(12) small arms mortars bombs
Europe												
1. Greece	r				l	d[a]		r				l
2. Portugal					l	l						l
3. Turkey	r		l[a]	l	l	n	l					l
4. Yugoslavia	d	d	l	l	l d	d	d		l d	d	n	l d
Latin America												
1. Argentina		l d	l d	l d	l	d	l	l	l d	l		l d
2. Brazil	c l[a]	l d	l	l d	l	d		d	l	l d	l	l d
3. Chile						l						n
4. Colombia						l d						l
5. Dominican Republic						n						d
6. Mexico						l						l
7. Peru					l d							
8. Venezuela		l[a]	l[a]			l						n
Africa												
1. Nigeria			l							l[a]		l
2. South Africa	l	l	l	l d	l	d			l	l	l d	l d
3. Sudan												l
4. Zimbabwe												l

Near/Middle East																
1. Egypt	l[a]		l[a]	l								l		l[a]		l
2. Iran	(r)		(l)	(l)								(l)		(l)		l
3. Israel	d	l d			d	l	d	l d		l d	d	d		l	d	l d
4. Libya	l[a]															
5. Saudi Arabia	l[a]			l[a]												l
Asia																
1. Burma								l							l	
2. India	l d	l d	l	l		l	d	l d	l	ld[a]	d	l	l	d	l	d
3. Indonesia		l d	l					d							l	
4. Korea (North)	l					l		d	l						l	
5. Korea (South)	l		l	l				l d		rl[a]		l		l	l	
6. Malaysia						l									l	
7. Pakistan	l	l	l	l											l	
8. Philippines		l d	l		n			l						l	l	
9. Singapore	r					l		l						l	l	
10. Taiwan	l d[a]	d	l	l		l		n				l		l	l	
11. Thailand								l d								

Notes: c. components only; r. repair, maintenance; l. licence, know-how transfer, assembly; d. indigenously developed, produced; n. not known whether l or d; a. planned; ( ) = status uncertain

Source: IFSH-Study Group on Armaments and Underdevelopment, *Transnational Transfer of Arms Production Technology*, IFSH Forschungsbericht 19, Hamburg: 1980, appendix, updated.

Argentina. Argentina has had a long history of unsuccessfully attempting to establish an autonomous arm industry and production is currently directed towards licence agreements. Brazil's arms producers stress licence production as well and have been successful in exporting armoured fighting vehicles and light aircraft. The other Latin American arms producers – Colombia, Chile, Dominican Republic, Mexico, Peru and Venezuela – concentrate on the production of small arms and small fighting ships. It has been reported in the past that helicopters were produced under licence or assembled in Peru, but recent information on that project is not available. Plans for similar production exist in Venezuela.

African arms production is, except for South Africa, negligible: helicopters are assembled in Nigeria; small arms are produced in Nigeria, Sudan and Zimbabwe. The Republic of South Africa is one of the major arms producers of the developing countries. Despite the UN arms embargo, weapons imports continue to flow to South Africa. Furthermore, South Africa has been able to import technology and production machinery to produce arms locally.

In the Middle Eastern region, Israel has established the most sophisticated and probably in terms of turnover the largest arms production of all developing countries (see Chapter 9). Arab countries, partly in response to Israeli efforts, planned a major arms production complex in Egypt, Saudi Arabia, Kuwait and Qatar. As one result of the Arab boycott of Egypt following the Israeli-Egyptian peace talks and treaties, the latter three countries cancelled their financial commitment to this joint enterprise. Consequently, most projects for which licence agreements had already been fixed or planned with West European and US arms producers have been cancelled. New projects are under discussion between Egypt and the US and France. The Egyptian government wants to proceed on its own while discussions are going on among the other Arab countries (including Iraq) about setting up an Arab arms industry excluding Egypt. The status of the previously far-reaching Iranian arms production plans are uncertain at the moment: some projects for licence production have been cancelled while the status of others is unknown. It seems likely that production is currently limited to small arms and simple spare parts. The Libyan government is considering the production of an Italian-made fighter.

Of the eleven Asian countries producing arms, India has launched the most ambitious programme, trying to reach self-sufficiency not only in production but also in design and development. The government of India has initiated programmes in each of the twelve arms categories covered in Table 10.1, often relying on indigenous development. Although the import of foreign technology and components have been reduced to a minimum in small arms production, substantial supplies from abroad are needed for the major arms production programmes. It is interesting to note that India and Peru are the only countries outside the group of socialist countries that receive production licences from the USSR. Second and third on the list in Asia are Taiwan and South Korea. Both countries have considerably expanded their production capacities for a large variety of weapon systems, especially during the last few years. Hardly any domestic development has been undertaken, however, making both countries reliant on the import of both technology and components. Other countries in Asia with ambitious plans and some production in progress are the Philippines, Indonesia, Pakistan, Singapore and North Korea.

Indigenous and Licence Production

Modern arms production usually requires a capacity for high technology in several branches of industry. The production process is highly complex and requires inputs from a diversified industrial base. Since such a broad industrial base is atypical for developing countries and since they tend to lack adequate research and development facilities, most arms produced in developing countries are licence-produced and not developed indigenously. When projects are developed indigenously, technical assistance from foreign personnel is often required and, as a rule, the most sophisticated components are imported.

Table 10.2 summarizes the number of major arms systems produced according to region and type of production. The majority of projects undertaken fall into two categories: aircraft and major surface ships/ fast patrol boats. Of the 46 different types of aircraft produced in developing countries, half (23) are licence productions, with the majority located in Asia. Modern fighters are developed in three countries: India, Israel and Yugoslavia. Licence production is even more common for the construction of ships. Only five of the 32 different types of major surface fighting ships or fast patrol boats are indigenously designed, in India, Israel, Peru and Yugoslavia. Similarly, the production of helicopters is apparently extremely difficult without the assistance of an experienced arms producer from an industrialized country. As far as is known, only one helicopter has been developed (in Argentina) for experimental purposes; sixteen other helicopter projects all involve production under licence. The fabrication of tanks has been undertaken only five times; submarines, six times. Of the five different

Table 10.2: Number of Major Arms Produced: Indigenous and Licensed[a]

	Aircraft	Helicopters	Tanks	Major Surface Ships Fast Patrol Boats	Submarines	Guided Missiles
Europe						
Indigenous	4			1	2	1
Licences		2		5	1	1
not known	1	1		1		
Latin America						
Indigenous	6	1	1	1		2
Licences	6	3	1	8	1	4
not known	1			1		
Africa						
Indigenous						
Licences	4	2		1		1
not known						1
Near/Middle East						
Indigenous	3		2	1		3
Licences				2		1
not known				1		
Asia						
Indigenous	7			2		
Licences	13	8	1	10	3	7
not known						
Total						
Indigenous	20	1	3	5	2	6
Licences	23	15	2	26	5	14
not known	2	1		3		1

Note: a. Only those projects with at least prototype production, in production, or with production completed have been included.
Source: IFSH-Study Group on Armaments and Underdevelopment, *Transnational Transfer of Arms Production Technology*, IFSH Forschungsbericht 19, Hamburg: 1980, appendix, updated.

tanks produced, three were domestic developments undertaken in Israel and Brazil, while two of the six submarine projects are known to be of Yugoslavian design. Twenty-one different guided missiles, ranging from simple portable anti-tank missiles to complicated surface-to-air systems, are produced in developing countries; two-thirds are produced under licences from industrialized countries.

Table 10.1 illustrates that apart from the domestic development of light planes, trainers and transport aircraft (developed in eight countries), the development of major arms is extremely limited. Only ten countries have produced indigenously designed weapons systems of any sort. While most of them are in production, though usually in small production runs, it should be pointed out that some of these projects have so far not passed the stage of prototype production. To reach the phase of series production often requires heavy industrial investments and additional technology imports. Table 10.1 also shows that three countries – Israel, India and Yugoslavia – have developed weapon systems in five or six of eight different categories. This table is, however, somewhat misleading regarding South Africa's arms production. South Africa has, especially during the 1960s, started a number of licence production schemes, mainly with Italian and French cooperation. Due to the UN arms embargo, the South African arms industry has received top priority and has thus been able to increase the domestic content of arms production. For this reason, some of the projects started with licences from abroad are now classified by some analysts as 'indigenous' projects.

Since arms production in developing countries depends heavily on licences, a large share of these projects relies on technology imports from producer nations. More than two-thirds of all licences listed in this chapter originated in only four countries: the United States, France, Great Britain and the Federal Republic of Germany. Apart from considerations of industrial self-sufficiency, questions of political independence and possibilities for reducing dependence on arms imports immediately arise in regard to licence production.

It is interesting to note how restrictively Soviet licences are granted. Of those licences recorded in Table 10.1, several date back many years and are related to Soviet-Yugoslav cooperation in the production of arms. More recently, India and North Korea have received licences for the production of fighters, guided missiles, and transport planes and while the assembly of Soviet helicopters in Peru was reported in the past, recent information on the possible existence of the project is not available.

Major Producing Companies and Countries

Information about the major arms producers in developing countries is scarce. Not surprisingly the largest producers are to be found in Israel, India, Brazil, South Africa, Argentina and probably Yugoslavia. Some of these ventures (often public enterprises or government departments) are large conglomerates with numerous subsidiaries, quite comparable in the size of their labour-force and turnover to some well-known arms producers in Western Europe and the US. For example, the Ministry of Defence in India controls nine companies with more than 90,000 employees plus 31 ordnance factories with about 150,000 employees.[4] ARMSCOR, the government-owned Armaments Development and Production Corporation of South Africa, controls a wide range of weapons and war materiel production and is said to have a budget of almost $1.2 billion. ARMSCOR purchases parts, technology and so on from about 800 companies. In addition to the 19,000 employees at ARMSCOR, another 100,000 workers are said to be employed by arms subcontractors.[5] One of the large conglomerates is the Argentinian military-industrial complex, founded in 1941 and run by the army, the Fabricaciones Militares.[6]

It is interesting to observe that the two major aircraft producers in developing countries, Israel Aircraft Industries and Hindustan Aeronautics Ltd, are comparable in sales and employment to some of the major aircraft producers of the world. The largest French aircraft producer SNI Aérospatiale employs about 34,000 people and had a sales volume of about $1.2 billion in 1977; Dassault-Breguet, with a slightly lower sales volume, employs about 15,000 people. The largest West German aircraft producer, MBB, employs about the same number of people as Israel Aircraft Industries, namely 20,000. The Hindustan Aeronautics Ltd labour force is 100 per cent larger. Northrop in the US has about the same number of employees as Israel Aircraft Industries. Brazil's two largest arms producing comanpanies, Embraer (4,800 employees) and Engesa (4,500 employees), are comparatively small.*

It is possible to delineate a rank order of countries according to size and diversification of arms production on the basis of the list of arms production projects developed in Table 10.1 (see Table 10.3). In the

*To arrive at a more descriptive comparison of the size of the companies it would be necessary to consider turnover, productivity, and value-added for each case. Many of the arms producers in developing countries cannot purchase components locally due to the lack of qualified subcontractors. Components are therefore produced to a greater extent in the company or they have to be imported.

top group of countries, Israel is found in first place. India is second, Brazil third, and Yugoslavia fourth. In these countries, weapons are produced in almost all of the twelve weapons categories listed in Table 10.1 and several of these projects are indigenous developments. A second group of countries – South Africa, Argentina, Taiwan and South Korea – also produces weapons in most of the twelve categories but has fewer indigenous projects. The third group, with production in several weapon categories but much smaller in size and without any major capacity for indigenous development, consists of the Philippines,

Table 10.3: Rank Order of Major Arms Producing Countries

Rank Order	Country	Group of Countries
1	Israel	I
2	India	diversified and
3	Brazil	sizeable production
4	Yugoslavia	
5	South Africa	II
6	Argentina	production most of
7	Taiwan	the twelve weapons
8	Korea (South)	categories
9	Philippines	III
10	Turkey	production in several
11	Indonesia	weapons categories,
12	Egypt	without substantial
13	Korea (North)	capacity for indigenous
14	Pakistan	development
15	Singapore	
16	Iran	IV
17	Colombia	
18	Portugal	
19	Greece	
20	Peru	some isolated projects
21	Thailand	
22	Venezuela	
23	Domin. Republic	
24	Nigeria	
25	Mexico	
26	Malaysia	
27	Burma	
28	Chile	
29	Saudi Arabia	
30	Sudan	
31	Zimbabwe	
32	Libya	

Source: Table 10.1 and IFSH-Study Group on Armaments and Underdevelopment, *Transnational Transfer of Arms Production Technology*, IFSH Forschungsbericht 19, Hamburg: 1980, appendix.

Turkey, Indonesia, Egypt, North Korea, Pakistan and Singapore. Most of these countries have ambitious expansion plans and it can be expected that despite severe economic problems (particularly in Turkey, Pakistan and Indonesia) considerable investments will be allocated for setting up new arms industries. As of 1980, all other countries, from rank 16 to 32, have undertaken only isolated projects in two or three of the twelve weapon categories.*

If the expectations or claims by most governments in arms-producing countries regarding the effects of arms production are taken seriously, a strong inverse correlation between arms production and arms imports should exist. In other words, the major arms producers should not be the same as the major arms importers. Arms transfer data illustrate that such a clear negative correlation does not exist. A number of major arms importers are among the top developing-nation arms producers. Despite considerable efforts in arms production, weapons imports are required to meet the demands of the armed forces.

The Potential Arms Production Base

Sophisticated modern weapon systems are usually manufactured from hundreds of different kinds of industrial metals and materials; standardized components and parts demand a rigid uniformity in materials specifications and manufacturing tolerance. Literally thousands of different components have to be produced. As a consequence, arms production is dependent on supplies from a variety of industries: iron and steel, electronics, foundry, metallurgy, transportation equipment, and machine tools, to name only the most obvious. To establish an enclave-type of arms industry therefore, seems to be an unrealistic approach. The production of some types of major weapon systems appears to create greater difficulties than others. In an early study on the transfer of arms to developing countries, it was stated:

> At first glance it may seem that the construction of tanks would be relatively simpler than that of modern aircraft. However, tank construction involves not only very heavy industrial equipment but skills which are less likely to be found in the developing areas and are more difficult to bring into being than a capacity for aircraft

*Iran's arms production was larger in 1977; it is included in this group since many ambitious projects have since been cancelled.

> *assembly*. Most of the aircraft produced under licence have demanded only the development of light industry to fabricate parts and produce assemblies: the engines and more difficult electronic components were usually available from the original manufacturer or another major industrial power (emphasis added).[7]

Large-size foundry pieces used for tanks are particularly difficult to produce and this is probably the reason for the small number of tank production programmes in developing countries. The above-cited article also draws attention to the remaining import dependency in aircraft production, something that is confirmed by the large number of licence agreements mentioned above. Similarly, the production of large fighting ships and artillery depends on the existence of heavy industrial capacity and it is assumed that most electronic systems related to these have to be imported. Even the production of small arms requires considerable technological expertise. Gavin Kennedy has pointed out:

> Small arms production requires machine-shop skills such as are found in tool-makers, machinists, fitters, drillers, borers, lathe operators, forgers, reamers, press workers, heat treatment specialists and so on. The workers would need to work to fine tolerance and, for operational effectiveness, the parts would need to be interchangeable.[8]

A weak industrial infrastructure, typical of developing countries, imposes technological limitations on arms production. A low degree of industrial diversification and a shortage of qualified manpower, production equipment, and special raw materials inhibits plans for indigenous arms production. A production base of reasonable proportions is a precondition for the initiation of arms production. The smaller the industrial base and the manpower potential, the less likely the production of arms. If arms are to be produced in a particular country despite serious limitations in the industrial base and lack of skilled personnel, the greater the need to import licences, parts and other technology.

The increase in domestic arms production in developing countries can probably be explained in part by the growing manufacturing sector. Conversely, the desire to produce arms might also have enhanced the propensity to give priority to industrialization in general and to certain industries in particular. To get an idea of the size of the capacity for arms production in developing countries, an attempt has been made to identify what may be called the *potential arms production base*. To arrive at a rank order of countries according to this potential production base, which can in turn be compared with the actual rank order of arms producers, two basic indicators have been chosen: ***the industrial base and the manpower base.***

Since the United Nations system of International Standard Industrial Classification (ISIC) does not include a category 'arms industry', an attempt has been made here to identify those sub-categories which are most relevant for arms production. These are: ISIC no. 371, iron and steel; ISIC no. 372, non-ferrous metal; ISIC no. 381, metal products; ISIC no. 382, machinery (non-electrical); ISIC no. 383, electrical machinery; and ISIC no. 384, transportation equipment. This selection implies the exclusion of major divisions number 2, 4 and 5: mining, energy production and construction.

These six key industries, which will be called the ***'relevant industries'***, for arms production are some of the most advanced industries in terms of their ability to incorporate new technologies and to apply research and development. Their output has been taken as the first indicator of arms production capacity.[9] The second indicator, 'manpower base', consists of two sets of data: the employees or persons engaged in the 'relevant industries' and the total number of scientists, engineers and technicians involved in R&D during the latest year for which data are available.

From these two indicators, a combined rank order of countries has been constructed. Since no comprehensive data are available for all the countries of interest, some estimates have had to be made in order to arrive at this rank order. It should be pointed out that for the purpose of this discussion an exact knowledge of the value of output or the number of qualified personnel is not so important. Of particular interest is the magnitude of arms production potential and the rank order. It is, however, obvious that below a certain level of production potential the fabrication of weapons is hardly possible. It is estimated that those countries in Table 10.4 having a rank order of 26 or more would not possess the minimum industrial capacity necessary for arms production. The same is true of all developing countries not included in Table 10.4 and all of these countries can be expected to have only negligible domestic arms production, if any at all. Only those countries up to around rank order number 15 (plus or minus one or two previous or subsequent ones) would be in a position to initiate arms production programmes beyond the assembly of parts.

The first conclusion that can be drawn from Table 10.4 is that a strong positive correlation exists between the share of total manufacturing in gross domestic product (GDP) and the share of the six 'relevant industries' as a percentage of manufacturing. If the *rank order of actual arms production* is compared with the rank order of the potential arms production base, a strong positive correlation is evident. All but one of the top fifteen major arms producing countries (North Korea, for which no information is available) belong to those 25 countries having a substantial potential for arms production. There are, however, some deviant cases of particular interest for this chapter.

A few countries with a substantial industrial and/or manpower base are not among the major arms producers: Mexico (ranked fifth in arms production potential), Greece (ranked tenth), Hong Kong (ranked fifteenth), Chile (ranked sixteenth), and Algeria (ranked twenty-first). According to their industrial and manpower base, these five countries are in a position to produce more weapons than they actually do. That they do not produce up to 'capacity' is due either to a policy decision or to certain constraints (like scarce hard currency reserves or difficulties in obtaining licences). The opposite also occurs. Certain countries have apparently overburdened their industrial and manpower bases with ambitious arms-production programmes. A particular case in point is Israel, the largest arms producing country, ranked only twelfth on the list of arms production potential. (See Chapter 9 also.) The Philippines is ranked ninth in terms of actual arms production and eighteenth in terms of the potential to produce arms, Indonesia, eleventh and twenty-third, respectively; and Pakistan, fourteenth and twenty-fourth. It seems likely that these four countries will experience economic and technical difficulties since their arms production programmes are not based on an adequate industrial base.

Some Theoretical Foundations and Empirical Evidence

As argued in the preceding section, the size and diversification of an arms industry are dependent on the structure and capability of a country's manufacturing industry in general. They are also closely related to its *industrial strategy*. In simplifying the industrial concepts held by developing countries, two basic strategies of industrialization can be identified: import substitution and export orientation. Each affects the kind of arms produced. Similarly, two models of arms production can be developed. First is the self-sufficiency concept based on substituting production for arms imports, whenever possible using indigenous designs. Second comes the model of internationally inte-

Table 10.4: Potential Arms Production Base

Rank Order	Country	Industrial Base			Manpower Potential	
		Manufacturing as percentage of GDP	Relevant industries as percentage of manufacturing	Output of relevant industries in US $ million	Scientists, engineers technicians in research & development (thousands)	Employees/persons engaged in the relevant industries (thousands)
(1)	(2)	(3)	(4)	(5)	(6)	(7)
1	India	16	32	5,025	97[xy]	1,688
2	Brazil	25	36	17,025	8	1,194
3	Yugoslavia	31	40	4,800	32	578
4	South Africa	23	38[y]	3,925	. .	396[y]
5	Mexico	28	. . .	. . .	6[xy]	167
6	Argentina	37	. . .	. . .	19	112
7	Taiwan	37	38	3,375	. . .	263
8	South Korea	25	21	2,500	19	322
9	Turkey	20	21	2,050	9[xy]	218
10	Greece	19	23	1,375	4	114
11	Iran	13	35	3,500	6[y]	90
12	Israel	30	33	1,300	3[x]	97
13	Portugal	36	20	1,275	4	130
14	Egypt	24	20[y]	875	11[xy]	98[y]
15	Hong Kong	26	21[y]	650	. . .	151[y]
16	Chile	20	45	1,325	6[y]	76
17	Venezuela	15	22	1,300	4[y]	79
18	Philippines	25	15	900	. . .	80

19	Colombia	19	17	625	1[x]	88
20	Thailand	20	21[y]	900	6[x]	...
21	Algeria	11	28[y]	625	0.3[x]	...
22	Singapore	25	32	600	1	91
23	Indonesia	9	12	525	19	61
24	Pakistan	16	12[y]	325	9	78[y]
25	Peru	19	25	425	...	49[y]
26	Malaysia	18	15[y]	425	...	72[y]
27	Nigeria	9	17	465	3[y]	23
28	Saudi Arabia	5	...	...	...	...
29	Zimbabwe	21	30	225	...	47
30	Morocco	12	14	200	...	18[y]
31	Iraq	7	12	150	2	16
32	Kenya	12	11	75	1[y]	27
33	Sri Lanka	15	14	50	...	21
34	Syria	11	11	75	...	14

Notes: x. only scientists and engineers; y. no data available after 1974.
Sources: Column 3: World Bank, *World Development Report 1979*, Washington, DC: 1979.
Columns 4 and 7: United Nations, *Yearbook of Industrial Statistics*, vol. I, New York: various years.
Column 5: Columns 3 and 4 plus International Monetary Fund, *International Financial Statistics*, Washington, DC: November 1979, for GDP figures.
Column 6: United Nations, *Statistical Yearbook 1978*, New York: 1978.
In addition, some national government statistics have been used.

grated arms production carried out in developing countries in close cooperation with arms producers of industrialized countries.

Import-substitution Industrialization and Self-sufficient Arms Production

Import-substitution industrialization was adopted as a growth-oriented industrial strategy in a number of developing countries between the late 1940s and the early 1960s.[10] The primary goal was to produce locally manufactured goods which had previously been imported. Profound structural changes in the economy were essential to build up autonomous production capacity. Stagnating traditional exports, lack of employment opportunities and, in several cases, increasing balance-of-payment problems led to the wide acceptance of this industrial strategy. At the end of the 1960s when this strategy had already failed in countries like Brazil and Argentina, it was hailed in other countries, especially in Asia, as the main avenue to industrialization and an essential condition for development.

The positive growth effects experienced during the early phase of import-substitution were, however, not long-lasting. One of the major problems of this industrialization strategy was the small size of local markets. Due to extremely unequal income distribution in third-world countries, the demand for domestic production was restricted and heterogeneous. The technologies used for production were usually capital-intensive, resulting in little growth of industrial employment and only limited increases in wages. To meet domestic demand, heterogeneous industrial structures were installed, which worked behind protective tariff barriers at high costs. The inefficiency of the new industries was therefore perpetuated: production costs normally were significantly above world-market prices; production was confined to import substitution and could not compete on the world market. Underutilized capacities were nevertheless profitable since the protection of the nascent industries (initially only a temporary tactic) continued on a permanent basis to defend local capital against international competition.

An additional bottleneck was inherent in the import-substitution strategy. Import substitution which stimulated growth in some industries required even greater imports in previous production stages. As a result, the composition of the import bill changed and existing balance-of-payment problems were not solved. Much of the initial success of the import-substitution strategy depended on the importation of production technology and on the influx of foreign capital. The capital of large multinational concerns which was basically confined to the extractive sector diversified during the import-substitution phase and penetrated the economies of developing countries on a much larger scale. Instead of reducing external dependence, the domestic production process became more vulnerable to outside pressures and more dependent on external inputs.

Obviously parallels exist between *import-substitution industrialization and arms production* in several developing countries. Arms production was part of the industrialization strategy and meshed well with aspirations for maximum possible political independence. Consequently, arms production programmes aiming at national self-sufficiency were instituted in some of those countries which emphasized import-substitution strategies. This is particularly true for arms production in Argentina under Perón and for India since the beginning of the 1960s. The major exception is Mexico, with intensive import-substitution industrialization but only limited arms production to date. Self-sufficiency in arms production is given even higher priority than industrialization in Israel, South Africa and, more recently, Taiwan and South Korea, whereas large-scale domestic production of arms began in Brazil only after import-substitution industrialization had been abandoned.

Expectations similar to those for import-substitution industrialization are attached to domestic arms production: reducing the dependence on imports, decreasing production costs, saving foreign exchange, creating employment opportunities, training labour, stimulating R&D and so on. In the final stage of the arms production programme, governments hope to supply the complete range of weapon systems for their armed forces without relying on technology imports.

So far, the record of the self-sufficient arms production model is not very much different from that of the import-substitution industrialization strategy. Structural difficulties and bottlenecks in the economies of the developing countries hamstring a policy of self-sufficiency. As long as domestic arms production is based on a weak industrial base, very large investments are required to initiate the design and production of the numerous components of modern weapon systems. Suboptimal utilization of production capacity characterizes both late-comer industrialization oriented towards the domestic market and late-comer arms production. Technological specialization leads to investments in highly diverse and only partly integrated production capacities; the limited demand of the domestic market – in the case of arms production, a part of the budget of the armed forces – results in oversized factories and, eventually, substantial cost overruns. While foreign exchange requirements might be eased by producing a particular weapon system instead of importing it, it seems likely that the imports of production technology required to set up the industrial plants involve a drain on the balance of payments which might be higher than the original saving. Because of the lack of industrial structure and despite relatively low labour costs, producing at costs below world-market prices is hardly possible. The exact cost calculations on arms production are not publicly available, although it can be observed that the major arms production programmes in developing countries have usually been plagued by cost overruns (which is not unusual for arms production in industrialized countries as well). Whenever prices are quoted for the export of goods, producers in developing countries seem to ignore actual production costs and take world-market prices as a guideline.[11]

An indicator of the extent of self-sufficiency in arms production is the content of indigenous production in a particular weapon system. Usually, only a limited percentage can be produced locally. Reducing the import content below 30 per cent – even in production programmes marked as 'indigenous development' – is the exception rather than the rule. Even the figures on indigenous content or detailed lists of imports of components are not the final word on technological dependence. In the early phases of the ambitious 'self-reliant' arms production in Argentina and also for the 'indigenous' development of modern, supersonic fighters in India, the know-how of specialists from industrialized countries had to be obtained. A similar pattern can be observed in Israel and South Africa today: thousands of specialists have been recruited in Western Europe and the US. Other countries involved in arms production like South Korea, Indonesia, Taiwan, the Philippines, Turkey, Pakistan or Singapore are obliged to rely on technology imports as well as imports of patents, licences and so on in order to go ahead with their production plans. Given a certain industrial base, the initiation of weapon assembly and the production of simple parts can be successfully managed within a short span of time. Beyond basic assembly and simple production, the technical diseconomies of further import substitution become substantial. An attempt to increase the domestic content per unit by substituting for imports normally leads to a steep rise in cost. (This pattern is illustrated in Figure 10.1.) The higher the indigenous content, the lower the import of parts and so on but the more likely the need for intensified cooperation with foreign firms as far as high technology is concerned.

Figure 10.1: Indigenous Content/Cost Function

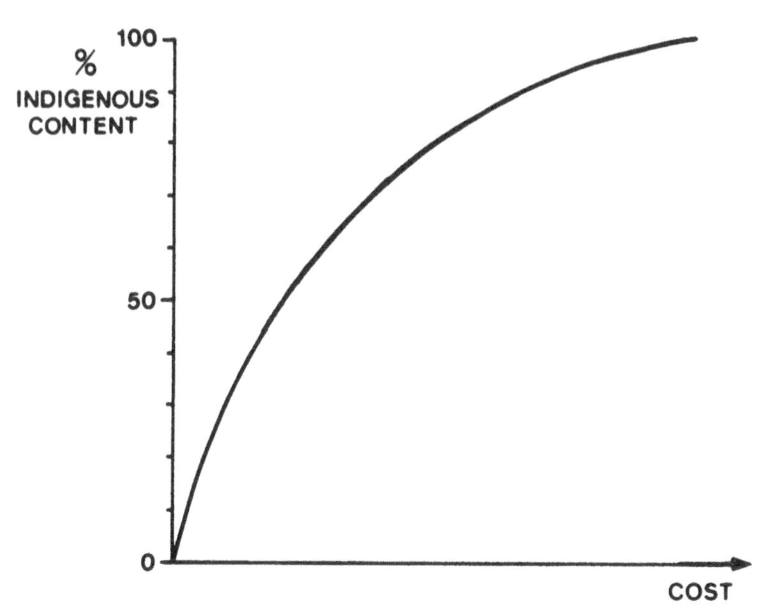

The anticipated training effect for the domestic work-force therefore seems doubtful, especially since the arms producers from industrialized countries are not likely to adapt their technologies to the particular requirements of the developing countries' industry and industrial labour. The Institute of Defence Studies and Analyses in India concludes that despite the long-term experience of arms production in India 'the implications are that, for modern and sophisticated weaponry, dependence upon the four major arms producing nations (USA, USSR, France, Great Britain) cannot be avoided.'[12]

Export-oriented and Internationally Integrated Arms Production

In contrast to direct foreign investment in the civilian sector, foreign investment in arms production companies in developing countries is rare. Regarding the general shift from import-substitution to export-oriented industrialization, Michael Sharpston has concluded:

> By the late fifties and early sixties, many countries had found import substitution an unsatisfactory method of industrialization, particularly for small countries where there were serious difficulties in obtaining adequate economies of scale; it is probably no coincidence that international subcontracting has been most in evidence

> in countries such as Hong Kong, Singapore, Taiwan, and Korea, where the small size of the domestic market made open economic policies particularly desirable. At the same time, there was a growing emphasis among development economists on export rather than import-substitution, and indeed this has now become the new orthodoxy, affecting government policy in quite a few developing countries. An increasing minority began to open up their economies, making importation and exportation easier, and often also providing certain direct or indirect subsidies to export industries.[13]

During the phase of import substitution, foreign capital was mainly concerned not to lose markets in developing countries and thus invested directly because of trade restrictions. The export-orientation in a number of developing countries, with subsidies for exports, tax benefits, low labour costs, restrictions on union activities and so on, leads foreign capital to invest in these countries beyond the immediate need of the domestic market.

The possibility of dividing the fabrication of goods into several elementary production processes (Taylorism) and the reduction of the time taken for transportation have encouraged the fabrication of components or their assembly in developing countries. The labour-intensity of the production process is the most typical reason for which this sort of production and assembly occurs in developing countries, and they tend to be intensive in the use of cheap unskilled or semi-skilled labour.

As this chapter has explained, numerous arms production programmes in developing countries can be described as an integral part of the import-substitution strategy. Probably the most obvious motive for producing arms domestically is the desire to reduce imports. There are several reasons why (at least until the end of the 1970s) the export-oriented industrialization strategy and arms production in developing countries coincided only partially. First, few industrial branches are under such close governmental control and as *dependent on public demand* as the arms industry. Even though the export of arms from most industrialized countries has been 'liberalized', export licences are usually required.

Second, in developing countries the major arms producing companies are *public enterprises*; only a limited number of foreign investments takes place in the arms industry in these countries. Export-oriented industrialization, however, is based on the removal of practically all political constraints and trade restrictions. Incentives must be offered to foreign investors to stimulate direct investment in developing countries. Third, arms technology and arms-production technology, especially for modern weapon systems, is often considered to be *'sensitive' or 'critical' technology* that should not fall into the hands of the perceived enemy. Governments in industrialized countries are therefore reluctant or unwilling to agree to the diffusion of the latest technology. A fourth reason for the limitations on arms production might be the *production process itself*: while export-oriented industrialization in the civilian sector is often based on mass production, using unskilled and semi-skilled labour, arms production typically consists of short production runs (with a few exceptions like small arms) requiring an over-proportionate number of highly skilled technicians, engineers, system managers and so on.

Nevertheless, in several countries like Brazil, Israel, Taiwan, South Korea and Singapore, governments have promoted the creation of an arms industry with the explicit intention of sales on the world market. Even so, many of the new suppliers remain themselves heavily dependent on imports from industrialized countries. It has been described above the extent to which arms production in developing countries requires licences from industrialized countries (see pages 313-20 and Tables 10.1 and 10.2). It was also stressed that in practically all so-called 'indigenous' projects, imports are also required. One important result of the export-oriented strategy is the inclusion of many arms producers of the developing world in an international division of labour in arms production.

What are the reasons that arms manufacturers in industrialized countries have for cooperating with arms producers in developing countries? The same motives encourage export-oriented civilian industrialization and export-oriented arms manufacturing. First of all, producers in industrialized countries do not want to lose a market. Rather than lose a customer to a competitor, a producer will be more accommodating (at times reluctantly) towards co-production and licensing arrangements. Second, in arms production the low labour costs and control of labour unions in developing countries might encourage industrialized-country producers to contract-out some components or the assembly of weapon systems.[14] Third, it is not an unusual practice for producers in industrialized countries (especially the Federal Republic of Germany) to evade export restrictions at home by producing abroad.

Different degrees of integration of developing countries into internationally organized arms production can be discerned. Usually, the first stage of production consists of *assembly work* only and is mainly directed at import substitution. Occasionally, however, it takes the form of a market expansion strategy of producers from industrialized countries. A typical feature of the division of labour between industrialized and developing countries in the civilian sector, namely the shipment of components to developing countries for assembly which are then sent back to the producer in an industrialized country, has so far not been observed in arms production. The afore-mentioned factors which limit export-oriented, internationally integrated arms production (short production runs, high transportation costs for heavy equipment, sensitive technology etc.) might be the reasons for restraint.

A second stage, usually in temporal sequence, is the *production of components*, starting with the least technologically sophisticated parts and gradually increasing the number of components and sub-systems produced domestically. There are two distinct forms of this sort of cooperation. The first is production for local 'demand'. Most of the projects fall into this category. When a particular weapon system is procured by the armed forces of a developing country, the government insists on domestic assembly, and when enough experience is collected in assembly work, additional components are produced locally. The declared intention of such projects usually includes self-reliant production as the final stage. The second form is when components are produced for the licenser in an industrialized country. This type of cooperation comes close to the often practised international sub-contracting of civilian industrial products. While this kind of cooperation in arms production is often carried out between producers in industrialized countries and is intended to be a compensation for the industry of the importing country, it is not so common between industrialized and developing countries. Few examples are known. Israel Aircraft Industries, for example, produces components for US-produced F-15 fighters. One can call this case a truly international division of labour under the guidance and control of a US company. Additional examples are the production of parts for the F-5 Northrop fighter in Brazil, or the L-70 Swiss-made Fire Control Racer in India. At present, a division of labour in aircraft production has been agreed upon between France and Brazil: an order for a Brazilian-made transport plane (Xingu) has been placed by the French while the Brazilian government has in turn ordered French fighters (Mirage) and helicopters (Puma).

A third stage of a different degree of export-oriented production of arms is fabrication *under the control of companies in developing countries.* It is probably most advanced in Israel and in Brazil. The

Brazilian company, Engesa, for example, exports a so-called indigenously developed armoured personnel carrier 'Cascavel' to several countries (Libya, Iraq, Qatar). In reality, the 'Cascavel' employs major parts (like the engine, transmission, electronic components, parts of the gun) imported from producers in industrialized countries or produced in Brazil by foreign-owned companies. Similarly, the Brazilian-made light fighter 'Bandeirante' consists of Brazilian-made as well as imported sub-systems. The sale of such products as 'Cascavel' and 'Bandeirante' or the Israeli-made 'Reshef' fast patrol boat or the 'Kfir' fighter on the world market is a rather new phenomenon of the late 1970s.*

A fourth stage, *direct investment of foreign companies* in arms production in developing countries, is rather the exception than the rule. The few cases for which information is available have been summarized in Table 10.5. Several of the few foreign investments are located in Israel, particularly in the electronics industry. It is also interesting to note that the original plans for building up an arms industry in Egypt were largely based on foreign investment. The status of these companies is, however, uncertain at present and it is likely that not all the five companies mentioned will come into operation. None of these foreign investors is expected to supply the capital shares in currency; it was planned to invest by granting licences for arms production. It is unlikely that all direct foreign investment in arms production companies has been covered in Table 10.5. It is known that some of the large arms producers from industrialized countries maintain service centres in developing countries and these may also be engaged in arms production.

A fifth and final stage of export-oriented arms production would be *fabrication in free-production zones*. So far, no information on such activities is available. One reason for the possible absence of arms production in free-production zones is probably the necessity of legitimizing the procurement of arms in a national context. Defending national security is the most obvious reason normally given for legitimizing arms procurement. A second reason might be the fact that arms are often sold below their actual costs because of government subsidies (like R&D reimbursement, inexpensive credits, pre-payments, and so on) that are granted to arms producers. These benefits might be more attractive than the economic incentives in free-production zones.

*See Chapter 9, pages 303-4 for a discussion of co-production in Israel.

Table 10.5: Direct Foreign Investment

Country	Company	Foreign Investor	Capital Share Holding (per cent)
Israel	Tadiran	G.T.E., USA	
	Elta Electronics	Carter Investment, UK	
	Motorola	Motorola, USA	
	Eljim	K.M.S., USA	
	Contahal	Control Data, USA	
	Turbochrom	Chromalloy, USA	
	Bet-Shemesh	Turbomeca, France	
Brazil	Helibras	Aérospatiale, France	45
Indonesia	Airtech	CASA, Spain	50
Egypt	5 companies planned	Rolls-Royce, UK	30
		SNECMA, France, UK	25
		Westland, UK	30
		Dassault, France	25
		BAC, UK	30
	Arab American Vehicle Comp.	American Motors, USA	49
Singapore	Vosper Thornycroft	Vosper Thornycroft, UK	100[a]
Malaysia	Hong Leong Lürssen	Lürssen Werft, FRG	40
Greece		Heckler & Koch, FRG	

Note a: This figure is uncertain.
Source: IFSH-Study Group on Armaments and Underdevelopment, *Transnational Transfer of Arms Production Technology*, IFSH Forschungsbericht 19, Hamburg: 1980, p. 58.

Some Empirical Evidence: Production Strategies of the Major Arms Producers

Self-reliant production in India and South Africa. Self-reliance or self-sufficiency has been the overriding principle of arms production in *India* for more than three decades. Conflicts with its neighbours, China and Pakistan, have stimulated the decision to decrease reliance on arms imports. The Minister of Defence stated in his 1976-7 *Report*, 'Past conflicts highlighted the need for Defence preparedness and therefore, of self-sufficiency and increasing self-reliance in the field of Defence equipment and weapon systems.'[15]

The volume of production in the 31 government-directed ordnance factories and nine public enterprises of the so-called 'defence production sector' has more than doubled during the last decade to more than $800 million. The arms production sector labour force has constantly been increasing, reaching – according to the annual *Report* of the

Ministry of Defence – a level of more than 190,000 employees in 1972 and over 240,000 in 1978.

One indication of the level of self-reliance is the indigenous content of major weapon systems. Official figures on the domestic value-added or the import content are only occasionally available; therefore, a definite answer cannot be given. Statements by the Indian government suggest that the indigenous content is constantly being increased. The frigate programme, for example, had an 82 per cent import-content for the first frigate while the last of a total of six units was down to 40 per cent. Occasionally, doubt is cast on such figures. For example, the journal, *Navy International*, pointed out in 1978 that Western companies 'have also been involved with the overall design and the main machinery for these ships'.[16] Nevertheless, such programmes are usually claimed to be indigenous. The fact that the government initiated a special 'indigenization programme' and referred to certain programme delays can be taken as evidence of continued difficulties in reaching self-sufficiency.

Probably the greatest failure of Indian arms production efforts was the development of the indigenous supersonic fighter-bomber HF-24 Marut, started in the early 1960s with a total production of about 140 units of different models. The aircraft never reached supersonic speed due to the failure to procure or develop an appropriate engine. A report in the Indian parliament criticized the protracted but eventually unsuccessful search by Hindustan Aeronautics Ltd to find more powerful jet engines and stated that by March 1975 development and production costs for the Marut Mk 1 were eight times higher than the original estimates.[17] Plans to produce an advanced version have apparently been abandoned in favour of producing the British strike-fighter Jaguar and possibly advanced Soviet fighters. (Both the MIG-23 and MIG-27 have been mentioned.)

The percentage of indigenous content in several projects reveal a high degree of autonomy in a number of these. A comparison with figures given for earlier periods suggests that dependence on imports has been decreasing over time. At the same time, it continues to be impossible for India to produce major modern weapon systems like fighters or sophisticated fighting ships entirely without foreign assistance.

ARMSCOR, the official procurement agent and major producer of armaments in the Republic of *South Africa*, increased its turnover from 32 million rand in 1968 to 979 million rand (almost $1.2 billion) in 1978. The basis for this tremendous growth is the frantic search for self-sufficiency which began in the 1960s and accelerated during the mid-1970s in expectation of a UN arms embargo. Government policy on arms production is described in the official White Paper:

> The voluntary arms embargo of the early sixties was really a blessing in disguise because it forced the RSA to take positive steps towards self-sufficiency. With the imposition of the mandatory arms embargo of 1977 the local armaments industry had already gained momentum. All those armaments disciplines concerning modern warfare had already been studied and, in many cases, fully mastered. The RSA is past the stage where it is only self-sufficient in respect of internal security. The conventional and advanced technological fields have been successfully entered and in many cases we have succeeded in moving through initial development and industrialized phases to line production . . .
>
> The mandatory arms embargo forced the RSA into those final phases of self-sufficiency that are typically the most difficult. But owing to the basis already established, the enthusiastic assistance of local industry and the innovative ability of our scientists, Armscor believes this challenge can be met. That the RSA has already achieved the number one position in the field of armaments in the Southern Hemisphere is no mean achievement.[18]

This statement is probably an appropriate description of the state of self-sufficiency. It is probably correct to say that all the weapons and equipment needed to control internal conflicts and to repress the black population, that is, small firearms, ammunition, simple communication equipment, military vehicles, armoured personnel carriers and light planes, are entirely domestically produced. The government policy statement also points to several likely loopholes in the UN embargo. First, the embargo was ignored by some companies and countries – especially Israel – which supplied weapons and components of weapons to South Africa. Second, numerous foreign companies have invested in South Africa; several of these subsidiaries supply equipment, parts, and technology to the South African arms industry.[19] Third, as has been pointed out by Peter Lock, international cooperation with South Africa's arms industry has increasingly shifted away from the supply of arms to the supply of machinery and patents, licences and so on.[20] Direct involvement has given way to a more indirect form of industrial assistance, expressed in constantly increasing imports of engineering products, especially from West Germany and the US.

Both India's and South Africa's arms exports are negligible and have been restricted, in the case of South Africa, to the former Rhodesian white minority regime and to Israel while India's exports have mainly concentrated on supplying small arms to some Asian countries and occasionally single aircraft, given away as military assistance.[21]

Export-oriented arms production in Brazil.[22] During the last two or three years, several military journals have reported enthusiastically about the success of *Brazil*'s arms industry on the world market. The export success story is basically that of two companies, Embraer and Engesa. Both companies are comparatively small, employing about 4,800 and 4,500 people respectively. The business concept of both companies is similar. Relatively simple, robust, and cheap light aircraft (at Embraer) and armoured vehicles (at Engesa) have been designed basically by using already existing and proven parts.

The characteristic feature of Brazil's economic sector has been an open-door policy leading to large-scale investment by transnational corporations and to an export-oriented industrial pattern. The entire international automobile industry is represented in Brazil; even the traditional subcontractors of US and European car manufacturers have established subsidiaries in Brazil. It is therefore not surprising to observe a fast growing arms industry that is based on the collaboration of the transportation and the machinery industries. Brazil's arms industry strategy is obviously, with some time lag, fairly well synchronized to the general export-oriented industrialization pattern. As in Israel, the Brazilian government does not restrict the export of arms on political grounds. Airplanes produced by Embraer have been exported to many countries in Latin America and Engesa's armoured vehicles are particularly popular in several Arabian countries.

A comparison of these two approaches of domestic arms production reveals that self-reliant arms production is quantitatively less dependent on imports but cannot be fully implemented without cooperation, technology supplies and licences from experienced arms producers in industrialized countries. On the other hand, the import of technology is part of the production strategy of internationally integrated domestic arms production. Transnational corporations are given an important role in the development and production process according to this second model.

Political-military Consequences: Self-sufficiency or Uninterrupted Dependence

Self-sufficiency in the supply of arms is the most common *raison d'etre* for arms production in developing countries. It is therefore legitimate to ask to what extent developing countries have, over time, developed the industrial base and the skills necessary to become self-reliant in arms production and large-scale military operations.

A qualitative change in the supplier-recipient relationships takes place when an importing country tries to diversify its supply sources. The direct leverage of a single supplier-country is reduced if arms are purchased in other supplier-countries and also if arms are produced domestically. There is virtually no major weapon system produced in developing countries that does not rely on the import of licences, production technology, knowledge, or components from industrialized countries. Given the underdeveloped character of the arms production base, the general paucity of skilled manpower and the low degree of industrial diversification, high-technology transfers and substantial cooperation from industrialized countries may well be required for the foreseeable future.

An examination of Indian, Israeli and Brazilian arms production programmes demonstrates that complete self-sufficiency has so far been reached only for some less-sophisticated weapon systems. Other countries, less advanced than Brazil or India both in general industrialization and in arms production, are even more dependent on foreign collaboration. The South Korean arms industry is almost totally oriented toward US licences and high-technology imports, even though the general industrial level is fairly advanced. Taiwan's efforts could be described in similar terms and the dependence of Egypt's plans to revise its arms production can best be demonstrated by pointing to the long list of West European and US collaborative projects which were discussed, then partly abandoned and are now once again in the planning stage, mainly in conjunction with US companies.

The defunct Arab Organization of Industrialization's excellent top-level management and good technological capabilities were praised by a US military journal. At the same time, the readers' attention was directed to the extensive training programme which was needed

> . . . in the US, England, and France to update Arab managerial, technical and production personnel in Western management, finance, design, engineering and production techniques. This year alone

> [1978] it was said some 2,500 Arabs are expected to complete their training in the US, England, France, Italy and West Germany including finance, airframe design, gas turbine design, propulsion, flight controls, metallurgy and structure.[23]

From the evidence collected in this chapter, it can be concluded that for the time being there is no short-term or even medium-term fulfilment of the desire of developing countries to reach a high degree of self-sufficiency in arms production. On the surface, a more equal partnership between the industrialized countries and those developing countries that have invested in arms production has emerged; in reality, basic asymmetries remain. As long as developing countries continue to formulate military scenarios akin to those prescribed for the East-West conflict, they are also bound to rely on the arms technology of industrialized countries, either imported as complete weapons or as production technology.

For developing countries the implications of the technological lead of the major industrial countries in arms production are that, for modern and sophisticated weaponry, dependence upon one or more of the major arms producing countries cannot be avoided. The implications of arms production in developing countries on the armaments dynamic and on disarmament efforts are manifold. More nations and producers are offering arms on the world market. The structure of the supplier market has therefore been directly affected. Effective control of arms transfers is apparently increasingly difficult. Concerted supplier action to reduce the transfer of arms seems ever more unlikely as the number of producers increases. In particular, the rugged, simple, cheap type of weaponry, like light aircraft, armoured vehicles, small arms and communication equipment, much in demand especially for internal control in many countries, is now offered for foreign sale by developing nation suppliers themselves. Several arms producers in developing countries have demonstrated their ability to produce these kinds of arms on the basis of a combination of domestic and imported technology.

Notes

1. This article is based on research undertaken at the IFSH-Study Group on Armaments and Underdevelopment. Due to lack of space, numerous details about the size of production, turnover and employment of companies cannot be presented here. For further details, see IFSH-Study Group on Armaments and Underdevelopment, *Transnational Transfer of Arms Production Technology*,

IFSH Forschungsbericht 19, Hamburg: 1980, and Herbert Wulf, *Rüstung als Technologietransfer*, Munich and London: Weltforumsverlag, 1980. Material on Israel (see Chapter 9) has also been omitted due to space considerations.

2. There are only a few studies available. A pioneering study was Stockholm International Peace Research Institute (SIPRI), *The Arms with the Third World*, Stockholm: Almqvist & Wiksell, 1971, Chapter 22. Updated information on domestic arms production is given in the *SIPRI Yearbooks*. See also, Raimo Väyrynen, 'The Role of Transnational Corporations in the Military Sector of South Africa', prepared for the United Nations Centre on Transnational Corporations, Helsinki: University of Helsinki, December 1978, mimeo; Helena Tuomi and Raimo Väyrynen, *Transnational Corporations, Armaments and Development*, Tampere: Peace Research Institute, 1980, Chapter V, pp. 144-207; Steven E. Miller, *Arms and the Third World: Indigenous Weapons Production*, PSIS Occasional Papers, no. 3, Geneva: Programme for Strategic and International Security Studies, Graduate Institute of International Studies, December 1980; Michael Moodie, 'Defense Industries in the Third World: Problems and Promises,' pp. 294-312, in *Arms Transfer in the Modern World*, eds. Stephanie G. Neuman and Robert E. Harkavy, New York: Praeger, 1979; and Andrew L. Ross, *Arms Production in Developing Countries: The Continuing Proliferation of Conventional Weapons*, Santa Monica, Calif.: Rand Corporation, 1981. For US co-production in developing countries, see Michael Klare, 'Arms, Technology, Dependency – US Military Co-Production Abroad', *NACLA's Latin America Report*, January 1977, pp. 25-32; Gavin Kennedy, *The Military in the Third World*, London: Duckworth, 1974, Chapters 15, 16; Ulrich Albrecht, Dieter Ernst, Peter Lock and Herbert Wulf, *Rüstung und Unterentwicklung. Iran, Indien, Griechenland/Türkei: Die verschärfte Militarisierung*, Reinbek: Rowohlt, 1976, Chapter 2; Peter Lock and Herbert Wulf, 'Register of Arms Production in Developing Countries', Hamburg: IFSH-Study Group on Armament and Underdevelopment, 1977, mimeo; Peter Lock and Herbert Wulf, 'Consequences of the Transfer of Military-Oriented Technology on the Development Process', *Bulletin of Peace Proposals* 8:2 (1977): 127-136; and Wulf, *Rüstung als Technologietransfer*.

3. See Moodie, 'Defense Industries in the Third World', p. 298. This chapter will not attempt to assess the validity of such arguments.

4. Government of India, Ministry of Defence, *Report 1977-1978*, New Delhi: 1978, Chapters VIII, IX. This is an annual publication.

5. Republic of South Africa, Department of Defence, *White Paper on Defence and Armaments Supply, 1979*, Capetown: 1979, p. 25.

6. See *Latin America Weekly Report*, 30 November 1979.

7. John L. Sutton and Geoffrey Kemp, *Arms to Developing Countries 1945-1965*, Adelphi Papers, no. 28, London: Institute of Strategic Studies, October 1966, p. 26.

8. Kennedy, *The Military in the Third World*, p. 295.

9. The output of the six industrial branches is also given in the United Nations *Yearbook of Industrial Statistics*, (vol. I), an annual publication. The precision of reporting varies to some extent, however, from country to country. To arrive at comparable data, the share of the six industries as a percentage of manufacturing output was computed. Using the share of manufacturing as a percentage of gross domestic product it was possible to arrive at a comparable output figure for more than 50 developing countries.

10. It is impossible to refer to the voluminous literature on the subject here. For a discussion of some of the basic problems in Latin America, see Werner Baer and Larry Samuelson, eds., 'Post Import-Substitution Industrialization in Latin America' (Issue Title), *World Development* 5 (January-February 1977): 1-168.

11. Often prices for arms are 'political' prices unrelated to the actual produc-

tion costs. A comparison of prices for light planes illustrates that those planes produced in Israel and Brazil were not the cheapest planes available. See *Interavia* (German edition), July 1974, p. 662. Prices for Israeli-produced fast patrol boats are, however, quoted at prices about 35-40 per cent below world market prices. See *The Economist*, 28 August 1976.

12. P.R. Chari, 'Indo-Soviet Military Cooperation: A Review', *Strategic Digest*, May 1979, p. 303.

13. Michael Sharpston, 'International Sub-contracting', *Oxford Economic Papers* 27:1 (1975): 99.

14. In an advertisement in *Internationale Wehrrevue*, no. 3 (1979): 365, the British fighting-ship producer, Vosper, poses the question: Why are fighting ships produced by a subsidiary in Singapore? The answer regarding labour relations: 'Difficulties in industry are unknown.'

15. Government of India, Ministry of Defence, *Report 1976-77*, New Delhi: 1977, p. 28.

16. *Navy International*, July 1978, p. 25.

17. This is according to *Milavnews*, no. 1 (1978).

18. Republic of South Africa, *White Paper on Defence*, pp. 25-6.

19. For a description of the involvement of transnational companies in the military sector, see Väyrynen, 'The Role of Transnational Corporations'.

20. Peter Lock, 'Rüstungstransfer und Weltmarktintegration', Hamburg: Free University of Berlin, dissertation manuscript, 1979.

21. For a list of exports and the recipient countries, see Wulf, *Rüstung als Technologietransfer*, Table 19.

22. The arms industry in Brazil is described in Lock, 'Rüstungstransfer', Chapter 7. Israel also follows the export-oriented arms production strategy. Israel is accorded a separate chapter in this volume and is therefore not discussed here.

23. Quoted in Raimo Väyrynen, 'The Arab Organization of Industrialization: A Case Study in the Multinational Production of Arms', *Current Research on Peace and Violence*, no. 2 (1979): pp. 66-79.

APPENDIX 1: THE UNITED KINGDOM

Compiled by Nicole Ball

The British defence sector exhibits both similarities and differences when compared with the defence sectors of other major Western countries. Military expenditure expressed as a percentage of gross national product has tended since 1950 to decrease in all the major NATO countries, the US, the UK, France and West Germany. British military spending expressed both in current pounds sterling and as a percentage of gross national product is shown in Table 11.1

Tables 1.2-4 list some of the major defence producers (divided according to ownership), their functions, and – where data are available – the number of employees and the location of production facilities.

Table 11.5 lists the twenty largest arms producers in the UK at the end of the 1970s. The degree of dependence on arms production (measured as a share of arms production in total sales) varies widely among these top twenty firms. Some, notably General Electric, British Leyland, EMI, Lucas Industries and Imperial Metal Industries, have diversified production and are not, as a corporate whole, dependent on arms production. Individual units within these corporations may, of course, be heavily reliant on the production of arms. Table 11.6 shows that for ten of the top 24 arms producers (42 per cent) with over £25 million in sales in 1977, arms production accounted for 50 per cent or more of total sales. In France and West Germany, a similar percentage of major defence contracts is dependent on the defence sector for 50 per cent or more of their sales. In Italy, however, more than 70 per cent of the major contractors show this degree of dependence on arms sales.

Figure 11.1 shows the geographic distribution of the fifteen largest defence contractors in 1979/80. The differences in the firms included in Figure 11.1 and in Table 11.5 arise partly from the different years from which data has been used but primarily from the different definitions of 'major contractor'. In Table 11.5, major contractors are firms which have sold more than £25 million in arms to any source. In Figure 11.1, major contractors are firms which have sold more than £25 million to the British Ministry of Defence. As for most of the other major weapon-producing countries, arms production in the UK is concentrated in particular geographic regions. The effects of converting arms production facilities to civilian uses are likely, therefore, to differ from region to region. There is no reason, however, to anticipate either that the effect of conversion on the economy as a whole would be negative or that individual areas could not adjust to cutbacks in military industries provided that industrial replacement took place.

UK arms exports rose rapidly in the late 1950s, dropped during the early 1960s and then rose again during the mid- to late-1960s. Between 1970 and 1979, the UK was ranked fourth by SIPRI as an exporter of major weapons. Like France, the export of arms from the UK accounts for a relatively large proportion of total exports of machinery and transport goods (6.8 per cent in 1978). For West Germany and Italy, the proportion is much lower (1.3 and 3.3 per cent in 1978, respectively). Table 11.7 provides data on the export of military equipment from the UK between 1975 and 1980.

Tables 11.8 and 11.9 provide information on UK defence-related research and development expenditure. The percentage of public R&D devoted to defence in the UK is rather high compared to other European states. Only France comes close with an average of 30 per cent of public R&D devoted to defence. Of the major Western countries, only the US spends more of its public R&D funds on defence.

Table 11.10 and Figure 11.2 give some indication of the way in which production costs for weapons have increased since the end of World War II. Table 11.10 provides data for selected naval vessels while Figure 11.2 shows relative cost increases for successive generations of shells, guided missiles, helicopters and aircraft. Some of these increases can be explained by inflation but by no means all. It must be noted that these comparisons are based only on production costs, not total life-cycle costs. It is likely that if life-cycle costs were available, the cost differentials between subsequent generations of military equipment would be even larger. With regard to naval vessels, for example, British naval sources estimate that production costs account for only 25 per cent of life-cycle costs.

Tables 11.11 and 11.12 deal with concentration in the UK arms industry. The trend in all major Western arms industries is towards increasing concentration and Britain is no exception.

Table 11.1: Defence Expenditure in the UK, 1950-80 (millions current £ and percentage GNP)

Year	Defence expenditure (1)	GNP[a] (2)	(1) as a percentage of (2)
1950	849	13,325	6.4
1951	1,149		
1952	1,561		
1953	1,681		
1954	1,571		
1955	1,567	19,281	8.1
1956	1,615		
1957	1,574		
1958	1,593		
1959	1,595		
1960	1,657	25,709	6.4
1961	1,709	27,426	6.2
1962	1,814	28,756	6.3
1963	1,870	30,548	6.1
1964	2,000	33,276	6.0
1965	2,091	35,922	5.8
1966	2,153	38,230	5.6
1967	2,276	40,353	5.6
1968	2,332	43,551	5.4
1969	2,303	46,718	4.9
1970	2,444	51,462	4.7
1971	2,815	57,528	4.9
1972	3,258	63,540	5.1
1973	3,512	73,565	4.8
1974	4,160	83,092	5.0
1975	5,165	103,237	5.0
1976	6,132	123,235	5.0
1977	6,810	140,534	4.8
1978	7,616		
1979	9,029		
1980	11,306		

Note: a. market prices

Source: For military expenditure — Stockholm International Peace Research Institute (SIPRI), *SIPRI Yearbook of World Armaments and Disarmament, 1968/69*, Stockholm: Almqvist & Wiksell, 1969, pp. 202-3; SIPRI, *World Armaments and Disarmament, SIPRI Yearbook, 1979*, London: Taylor & Francis, 1979, pp. 36-7; and SIPRI, *World Armaments and Disarmament, SIPRI Yearbook, 1981*, London: Taylor & Francis, Ltd., 1981, p. 163.

For GNP — 1950/1964: World Bank, *World Tables, 1976*, Baltimore, Md. & London: Johns Hopkins University Press, 1976, p. 280.

1965/1977: World Bank, *World Tables*, 2nd edn. Baltimore, Md. & London: Johns Hopkins University Press, 1981, pp. 266-7.

Table 11.2: British Government Defence Establishments

	Name	Function	Employees	Facilities
1.	Admiralty Marine Technology Establishment (AMTE)	R&D, for example structural design of ships, submarines; hydrodynamics		
2.	Admiralty Underwater Weapons Establishment (AUWE)	R&D, undersea warfare activities, for example, torpedo research, sonar		
3.	Admiralty Surface Weapons Establishment (ASWE)	R&D, on guided weapons, guns, surveillance and tracking systems, communications, navigation		
4.	Royal Signals and Radar Establishment (RSRE)	R&D, applied electronics, for radar, guided weapons, computing etc.		
5.	Atomic Weapons Research Establishment (AWRE)	R&D, primarily but not exclusively for nuclear programme		
6.	Chemical Defence Establishment (CDE)	research into and evaluation of threats from chemical and biological weapons		
7.	Military Vehicles and Engineering Establishments (MVEE)	R&D, combat and logistics vehicles, engineering equipment, military bridges		
8.	Propellants, Explosives and Rocket Motor Establishment (PERME)	R&D, propellant formulation/performance, motor materials, gun propellants, primary explosives		
9.	Royal Armament Research Establishment (RARDE)	R&D, guns, mortars, mines, warheads, antitank guided weapons, etc.		
10.	Aeroplane and Armament Experimental Establishment (A&AEE)	testing of military aircraft and associated weapons systems		
11.	Royal Aircraft Establishment (RAE)	R&D, aerospace (other than engines and radar)		

	Name	Function	Employees	Facilities
12.	National Gas Turbine Establishment (NGTE)	R&D, gas turbines, diesel engines, electrical systems, engineering equipment (for ships)		
13.	Royal Ordnance Factories (ROF)	major manufacturer of guns, armoured vehicles, small arms, munitions, some R&D	23,000[a]	11 plants[b] (Glascoed, Chorley, Nottingham, Leeds, Radway Green, Blackburn, Birtley, Bishopton, Enfield, Bridgewater, and Patricroft)
14.	Royal Dockyards	ship construction (UK and foreign navies)	30,000[b]	5 dockyards[b] (Portsmouth, Devonport, Chatham, Rosyth in UK, and Gibraltar)
15.	Defence Sales Organisation (DSO)	provides sales support for Royal Ordnance Factories, nationalized industries, and private firms		

Notes: a. data from 1977.
b. Data from 1979.
Source: Derived from Rae Angus, *The Organisation of Defence Procurement and Production in the United Kingdom*, Aberdeen Studies in Defence Economics, no. 13, Aberdeen: Centre for Defence Studies, Dec. 1979, pp. 24-7.

Table 11.3: Major British Nationalized and Public Sector Defence Industries

	Name	Function	Employees	Facilities
1.	British Aerospace, Ltd. (BAe)[a]	produces airframes, missiles	69,000[b]	Aircraft Group: 6 plants (Hatfield-Chester, Kingston-Brough, Manchester, Scottish, Warton, Weybridge-Bristol); BAe Dynamics: headquarters at Stevenage
2.	British Shipbuilders	warship production	27,000[b] of total 78,000 workers engaged in defence production	Vickers, Vosper Thorneycroft, Yarrows, Swan Hunters, Cammell-Laird, Scotts, Brooke Marine, Robb Caledon, and Hall, Russell
3.	Rolls-Royce[c]	produces aero- and marine engines	61,700[d]	Bristol Division, Derby Engine Division, Small Engine Division, produce military goods
4.	Short Brothers[c]	aircraft and missile production	7,000[e]	
5.	British Leyland (BL)[c]	armoured and other vehicles	194,600[d]	Alvis facilities, Daimler/Jaguar plants, Rover Division, Scammell Division

Notes: a. In February 1981, BAe was partially de-nationalized. The government now own 49 per cent, shareholders, 48 per cent and employees, 3 per cent.
b. Data from 1979.
c. The primary purpose of this firm is not to produce for the military market but its contribution to the British defence sector is nonetheless important.
d. Data from 1977.
e. Data from 1981.
Sources: Derived from, Rae Angus, *The Organisation of Defence Procurement and Production in the United Kingdom*, Aberdeen Studies in Defence Economics, no. 13, Aberdeen: Centre for Defence Studies, December 1979, pp. 28-32; Michael Brzoska, 'Economic Problems of Arms Production in Western Europe – Diagnoses and Alternatives', in *Militarization and Arms Production*, eds. Helena Tuomi and Raimo Väyrynen, London: Croom Helm, 1983; and personal communication with Harry Dean, University of Sussex, May 1982.

Table 11.4: Private Sector Military Industries in the UK, Major Contractors[a]

Name	Function	Employees[b]	Facilities
Aerospace			
1. Westland	produces helicopters (and in its British Hovercraft facility it produces military air cushion vehicles)	12,900	
2. Hunting Engineering	produces, develops bombs, JP 233 airfield attack weapon	7,000	
3. Dowty	hydraulics, control systems	13,900	
4. Marshall of Cambridge	conversion, modification, repair work		
5. Lucas Aerospace			
Armoured Vehicles/Land Systems			
1. Vickers	development and production of medium tank, other armoured fighting vehicles and artillery pieces		Elswick
2. GKN Sankey	FV 432 and MCV 80		
3. Vauxhall Motors	trucks/transporters		Luton
4. David Brown Gear Industries	produces gearboxes		
5. Rolls Royce Motors	engines for heavy armoured military vehicles		
6. Sterling Armament Company	small arms, ammunition, esp. light machine guns		Dagenham
7. Fairey Engineering	manufactures bridges and trackways		
Electronics			
1. General Electric Company (GEC)	produces communications, radar and sonor equipment, surveillance equipment, electronic countermeasures equipment, fire control systems	192,000	Elliotts; Marconi Avionics; Marconi Radar Systems; Marconi Space and Defence Systems
2. Plessey Company	produces torpedoes, naval sonar and radar, ground-force radar and communications equipment, airfield control and air defence radar, ECM and ECCM systems	63,900	

3.	EMI Electronics	produces and develops sonar systems, mortar locating radar, thermal imaging equipment, anti-personnel mines	51,300
4.	Ferranti	airborne radar and target acquisition/designation equipment, action data automation systems for naval weapons, trainers and simulators	16,800
5.	Lucas Aerospace	produces a wide range of electronics	
6.	Philips	produces a wide range of electronics, especially communications and radar equipment, image intensifiers and sights	
7.	Racal	produces communications equipment	5,800
8.	Smiths Industries	produces flight instruments, flight control systems, air data computers, shipborne radar and sonar systems	20,400
9.	Sperry Rand	produces Ships Inertial Navigation Systems (SINS), ship- and airborne control systems	7,000[c]

Notes: a. Major contractors are firms which received more than £10 million in contracts from the British government in 1977-8.
b. Data from 1977.
c. Data from 1975.
Source: Derived from, Rae Angus, *The Organisation of Defence Procurement and Production in the United Kingdom*, Aberdeen Studies in Defence Economics, No. 13, Aberdeen: Centre for Defence Studies, December 1979, pp. 32-7, and Michael Brzoska, 'Economic Problems of Arms Production in Western Europe – Diagnoses and Alternatives', in *Militarization and Arms Production*, eds. Helena Tuomi and Raimo Väyrynen, London: Croom Helm, 1983.

Table 11.5: 20 Largest Arms Producers in the UK, 1977

	Name	Owner	Arms production (million £)[a]	Total sales (million £)	Arms production as a percentage of total sales
1.	British Aerospace	UK state	688	860	80
2.	Rolls-Royce Ltd (state-owned)	UK state	493	704	70
3.	General Electric Co.	various	351	2,343	15
4.	Vickers	various	159	409	40[b]
5.	Royal Ordnance Factories	UK state	150	150[b]	100[c]
6.	Plessey	various	134	611	22
7.	British Leyland	UK state	130	2,602	5[b]
8.	Westland	various	121	139	87
9.	Racal	family	109	182	60
10.	Vosper	D. Brown	98	98	100
11.	EMI	various	94	851	11
12.	Ferranti	UK state	69	125	55
13.	Decca	various	69	181	38
14.	Swan Hunter	Hunter	52	156	33[b]
15.	Dowty	various	50	136	37
16.	Hunting	various	47	88	53
17.	Lucas Industries	various	44	886	5[b]
18.	Cable Wireless	UK state	41	156	26[b]
19.	Imperial Metal Industries	ICI, GB	41	467	9[c]
20.	Sperry Rand	Sperry Rand USA	36	88[d]	41

Notes: a. Approximate figures

b. Maximum figure. Actual figure is probably a bit lower.

c. Minimum figure. Actual figure is probably a bit higher.

d. Figure for 1975.

Source: Derived from Michael Brzoska, 'Economic Problems of Arms Production in Western Europe – Diagnosis and Alternatives', in *Militarization and Arms Production*, eds. Helena Tuomi and Raimo Väyrynen, London: Croom Helm, 1983.

Table 11.6: Dependence of 24 British Arms Producers with over £25 Million in Sales, 1977, on Arms Production

Arms production as share of total sales, percentage	Number of companies
75 – 100%	6
50 – 75%	4
25 – 50%	6
10 – 25%	3
< 10%	5

Source: Michael Brzoska, 'Economic Problems of Arms Production in Western Europe – Diagnoses and Alternatives', in *Militarization and Arms Production*, eds. Helena Tuomi and Raimo Väyrynen, London: Croom Helm, 1983.

Figure 11.1: Geographic Distribution of UK Major Contractors, 1979-80[a]

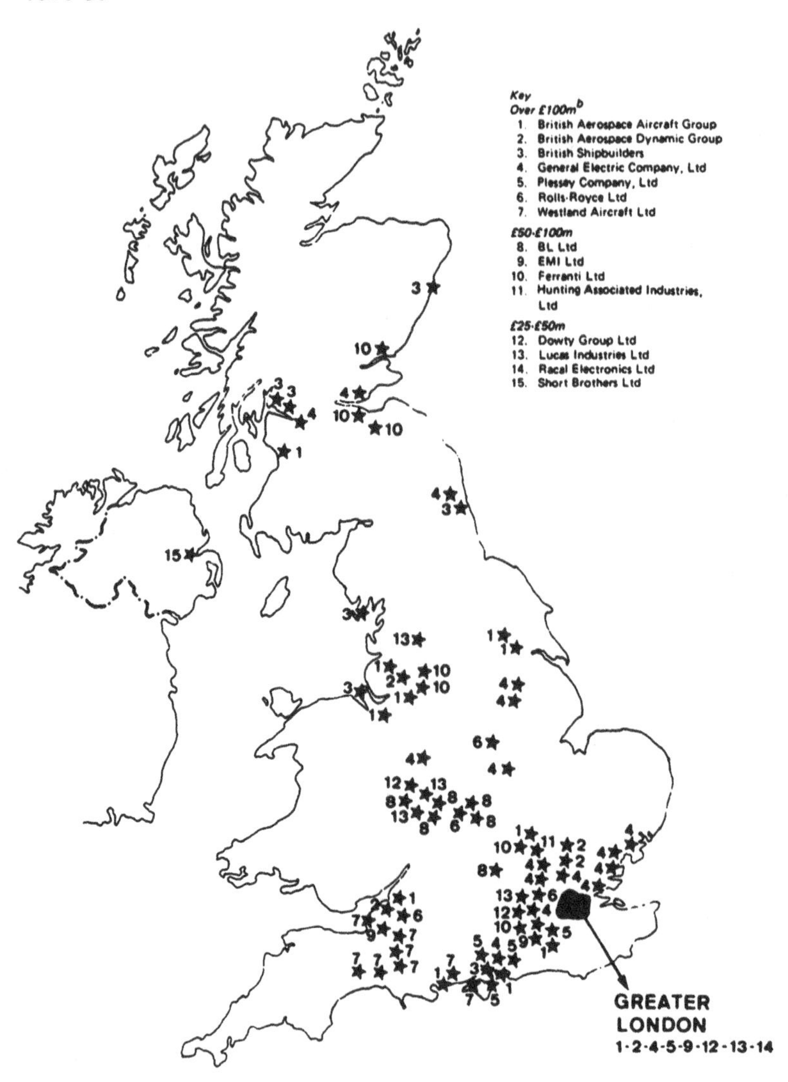

Notes: a. Major contractors are considered to be firms paid £5 million or more in 1979/80 by the Ministry of Defence. Here only contractors earning in excess of £25 million from MOD are included.

b. The Royal Ordnance Factories also received over £100 million from MOD in 1979/80 but are not shown on this map.

Source: *Statement on the Defence Estimates, 1981*, vol. I, Cmnd. 8212-I, London: HMSO, April 1981, Figure 13, p. 44.

Table 11.7: British Exports of Defence Equipment, 1975-80 (million current £)

Equipment category	1975	1976	1977	1978	1979	1980
Identified Equipment[a]						
Armoured fighting vehicles and parts	42	50	52	61	53	50
Military aircraft, combat and non-combat, including helicopters	43	42	63	59	52	170
Warships, including air cushion vehicles	42	36	123	39	82	59
Guns, small arms and parts	15	19	32	36	52	64
Guided weapons and missiles	13	18	27	24	24	25
Ammunition	43	53	59	97	76	102
Radio communication and radar apparatus	–	–	–	56	42	55
Optical equipment and training simulators	–	–	–	20	12	12
Identified equipment, total	198	218	356	392	393	537
Other Equipment[b]						
Military airframes, engines and parts[c]	199	241	323	338	286	na
Military aircraft equipment[d]	80	78	88	108	140	na
Military space equipment	–	–	10	2	11	na
Other military electronics	–	–	–	140	155	na
Other military road vehicles	–	–	–	90	90	na
Other equipment, total	279	319	421	678	682	na
Grand Total	477	537	777	1,070	1,075	na

Notes: a. Equipment which can be identified by the British Customs and Excise Tariff.
b. Equipment which cannot be distinguished from similar civilian items in the Customs and Excise Tariff. Information on these has been provided by the Society of British Aerospace Companies and individual electronics and motor vehicle manufacturing companies. These figures are therefore estimates.
c. Does not include export of aircraft parts, aeroengines and engine parts which are for international collaborative projects.
d. Excludes military airborne radars and ground flying trainers which are included in 'Identified Equipment'.
Source: Derived from ***Statement on the Defence Estimates 1981, vol. 2: Defence Statistics***. London: HMSO, April 1981. Table 2.8, p. 21.

Table 11.8: Public Defence-related R&D in the UK, 1961-80 (million £ and percentages)

Year	R&D	Defence-related of total R&D	Defence-related as percentage of total
1961	383.9	248.6	64.8
1962	393.7	243.1	61.7
1963	405.1	240.5	59.4
1964	452.8	261.9	57.8
1965	465.8	262.2	56.3
1966	500.4	260.6	52.1
1967	519.5	241.6	46.5
1968	549.2	236.7	43.1
1969	581.6	238.9	41.1
1970	605.8	258.2	42.6
1971	699.3	303.4	43.4
1972	776.6	351.8	45.3
1973	871.3	406.6	46.7
1974	1,074.4	500.2	46.6
1975	1,381.7	668.1	48.4
1976	1,515.2	756.1	49.9
1977[a]	1,647.2	862.9	52.4
1978	1,702.9	877.2	51.5
1979	2,026.3	1,083.2	53.5
1980	2,689.5	1,495.2	55.6

Note: a. Provisional figures.
Sources: 1961-69: OECD, *Changing Priorities for Government R&D: An Experimental Study of Trends in the Objectives of Government R&D Funding in 12 OECD Member Countries, 1961-1972*, Paris: pp. 140, 141, 320.
1970-3: United Kingdom, Central Statistical Office, *Research and Development, Expenditure and Employment*, Studies in Official Statistics, no. 27, London: HMSO, n.d., Table 7.
1974-7: United Kingdom, Central Statistical Office, *Economic Trends*, no. 309, London: HMSO, July 1979, Table 7.
1978-80: Statistical Office of the European Communities, *Government Financing of Research and Development, 1970-1980*, Luxembourg: 1981, pp. 150, 154.

Table 11.9: Composition of British Military R&D, 1970s[a] (percentages)

Category	1970/71	1971/72	1972/73	1973/74	1974/75	1975/76	1976/77	1977/78	1978/79
Military aircraft	32	35	37	36	35	33	37	39	38
Guided weapons	18	17	17	16	13	13	11	12	12
Other electronics	14	16	16	16	17	16	14	13	14
Ship construction/ Underwater warfare	8	8	9	10	9	10	10	10	9
Ordnance and other Army	11	11	10	10	9	8	8	7	8
Other R&D[b]	17	13	11	12	17	20	20	19	19
Military R&D as percentage of total budget	10.1	11.2	11.6	12.4	12.6	12.2	12.4	13.0	12.6

Notes: a. The R&D figures upon which these percentages are based are estimates, rather than actual expenditure. The percentages presented here should therefore be considered illustrative of trends only.
b. The category 'Other R&D' includes R&D related to Britain's strategic nuclear force.
Source: Derived from Lawrence Freedman, *Arms Production in the United Kingdom*, London: Royal Institute of International Affairs, 1978, Table 5, p. 13.

Figure 11.2: Relative Production Costs of Successive Generations of Selected UK Armaments

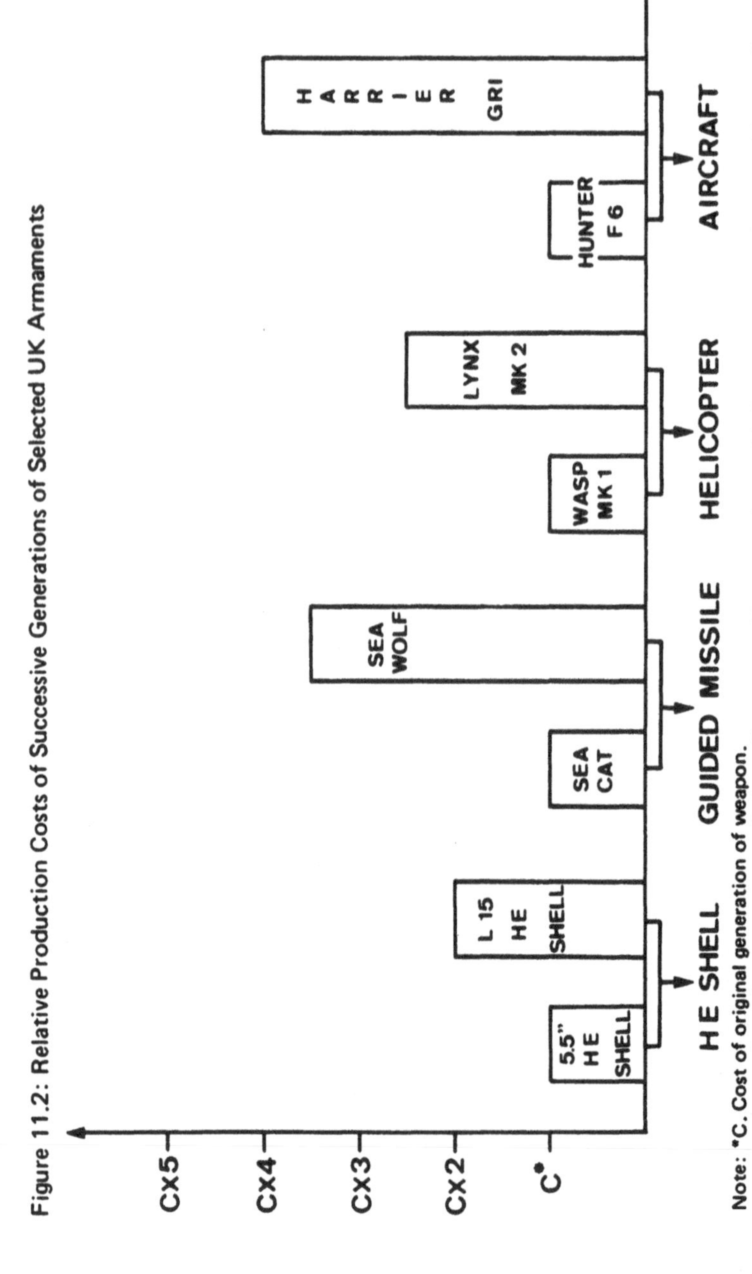

Note: *C. Cost of original generation of weapon.
Source: Derived from *Statement on the Defence Estimates, 1981*, vol. I, Cmnd. 8212-I, London: HMSO, April 1981, p. 45.

Table 11.10: Production Costs of Selected UK Naval Vessels

Name	Type	Completed	Cost per ton (£)
Battleships and Major Surface Vessels			
Vanguard	Battleship	1946	202.2
Centaur	Commando carrier	1953	476.5
Ark Royal	Aircraft carrier	1955	506.1
Blake	*Tiger* cruiser	1961	1,572.6
Bristol	Type 82 cruiser	1973	4,350.8
Invincible	Anti-submarine warfare cruiser	1978[a]	5,000.0[a]
Submarines			
Ambush	'A' class	1947	407.3
Ocelot	Patrol submarine	1961	1,813.7
Churchill	*Valiant* nuclear-powered submarine	1970	7,331.7
Sovereign	*Swiftsure* nuclear-powered submarine	1974	9,113.7
Frigates			
Morecambe Bay	*Bay* class	1949	312.5
Salisbury	Type 61	1957	1,343.3
Juno	*Leander* class	1967	2,040.8
Amazon	Type 21	1973	6,864.4
Destroyers			
Matapan	*Battle* class	1945	396.8
Defender	*Daring* class	1952	814.3
Sheffield	Type 42	1975	8,261.9

Note: a. estimated.
Source: Derived from Mary Kaldor, 'Defence Cuts and the Defence Industry', in *Alternative Work for Military Industries*, Dave Elliott, Mary Kaldor, Dan Smith and Ron Smith, London: Richardson Institute for Conflict and Peace Research, 1977, Table 2, p. 15.

Table 11.11: Concentration in the British Arms Industry, 1977

Largest companies	Share of total arms output (percentage)
1	10.4
3	24.0
6	31.0
10	38.4

Source: Michael Brzoska, 'Economic Problems of Arms Production in Western Europe – Diagnoses and Alternatives', in *Militarization and Arms Production*, eds. Helena Tuomi and Raimo Väyrynen, London: Croom Helm, 1983.

Table 11.12: Concentration in the British Aerospace Industry[a]

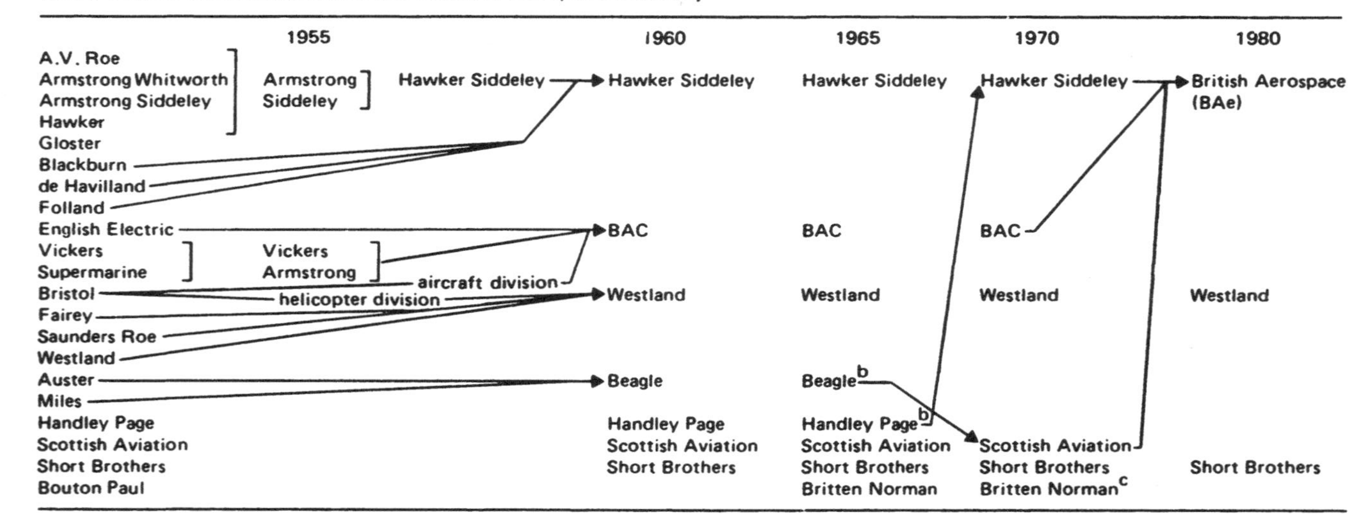

Notes: a. Aircraft producers only.
b. Went bankrupt. Production capacity absorbed by Hawker Siddeley and Scottish Aviation.
c. Assets acquired in 1978 by Pilatus Aircraft Ltd, of Switzerland.
Sources: 1955-70: Mary Kaldor, *European Defence Industries – National and International Implications*, ISIO Monograph no. 8, Brighton: Institute for the Study of International Organization, University of Sussex, 1972, Table 14.
1980: John W.R. Taylor, ed. *Jane's All the World's Aircraft, 1981-82*, London: Jane's, 1981.

APPENDIX 2: USE OF RAW MATERIALS FOR MILITARY PURPOSES

Compiled by Milton Leitenberg

One question for which very little information is available concerns the use of raw materials by the defence industrial sector. The United States is the only nation for which there appears to be any such data. An estimate of the combined consumption of 24 strategic raw materials directly absorbed by weapons production, as well as secondarily in the industrial machinery required to produce the weapons, showed a range of 1 to 13 per cent in 1973.[1] A comparable estimate for 1970 of the ten largest raw material requirements by the US defence sector showed a range of 7 to 13 per cent.[2] Due to the magnitude of the USSR's defence industrial sector, the size of its annual output, and knowledge of its weapon design as well as machinery design practices, it is quite reasonable to assume that these percentages of raw material use, as well as absolute amounts, may be somewhat higher for the USSR.

The use of petroleum by the US defence sector, including the direct operational use by military forces (ships, aircraft and so on) in 1970 – at the height of the Vietnam war – was 5 per cent.[3] Since operational and training practices of US and USSR military forces differ substantially, the use of petroleum may be lower for the USSR. Estimates of the use of other energy sources by the defence industrial sector are not available.

Notes

1. *Annual Report of the Joint Committee on Defense Production, 1975*, Washington, DC: US Govt. Printing Office, 1976, Appendix 3, pp. 70-6. See also, E.E. Hughes *et al.*, *Strategic Resources and National Security: An Initial Assessment*, Menlo Park, Calif.: Stanford Research Institute, 1975.
2. Stephen P. Dresch, *Disarmament: Economic Consequences and Developmental Potential*, prepared for UN, Department of Economic and Social Affairs, Centre for Development Planning, December 1972, Table 3.5.
3. C. Vansant, *Strategic Energy Supply and National Security*, New York: Praeger, 1971; R.W. Sullivan *et al.*, *A Brief Overview of the Energy Requirements of the Department of Defense*, Battelle-Columbus Laboratories, August 1972; United States, Department of Defense, *Management of Defense Energy Resources*, Report of the Defense Energy Task Group, Washington, DC: 15

November 1973; C.C. Mow and J.K. Ives, *Energy Consumption by Industries in Support of National Defense: An Energy Demand Model*, R-1448-ARPA, Santa Monica, Calif.: Rand Corporation, August 1974; and W.D. Gosch and W.E. Mooz, *A USAF Energy Consumption Projection Model*, R-1553-ARPA, Santa Monica, Calif.: Rand Corporation, November 1974. See also, Dresch, *Disarmament*.

NOTES ON CONTRIBUTORS

Nicole Ball is Visiting Research Associate at the Swedish Institute of International Affairs in Stockholm. Since 1979 she has been studying the role of the military in third-world development. Her most recent publications include *World Hunger: A Guide to the Economic and Political Dimensions* (ABC-Clio, 1981) and *The Military in the Development Process: A Guide to Issues* (Regina Books, 1981).

Frank Blackaby, an economist, is director of the Stockholm International Peace Research Institute. In addition to his work on military expenditure, he has written on a variety of economic issues, particularly British economic policy. He was formerly Deputy Director of the National Institute for Economic and Social Research in London.

Michael Brzoska, an economist by training, is a researcher at the Institute for Peace Research and Security Studies at the University of Hamburg. He has written on data problems in the socio-military field, the effects of military expenditures in the third world, and arms production in Western Europe.

David Holloway is Lecturer in Politics at the University of Edinburgh. His most recent publications include 'Innovation in the Defence Sector', and 'Innovation in the Defence Sector: Two Case Studies' in R. Amann and J. Cooper (eds.), *Industrial Innovation in the Soviet Union* (Yale University Press, 1982) and *The Soviet Union and the Arms Race* (Yale University Press, 1983).

Per Holmström is a researcher at the Department of Economic History at the University of Umeå in Sweden. At present he is working on a project concerning decision-making in private companies.

Sydney Jammes is an analyst with the Central Intelligence Agency in Washington, DC. His most recent unclassified publications have been in support of US Congressional Committees including 'China's Military Capabilities' for the Senate Foreign Relations Committee and joint authorship of 'China's Military Strategic Requirements' for the Joint Economic Committee.

Edward Kolodziej is Professor of Political Science and Co-director of the Office of Arms Control, Disarmament, and International Security at the University of Illinois, Champaign-Urbana. He has recently edited (with Robert Harkavy) two books, *American Security and Security Policy-Making* and *Security Policies of Developing Countries*. He is completing a manuscript on French arms transfer policy and the global arms transfer system.

Milton Leitenberg is Visiting Research Associate at the Swedish Institute of International Affairs. He has edited several volumes and written some 90 papers and monographs since beginning work in arms control and strategic studies in 1966. Among these are the major portion of, *Tactical Nuclear Weapons, European Perspectives*, SIPRI (Taylor & Francis, 1978) and *Great-Power Intervention in the Middle East* (edited) (Pergamon Press, 1979). He is presently completing a book-length study on long-range theatre nuclear weapons in Europe.

Ulf Olsson is Professor of Economic History at the University of Umeå in Sweden. His publications include *L.M. Ericsson 100 Years* (1976), *The Creation of a Modern Arms Industry. Sweden 1939-1974* (University of Gothenburg, 1977), and 'The State and Industry in Swedish Rearmament' in *The Adaptable Nation. Essays in Swedish Economy During the Second World War* (1982).

Judith Reppy is with the Peace Studies Program at Cornell University. Her current research interests centre on the long-term effects of military spending on the structure of the US economy. She is author of two recent case studies of military influence on civilian technology in the semiconductor and machine tool industries (Harvard Business School Case Study numbers 9-382-760 and 9-682-624).

Sergio Rossi is Deputy Director of CESDI (International Research Studies and Documentation Centre) in Turin and foreign and defence correspondent for *Il Sole-24 Ore* of Milan. His book, *Rischio atomico ed equilibri mondiali: Salt, euromissili, crisi afghana* (SEI, 1980) is the first comprehensive Italian survey of the SALT and INF negotiations. Since 1982 he has directed the Seminar on Strategic Studies at the University Institute for European Studies in Turin. He is currently researching Italian defence policy in the 1980s with special reference to policy relating to NATO's nuclear forces.

Gerald Steinberg teaches in the Political Science Department of Hebrew University in Jerusalem. He has an MSc. in physics from the University of California, a PhD in government from Cornell University and was recently a fellow at the Massachusetts Institute of Technology. His research deals with science, technology and public policy. His book, *Satellite Reconnaissance: The Role of Informal Bargaining* is forthcoming from Praeger (New York).

Stephen Tiedtke is Research Associate at the Peace Research Institute Frankfurt (PRIF). His publications include *Die Warschauer Vertragsorganisation zum Verhältnis von Militär- und Entspannungspolitik in Osteuropa* [The Warsaw Treaty Organisation. Relations between Military and Détente Policy in Eastern Europe] (Oldenbourg-Verlag, 1980) and *Rüstungskontrolle aus sowjetischer Sicht. Die Rahmenbedingungen der sowjetischen MBFR-Politik* [Armament Control from a Soviet Point of View. The Conditions of the Soviet MBFR Policy] (Campus-Verlag, 1980).

Herbert Wulf is a staff member of the Institute of Peace Research and Security Policy, University of Hamburg. He is co-author of *Rüstung und Unterentwicklung* [Armament and Underdevelopment] (Reinbek, 1976), *Arbeitsplätze durch Rüstung?* [Employment through Armament?] (Reinbek, 1978) and *Sicherheitspolitik, Rüstung und Abrüstung* [Security Policy, Armament and Disarmament] (Frankfurt, 1982). He is the author of *Rüstungsimport als Technologietransfer* [Weapon Imports as Technology Transfer] (München, 1979).

INDEX